Practical Law of Architecture, Engineering, and Geoscience

Second Canadian Edition

Brian M. Samuels
Samuels & Co., Barristers and Solicitors

Doug R. Sanders
Borden Ladner Gervais LLP

Pearson Canada
Toronto

Library and Archives Canada Cataloguing in Publication

Samuels, Brian M.
Practical law of architecture, engineering and geoscience / Brian M. Samuels, Doug R. Sanders.—
2nd Canadian ed.

Includes bibliographical references and index.
ISBN 978-0-13-700408-9

1. Architects—Legal status, laws, etc.—Canada. 2. Engineers—Legal status, laws, etc.—Canada.
3. Earth scientists—Legal status, laws, etc.—Canada. 4. Contracts—Canada. 5. Negligence—
Canada.
I. Sanders, Doug R. (Doug Ronald), 1964– II. Title.

| KE2727.S26 2011 | 349.7102'462 | C2010-901336-0 |
| KF2925.S26 2011 | | |

ISBN 978-0-13-700408-9

Vice-President, Editorial Director: Gary Bennett
Sponsoring Editor: Carolin Sweig
Marketing Manager: Michelle Bish
Senior Developmental Editor: John Polanszky
Project Manager: Imee Salumbides
Copy Editor: Susan Broadhurst
Indexer: Sheila Flavel
Compositors: Debbie Kumpf, Christine Velakis
Art Director: Julia Hall
Cover and Interior Designer: Miguel Acevedo

For permission to reproduce copyrighted material, the publisher gratefully acknowledges the
copyright holders listed beneath figures and tables throughout the text, which are considered an
extension of this copyright page.

The information in this text is intended to be current and accurate. It is not, however, intended
to be comprehensive or complete, and therefore should not be relied upon in making decisions
on particular legal problems. In such cases, the services of a competent professional should be
sought. The authors and publishers expressly disclaim any responsibility for any liability, loss,
or risk, personal or otherwise, that is incurred as a consequence, directly or indirectly, of the use
and application of any of the contents of this book.

1 2 3 4 5   14 13 12 11 10

Printed and bound in the United States of America.

With love to Lacey, Elizabeth, and Katie

With love to Kali, Erin, and Anna

ABOUT THE AUTHORS

Brian M. Samuels

Brian Samuels is the principal of Samuels & Co., a Vancouver construction law firm, and is counsel to the firm of Lee & Associates in Denver, Colorado. Mr. Samuels was admitted to the Bar in British Columbia in 1988 and in Colorado in 1993. He holds a Bachelor of Laws (1987) from the University of Victoria, an MBA (1982) from the University of British Columbia, and a Bachelor of Engineering (Civil, 1978) from McGill University. He has been registered as a Professional Engineer in British Columbia since 1981.

Mr. Samuels is a construction lawyer, mediator, and arbitrator, as well as the author of the textbook *Construction Law* (Prentice Hall, 1996) and the co-editor of the textbook *Expert Evidence in British Columbia Civil Proceedings* (CLE, 2001). He is an Adjunct Professor in the Civil Engineering department at the University of British Columbia, where he teaches Construction Law.

Doug R. Sanders

Doug Sanders is a partner at the Vancouver office of Borden Ladner Gervais LLP. Mr. Sanders was admitted to the Saskatchewan Bar in 1990 and the British Columbia Bar in 1999. He is a graduate of the University of Saskatchewan with a Bachelor of Laws (1989) and a Bachelor of Science in Civil Engineering (1986). Mr. Sanders is a Professional Engineer licensed in the provinces of Saskatchewan and British Columbia.

Mr. Sanders practises in the areas of construction contracts (drafting and review), public-private partnerships, and construction litigation. He is the author of numerous articles and a presenter at construction and engineering law seminars. Mr. Sanders has been ranked by his peers for inclusion in The Best Lawyers in Canada (Construction Law) and with Practical Law Company.

Mr. Sanders is also a volunteer with provincial and national engineering organizations. He is a past board member of the Canadian Engineering Qualifications Board, and the Canadian Engineering International Board. He is past chair of the Association of Professional Engineers of Saskatchewan (APEGS) Registration Committee, a current member of the Association of Professional Engineers of British Columbia (APEGBC) Registration Task Force, and a past member of the APEGBC Software Engineering Task Force. He originated the Law and Ethics seminars in Saskatchewan and British Columbia and has been an instructor for both seminars. In 2000, APEGS presented him with the McCannel Award for exemplary service to the engineering profession. He is also a long-time member of Construction Specifications Canada.

BRIEF CONTENTS

List of Abbreviations   xiv

Preface   xvi

CHAPTER 1   The Canadian Legal System   1

CHAPTER 2   Regulation of the Professions   8

CHAPTER 3   Ethical Considerations   14

CHAPTER 4   Property Law   23

CHAPTER 5   Business Organizations   34

CHAPTER 6   Contracts   44

CHAPTER 7   Breach of Contract   57

CHAPTER 8   Interpreting and Drafting Contracts   64

CHAPTER 9   Selected Contract Issues   68

CHAPTER 10   Getting to a Contract   84

CHAPTER 11   Specific Contracts and Clauses   113

CHAPTER 12   Torts   131

CHAPTER 13   Common Issues in Contract and Tort   148

CHAPTER 14   Dispute Resolution and Expert Evidence   155

CHAPTER 15   Risk, Responsibility, and Dispute Avoidance   166

CHAPTER 16   Insurance   175

CHAPTER 17   Bonds   191

CHAPTER 18   Construction Liens   202

CHAPTER 19   Delay and Impact Claims   216

CHAPTER 20   Labour Law   226

CHAPTER 21   Employment Law   237

CHAPTER 22   Health and Safety Law   247

CHAPTER 23   Environmental Law   256

CHAPTER 24   Aboriginal Law   265

CHAPTER 25   Securities Law   268

CHAPTER 26   Privacy Law   271

CHAPTER 27   Internet Law   273

Appendix   279

Glossary   342

Selected References   351

Index   353

CONTENTS

List of Abbreviations xiv

Preface xvi

CHAPTER 1   The Canadian Legal System 1

1.1   The Canadian Constitution   1

1.2   The Canadian Court System   2

1.3   The Creation of Law   3

1.4   Quebec Law   4

1.5   Claims and Disputes   4

1.6   International Law   5

1.7   Subject Areas and Principles   6

1.8   Case Citations   7

CHAPTER 2   Regulation of the Professions   8

Overview   8

2.1   Right to Title   9

2.2   Scope of Practice   9

2.3   Registration   11

2.4   The Obligations of a Professional   11

2.5   Discipline and Enforcement   12

2.6   Professional Seals and Letters of Assurance   12

Chapter Questions   12

Answers   13

CHAPTER 3   Ethical Considerations   14

Overview   14

3.1   The Relationship between Ethics and the Law   14

3.2   Codes of Ethics   16

3.3   The Duty to the Public   17

3.4   The Client's Interest   18

3.5   The Employer's Interest   18

3.6   Duty to the Profession   19

3.7   The Dual Role of the Consultant as Owner's Agent and Impartial Arbiter   20

Chapter Questions   21

Answers   22

CHAPTER 4   Property Law   23

Overview   23
4.1   Real Property   24
4.2   Chattels   28
4.3   Intellectual Property   28
Chapter Questions   *32*
Answers   *33*

CHAPTER 5   Business Organizations   34

Overview   34
5.1   Sole Proprietorships   34
5.2   Partnerships   35
5.3   Corporations   37
5.4   Advantages and Disadvantages of Organizational Structures   41
Chapter Questions   *41*
Answers   *43*

CHAPTER 6   Contracts   44

Overview   44
6.1   Contract Formation, Offer, and Acceptance   45
6.2   Consideration   46
6.3   Agreements to Agree   49
6.4   Voiding a Contract   50
6.5   Amendment of Contracts   52
6.6   Waiver and Estoppel   52
6.7   Quasi-contract   53
Chapter Questions   *54*
Answers   *55*

CHAPTER 7   Breach of Contract   57

Overview   57
7.1   Breach   57
7.2   Damages   59
7.3   Contract Termination   61
Chapter Questions   *62*
Answers   *63*

CHAPTER 8   Interpreting and Drafting Contracts   64

8.1   Interpretation   64

8.2   Drafting Contracts   66

Chapter Questions   67

Answers   67

CHAPTER 9   Selected Contract Issues   68

Overview   68

9.1   Agency and Authority   68

9.2   Indemnities   70

9.3   Change Orders   71

9.4   Subcontract Issues   74

9.5   Unforeseen Conditions   76

9.6   Specifications and Drawings   78

9.7   Contract Administration   79

Chapter Questions   82

Answers   83

CHAPTER 10   Getting to a Contract   84

Overview   84

10.1   Delivering a Project   85

10.2   Transfer of Risk and Obligation   86

10.3   The Processes for Choosing Project Participants   87

10.4   Tendering   92

10.5   Choosing the Best Process and Delivery Method for a Project   105

10.6   Effects of Interprovincial and International Trade Agreements   110

Chapter Questions   111

Answers   112

CHAPTER 11   Specific Contracts and Clauses   113

Overview   113

11.1   Standard Form Contracts   113

11.2   Construction Contracts   114

11.3   Professional Service Agreements   120

11.4   Licensing Agreements   122

11.5   Geoscience Agreements   122

11.6   Standard Clauses   123

11.7   Project Finance   128

Chapter Questions   129

Answers   130

CHAPTER 12   **Torts**   131

Overview   131

12.1 Duty of Care   132

12.2 Breach of Duty   136

12.3 Proximate Cause   137

12.4 Loss Caused by the Breach   139

12.5 Tortious Misrepresentation   139

12.6 Fraud   140

12.7 Fiduciary Duty   141

12.8 Trespass   142

12.9 Duty to Warn   143

12.10 Product Liability   143

Chapter Questions   144

Answers   145

CHAPTER 13   **Common Issues in Contract and Tort**   148

13.1 Concurrent Liability in Contract and Tort   148

13.2 Limitation Periods   148

13.3 Joint and Several Liability   150

13.4 Vicarious Liability   151

13.5 Codes and Standards   151

Chapter Questions   153

Answers   154

CHAPTER 14   **Dispute Resolution and Expert Evidence**   155

Overview   155

14.1 Litigation   155

14.2 Arbitration   159

14.3 Negotiation   161

14.4 Mediation   161

14.5 Other Dispute Resolution Methods   162

14.6 Expert Witnesses   162

Chapter Questions   164

Answers   165

CHAPTER 15   **Risk, Responsibility, and Dispute Avoidance**   166

Overview   166

15.1 Risk Assessment   166

15.2 Common Law Presumptions   167

15.3 Shifting Risk   168

15.4 Disputes Caused by Client Dissatisfaction   169

15.5 Disclaimers   171

15.6 Record Keeping   173

15.7 Problem Solving   174

Chapter Questions   174

Answers   174

CHAPTER 16   **Insurance**   175

Overview   175

16.1 Operating Without Insurance   176

16.2 The Duty to Defend   177

16.3 Subrogation   177

16.4 Insurable Interest   178

16.5 Claims-Made and Occurrence Policies   180

16.6 Material Non-Disclosure and Prejudice to Third Parties   181

16.7 Cooperation and Conflict between Insurer and Insured   182

16.8 Types of Policies and Their Common Exclusions   183

Chapter Questions   187

Answers   188

CHAPTER 17   **Bonds**   191

Overview   191

17.1 Roles and Responsibilities in Bonds   192

17.2 Indemnities and Other Surety Recourses   192

17.3 Bid Bonds   193

17.4 Performance Bonds   194

17.5 Defences under a Performance Bond   197

17.6 Payment Bonds   198

Chapter Questions   199

Answers   201

CHAPTER 18 Construction Liens 202

Overview 202
18.1 Making and Proving a Lien Claim 203
18.2 Who May Claim a Lien 204
18.3 Substitute Lien Security 205
18.4 Trust Provisions 207
18.5 Holdback 209
18.6 Risk to the Contractor and Owner 213
Chapter Questions 213
Answers 215

CHAPTER 19 Delay and Impact Claims 216

Overview 216
19.1 Scheduling Principles 217
19.2 Compensable and Excusable Delays 218
19.3 Concurrent Delays 219
19.4 No-Damages-for-Delay Clauses 220
19.5 Acceleration 220
19.6 Impact Claims 221
19.7 Proving a Delay, Impact, or Acceleration Claim 221
Chapter Questions 224
Answers 225

CHAPTER 20 Labour Law 226

Overview 226
20.1 Establishing Union Representation 228
20.2 Trade Unions and Jurisdictional Disputes 229
20.3 Union Security and Right-To-Work 230
20.4 Work Stoppages 230
20.5 Secondary Activity 231
20.6 Successor Employers and Common Employers 231
20.7 Enforcement of Collective Agreements 233
20.8 Layoffs and Seniority 234
Chapter Questions 234
Answers 235

CHAPTER 21 Employment Law 237

Overview 237

21.1 Implied Terms in the Common Law Employment Contract 237

21.2 Independent Contractor or Employee 242

21.3 Employment Standards Legislation 243

21.4 Human Rights 244

21.5 Employees Facing Termination 244

21.6 The Charter of Rights and Freedoms 244

Chapter Questions 245

Answers 246

CHAPTER 22 Health and Safety Law 247

Overview 247

22.1 Occupational Health and Safety 248

22.2 Contracts 252

22.3 Torts and Workers' Compensation Legislation 252

22.4 Ethical Considerations 253

Chapter Questions 254

Answers 255

CHAPTER 23 Environmental Law 256

Overview 256

23.1 Environmental Site Assessments and Audits 257

23.2 Remedies for Private Landowners 258

23.3 Governmental Regulation 260

23.4 The Environmental Assessment Process 263

Chapter Questions 263

Answers 264

CHAPTER 24 Aboriginal Law 265

Overview 265

24.1 The Duty to Consult 266

24.2 Nature of Reserve Property 266

24.3 Aboriginal Participation in Projects 266

24.4 Contracts with Band Councils 266

Chapter Questions 267

Answers 267

CHAPTER 25 Securities Law 268

Overview   268

25.1 Information Disclosure Requirements   268

25.2 Technical Disclosure Guidelines   269

25.3 Common Law and Statutory Liability   269

Chapter Questions   *270*

Answers   *270*

CHAPTER 26 Privacy Law 271

Overview   271

26.1 Federal Legislation   271

26.2 Provincial Legislation   272

Chapter Questions   *272*

Answers   *272*

CHAPTER 27 Internet Law 273

Overview   273

27.1 Jurisdiction   274

27.2 Torts   274

27.3 Copyright   274

27.4 Trademarks   275

27.5 Privacy and Security   275

27.6 Securities Regulation   275

27.7 Electronic Contracts   275

27.8 Websites   276

27.9 Communications System Risk Management   277

Chapter Questions   *278*

Answers   *278*

Appendix   279

Glossary   342

Selected References   351

Index   353

LIST OF ABBREVIATIONS

AAA	American Arbitration Association
ACEC	Association of Consulting Engineering Companies
ACEC 31	ACEC Prime Agreement between Client and Engineer
AIA	American Institute of Architects
AIT	Agreement on Internal Trade
AOD	Association of Owners and Developers
APEGBC	Association of Professional Engineers and Geoscientists of British Columbia
APEGGA	Association of Professional Engineers, Geologists, and Geoscientists of Alberta
APEGS	Association of Professional Engineers and Geoscientists of Saskatchewan
ASCE	American Society of Civil Engineers
BCCA	British Columbia Construction Association
BCCA 200	BCCA Subcontract
CCA	Canadian Construction Association
CCA 14	CCA Design-Build Stipulated Price Contract
CCA 17	CCA Stipulated Price Contract for Trade Contractors on Construction Management Projects
CCA 5	CCA Construction Management Contract
CCDC	Canadian Construction Documents Committee
CCDC 2	CCDC Stipulated Price Contract
CCDC 23	CCDC Guide to Calling Bids and Awarding Contracts
CCDC 220	CCDC Performance Bond
CCDC 221	CCDC Bid Bond
CCDC 222	CCDC Labour and Material Payment Bond
CCDC 3	CCDC Cost Plus Contract
CCDC 4	CCDC Unit Price Contract
CCPG	Canadian Council of Professional Geoscientists
CEAA	Canadian Environmental Assessment Act
CEPA	Canadian Environmental Protection Act
CGL Insurance	Commercial General Liability Insurance
CGS	The Canadian Geotechnical Society
CHSR	Canada Health and Safety Regulations
CIM	Canadian Institute of Mining
CITT	Canadian International Trade Tribunal
CLC	Canada Labour Code
CM	Construction Management
CPCA	Canadian Portland Cement Association
CPD	Continuing Professional Development
CPM	Critical Path Method
CSA	Canadian Standards Association
CSA	Canadian Securities Administrators
CSC	Construction Specifications Canada
CSCE	Canadian Society for Civil Engineering
CSME	Canadian Society of Mechanical Engineering
DBIA	Design-Build Institute of America
DMCA	Digital Millennium Copyright Act (U.S.)
DRP	Dispute Resolution Process
E&O Insurance	Errors and Omissions Insurance

EIT	Engineer-in-Training	OH&S	Occupational Health and Safety
ESA	Environmental Site Assessment	P.Eng.	Professional Engineer
FIDIC	International Federation of Consulting Engineers	PFI	Privately Financed Initiative
		PIPEDA	Personal Information Protection and Electronic Documents Act
GC	General Condition		
GMP	Guaranteed Maximum Price	PPP or P3	Public Private Partnership
GSC	Geological Survey of Canada	RAIC	Royal Architectural Institute of Canada
GST	Goods and Services Tax		
ICE	Institute of Civil Engineers	RAIC	
IEEE	Institute of Electrical and Electronics Engineers	Document 6	Canadian Standard Form of Contract for Architectural Services
ISO	International Organization for Standardization		
		RAIC	
ISP	Internet Service Provider	Document 7	Canadian Standard Form of Contract for Architectural Services (Abbreviated Version)
L&M Bond	Labour and Material Payment Bond		
NAFTA	North American Free Trade Agreement	RFP	Request for Proposals
		RFQ	Request for Qualifications
NCARB	National Council of Architectural Registration Boards (U.S.)	RRSP	Registered Retirement Savings Plan
NCEES	National Council of Examiners for Engineering and Surveying (U.S.)	Vmail	Voice Mail
		WCB	Workers' Compensation Board
		WORM	Write Once Read Many
NI	National Instrument	WTO	World Trade Organization

PREFACE

New to This Edition

This second Canadian edition has been updated and enhanced in the following respects:

- Case law has been updated to include recent leading case authority in several fields, including employment law, tendering, pure economic loss, contract enforceability, disclaimers, and other areas. For example, the recent Supreme Court of Canada decisions in *Tercon, Greater Fredericton Airport Authority, and Design Services* are discussed and explained.
- The most recent (2008) edition of the CCDC 2 Stipulated Price Contract has been included as an appendix.
- The section on Quebec has been expanded.
- The text contains a new section discussing the relationship between law and ethics.

About This Text

This text is intended to provide the following:

a) a broad overview of areas of the law relevant to the practice of architecture, engineering, and geoscience

b) practical, rather than theoretical, information

c) sufficient background to allow the reader to identify legal issues

d) simple, easy-to-follow language

This text is *not* intended to do the following:

a) make lawyers out of and thereby ruin perfectly good architects, engineers, and geoscientists

b) be a master's level thesis on any of the subjects, since most if not all of the topics have multiple full-length texts written about them

c) eliminate the need to seek appropriate legal advice

The title of this text is *Practical Law of Architecture, Engineering, and Geoscience*. Since the publication of the first edition, this practical guide to the law for design professionals has been adopted as the primary text for the Professional Practice Exams and the entrance exam for Professional Engineers and Geoscientists in various provinces. However, the intended audience of this text is not limited to architects, engineers, and geoscientists. Contractors, technicians and technologists, lawyers, suppliers, project managers, construction managers, software professionals, and others also may benefit from reading it.

One of the risks of creating a text that provides a broad overview is that the explanations in most areas are necessarily brief and do not contain the level of detail that would be found in comprehensive legal texts. Therefore, readers must recognize that this text is not intended as a substitute for legal advice. Legal problems are fact-specific, meaning that a slight change in the facts often can lead to a different conclusion. Furthermore, the law differs from one jurisdiction to another and changes over time. Non-lawyers should obtain specific advice for specific legal problems and should not attempt to act as their own counsel.

It has been said that ignorance of the law is no excuse.[1] The time has passed when architects, engineers, and geoscientists could rely on technical competence in their fields and ignore

---

[1] It has also been said that this statement does not apply to trial court judges, who have court of appeal judges to correct their errors for them.

the law. For these professionals, the likelihood of being involved in a lawsuit, whether as plaintiff or defendant, is much greater now than in the past. Since the cost of prosecuting, defending, and settling claims can be high, recognizing and preventing potential legal problems early is important. Moreover, all professionals need to know the basics of contract law so that they can negotiate appropriate agreements.

Many chapters in this text apply equally to all professions. For example, the basic principles of contract law and negligence, and property law and business organizations are as applicable to architecture as they are to engineering and geoscience. However, other portions of the text will be more useful to some professions than to others. For example, geoscientists may want to pay particular attention to section 11.5 on geoscience agreements and to Chapter 25, which covers the disclosure requirements for professionals involved in mining and oil and gas exploration.

Supplements

Practical Law of Architecture, Engineering, and Geoscience, Second Canadian Edition, is accompanied by a complete supplements package:

- *Instructor's Manual:* The Instructor's Manual features answers or notes to all of the questions in the textbook, brief case summaries, teaching suggestions, and additional questions for discussion along with suggested answers.

- *Test Item File:* A comprehensive testbank of various types of questions has been prepared in Word to accompany this Canadian edition.

- *PowerPoint® Slides:* Electronic slides are available in Microsoft PowerPoint. The slides illuminate and build on key concepts in the text.

The Instructor's Manual, Test Item File, and PowerPoint® Slides can be downloaded from Pearson Canada's online catalogue at http://vig.pearsoned.ca.

- *Companion Website: Practical Law of Architecture, Engineering, and Geoscience*, Second Canadian Edition, is supported by an excellent Companion Website that can be accessed by using the code that came packaged with this book. The Companion Website includes practice questions, key terms and concepts, sample contracts, and links to relevant websites.

- *CourseSmart:* CourseSmart goes beyond traditional expectations–providing instant, online access to the textbooks and course materials you need at a lower cost for students. And even as students save money, you can save time and hassle with a digital eTextbook that allows you to search for the most relevant content at the very moment you need it. See how when you visit www.coursesmart.com/instructors.

Acknowledgments

The manuscript for this book was reviewed at various stages of development by a number of academic and professional peers from across Canada. We wish to thank those who shared their insight and constructive criticism of this manuscript.

Dr. Michael Bennett, University of Ontario Institute of Technology

Dr. Milton W. Petruk, Association of Professional Engineers, Geologists, and Geophysicists of Alberta

Gillian Pichler, Association of Professional Engineers and Geoscientists of British Columbia

Bryan S. Shapiro, University of British Columbia

Jan Shepherd McKee, University of Western Ontario

At Pearson Canada, we are grateful for the support and dedicated efforts of Carolin Sweig, Michelle Bish, John Polanszky, Imee Salumbides, and Susan Broadhurst.

A Great Way to Learn and Instruct Online

The Pearson Canada Companion Website is easy to navigate and is organized to correspond to the chapters in this textbook. Whether you are a student in the classroom or a distance learner you will discover helpful resources for in-depth study and research that empower you in your quest for greater knowledge and maximize your potential for success in the course.

Companion Website

www.pearsoned.ca/samuelssanders

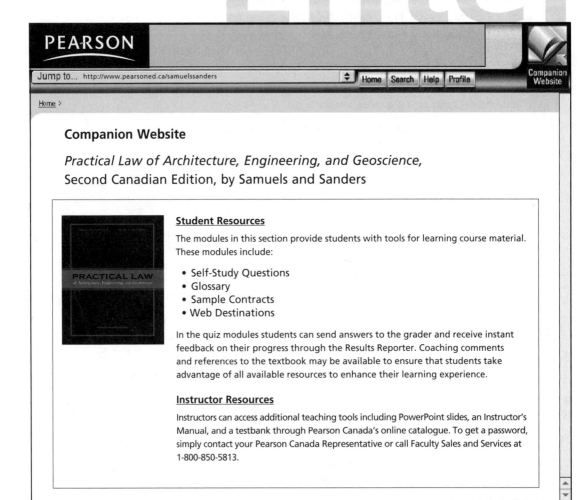

PEARSON

Jump to... http://www.pearsoned.ca/samuelssanders   Home | Search | Help | Profile

Companion Website

Home >

Companion Website

Practical Law of Architecture, Engineering, and Geoscience,
Second Canadian Edition, by Samuels and Sanders

PRACTICAL LAW
of Architecture, Engineering, and Geoscience

Student Resources

The modules in this section provide students with tools for learning course material. These modules include:

- Self-Study Questions
- Glossary
- Sample Contracts
- Web Destinations

In the quiz modules students can send answers to the grader and receive instant feedback on their progress through the Results Reporter. Coaching comments and references to the textbook may be available to ensure that students take advantage of all available resources to enhance their learning experience.

Instructor Resources

Instructors can access additional teaching tools including PowerPoint slides, an Instructor's Manual, and a testbank through Pearson Canada's online catalogue. To get a password, simply contact your Pearson Canada Representative or call Faculty Sales and Services at 1-800-850-5813.

CHAPTER 1

THE CANADIAN LEGAL SYSTEM

1.1 The Canadian Constitution

In Canada, as in many countries, the ability to create and enforce the law is derived from the Constitution. The Constitution is the source of all legal authority in Canada. The Constitution is made up of several statutes and conventions from both the British and the Canadian Parliaments. Although it is beyond the scope of this text to describe the intricacies of the Canadian Constitution, professionals in Canada must have a reasonable grasp of its content in order to work effectively.

The Constitution has three primary features: the division of powers, the creation of the courts, and the *Canadian Charter of Rights and Freedoms* (the **Charter**). The provisions relating to division of powers essentially assign rights to enact certain kinds of legislation to the federal Parliament and the provincial legislatures. In other words, the federal government has the right to enact laws relating to one set of areas, whereas the provincial governments have the right to do so over other areas. Matters of national importance, such as national defence, the postal service, criminal law, and specific fields such as patent law, all fall within federal jurisdiction. Most areas of private law, such as property rights, contracts, negligence law, and construction liens, fall under provincial jurisdiction. For the most part, laws affecting architecture, engineering, and geoscience practice fall under provincial jurisdiction. Note that wherever there are direct operational conflicts between a federal and a provincial statute, the Constitution holds that federal law prevails.

While private law falls under provincial jurisdiction, there are a few exceptions to this rule. Labour law is generally a matter of provincial jurisdiction; but industries that fall under federal jurisdiction, such as airlines, railroads, and the post office, are governed by federal labour laws. Another exception is competition law: matters such as bid rigging and unfair trade practices are governed by the *Competition Act*,[1] which is a federal statute.

Because much of the law relating to architecture, engineering, and geoscience falls within the provincial sphere, laws may differ from one province to another. For example, lien legislation in Alberta differs from

---

[1] R.S.C. 1985, c-34.

lien legislation in Ontario. The acts or statutes governing professions also differ from province to province. This text describes some of these legal differences. However, it focuses more on basic principles, which are in most cases either identical or very similar from one province to another.

This text generally refers to provinces, but in Canada there is another level of jurisdiction called a territory. Canada has three territories: Yukon, Northwest Territories, and Nunavut. Unlike provinces, territories are creations of federal legislation, and the federal government assigns powers to territorial governments. Despite this difference, in practice, territorial legislatures operate much like provincial legislatures and create legislation that is very similar to that of provinces. For example, the practice of engineering in Yukon is governed by Yukon legislation that has created the Association of Professional Engineers of Yukon as a regulatory body. References throughout the text to provinces should be interpreted to include territories.

1.2 The Canadian Court System

The Constitution also creates a system of courts, known as superior courts or *supreme courts*. Each province has a superior trial-level court as well as a court of appeal. The trial-level court is known by different names in each province, but each one has the same general powers as the Court of Queen's Bench in England. Those powers include "inherent jurisdiction," which is the power to administer justice even where there is no statutory basis for doing so. In British Columbia, the court is known as the Supreme Court of British Columbia. In Alberta, it is the Alberta Court of Queen's Bench. In Ontario and Quebec, it is called the Superior Court. While the names vary, the powers remain the same. All judges of these courts are appointed by the Governor General on the recommendation of the federal government and cannot be removed from the Bench, except for misconduct. Because the judiciary is an independent branch of government, judges must be able to function in an impartial and unbiased manner, despite being recommended by the federal government. In addition, federal courts known simply as the Federal Courts deal with federal matters such as patents and shipping disputes.

The court systems function independently of each other. Provincial litigants have the right to appeal a trial decision to the Court of Appeal of the province in which the decision was made. Litigants under the Federal Court may appeal to the Federal Court of Appeal. No court in any province has jurisdiction over courts in other provinces, and the Federal Courts have no jurisdiction over the provincial courts.

The ultimate level of appeal is the Supreme Court of Canada. But bringing a case to the Supreme Court is not fast or easy. In civil cases, litigants do not have an automatic right to bring an appeal to the Supreme Court of Canada. Instead, they must first obtain permission, or leave, from the Supreme Court of Canada to bring the appeal. In order to obtain leave, they must demonstrate that the issues in the case are of national importance. Thus, only rarely are cases involving architects, engineers, or geoscientists granted leave to appeal to the Supreme Court of Canada.

Each province also has a lower court level, known generally as the Provincial Court. This level includes provincial criminal courts and civil small claims courts. In some provinces, this level also includes family courts that deal with divorce and custody issues. Small claims courts have limited jurisdiction, both in terms of the dollar value of awards and the subject matter. The monetary jurisdiction may vary from one province to another. For example, in British Columbia, the small claims court award limit is $25 000. The small claims court in British Columbia also has no jurisdiction over construction liens.[2] Figure 1-1 shows the relationship between the various levels of court.

---

[2] See Chapter 18.

FIGURE 1-1 Canadian Court System

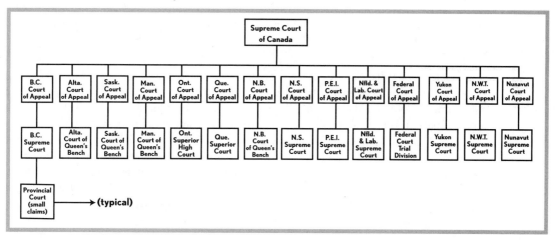

The **Charter** is a section of the Constitution that establishes individual rights that apply to all Canadians and in many cases to non-citizens in Canada. Some of these rights apply to criminal cases, such as the right to counsel, and the right not to self-incriminate. Other well-known Charter rights are freedom of religion, freedom of speech, freedom of association, and the right not to be discriminated against. The Charter lists prohibited grounds of discrimination, including colour, ethnic origin, religion, disability, and age. Note that the Charter applies to acts of the government and its branches, but not to interactions between private parties. Human rights statutes in each province prohibit discrimination by private individuals.

1.3 The Creation of Law

Law is an imprecise and somewhat unpredictable field. This unpredictability may be disconcerting to clients who present to their lawyer what they regard as a clear-cut legal problem, only to be advised, for example, that there is only a 60 percent likelihood of a favourable result, and the amount of recovery would be limited to $50 000.

One reason for the lack of predictability and precision is that the law is not static. Law is dynamic, changing over time and from place to place. What is considered settled law today might not be the rule tomorrow. Unlike laws of nature, human laws can be rewritten by legislatures and reinterpreted by courts.

(a) Statute Law

In Canada, laws can be created in several ways. Federal and provincial governments can enact statutes. Statute laws must be compatible with the Constitution in order to be enforceable. Statutes relevant to architecture, engineering, and geoscience include those governing joint and several liability in negligence, construction lien rights, formalities of contracts,[3] and workers' compensation requirements. Judges then interpret these statutes. In theory, judges do not make laws; they only interpret them. In practice, however, the act of interpretation unavoidably results in the creation of new rights and remedies.

---

[3] In many jurisdictions, statutes require certain types of contracts to be written contracts. For example, construction contracts generally do not have to be written, whereas in some provinces, legislation requires that surety contracts must be in writing in order to be enforceable.

Legislation frequently gives certain other agencies the power to create regulations that may be considered law. For example, a provincial Association of Engineers and Geoscientists may enact bylaws, a breach of which can give rise to liability for damages as well as other consequences.

b) Common Law

Large bodies of Canadian law are not based on statutes. These laws were not created by legislatures, nor were they the result of interpretation of statutes. Rather, they are laws created by judges based upon principles of law and equity established and modified over hundreds of years, tracing their roots to England. This body of judicial precedents is known as the **common law**.

Almost all jurisdictions in Canada and in the United States are common law jurisdictions. The only notable exceptions are Quebec and Louisiana, which have civil law systems based on the Napoleonic Code of France. In civil law jurisdictions, all law is statute-based. In other words, **civil law** is a more codified version of the common law. Professionals practising in Quebec should become familiar with the Civil Code. This text does not discuss the Civil Code, with the exception of a general overview in this chapter.

1.4 Quebec Law

As stated, the province of Quebec uses a civil law system that differs from the common law system in force throughout the rest of Canada. Canadian common law does in fact apply in Quebec with respect to certain matters, such as criminal law, patent law, and other areas of federal jurisdiction. However, relationships between private individuals are governed by the Civil Code of Quebec, for the most part. The law governing private relationships includes contract law, torts, employment law, business organizations, insurance, bonds, and many of the other subject matters covered in this text. Architects, engineers, and geoscientists practising in Quebec will find that many of their business dealings, as well as personal matters, will be governed by the Civil Code of Quebec.

One essential difference between a civil law jurisdiction, such as Quebec, and a common law jurisdiction, such as any other province of Canada, is that in a common law jurisdiction there are areas of law for which no statute exists, and the basis for the law is the body of case law decided by judges in the past. For example, contract law deals with how contracts are formed, what is required to make a contract enforceable, and in what circumstances a contract may be held to be void or unenforceable. In a common law province, these essential principles may be subject to little or no statute law. Enforceability of a contract, for example, depends on the existence of certain essential elements such as consideration, but there is no statute that sets out what constitutes consideration, or even the fact that it is required. As will be discussed in Chapter 6, the common law regarding consideration is evolving and is not as cut and dried as it once was. In Quebec, in contrast, the Civil Code of Quebec is an attempt to codify the basic principles of law in all areas.

In practice, the law does not result in much difference to a design or construction professional with respect to many aspects of practising one's profession in Quebec or elsewhere in Canada. Basic principles of negligence, contract formation, insurance, and other matters are fairly consistent across the country. There are, of course, some differences, but it is equally true that differences may exist in the law from one common law province to another.

1.5 Claims and Disputes

Claims and disputes are a fact of life in industry and business. These include claims for negligence, breach of contract, refusal to pay, and other causes of action. Disputes and claims are costly to resolve, not only in terms of the amounts paid in settlement, but also in terms of legal and expert fees, preparation time for employees, and the effects of uncertainty and contingent liability on the ability to carry on a business.

Many disputes involving architects and engineers arise from construction projects. Construction projects involve many parties: owners, architects, engineers (including structural, mechanical, electrical, and geotechnical engineers), contractors, subcontractors, suppliers, individual workers, estimators, construction managers, and others. When disputes develop, many if not all of these parties become involved. Thus, one of the characteristics of construction litigation is the large number of parties involved. Another is the large number of documents generated during the project that are relevant to the case. The combination of numerous parties and extensive documents renders litigation of these cases very expensive and time-consuming. In some instances, the costs of litigation can exceed the amount of money in dispute.

Yet despite the complexity, many claims and disputes are successfully resolved in accordance with a small number of legal principles, especially the law of contract and the law of negligence. Thus, a good understanding of contract and negligence law helps professionals not only in analyzing and resolving disputes but also in avoiding disputes in the first place. One of the purposes of this text is to provide the reader with a sufficient understanding of legal principles to avoid some of the pitfalls that lead to claims and disputes.

In some situations, claims and disputes are unavoidable. A party can become insolvent or bankrupt, for example, and be unable to pay. Alternatively, an accident can cause serious injury or property damage. In such instances, parties must be aware of the steps that must be taken, often quickly, in order to preserve their legal rights. For instance, in an insolvency dispute, an unpaid subcontractor needs to file a lien or assert a claim against a labour and material payment bond. Such steps must be done in accordance with the applicable legislation or in accordance with the terms of the bond. A lack of awareness that these legal remedies exist, or that **limitation periods** and procedural requirements must be met, can seriously prejudice the claimant's rights. For example, in the event of an accident, the insured party must comply with requirements of any insurance contracts or risk losing coverage. This text will help readers learn when to seek legal advice in order to avoid such consequences.

Allocation of risk is a common theme throughout this text. The professions of architecture, engineering, and geoscience are by their very nature risky. The risks of accident, failure, financial loss, and unforeseen results are always present. As a result, all parties in a construction project try to minimize their own risks. They may design contracts to allocate risks between the parties. They may also use insurance to shift risk to a third party, the insurer. In addition, they may use bonds to protect parties against risks of non-performance and non-payment. As well, they may include various types of contractual disclaimers in an attempt to shift risk. Readers should consider all principles discussed and explained in this text in the context of risks and keep in mind to which parties those risks are allocated.

1.6 International Law

As with each of the other areas covered in this text, international law is itself a specialized area. This section highlights a few basic principles of international law as they apply to architects, engineers, and geoscientists.

First, international treaties typically take precedence over domestic law where there are inconsistencies between the two. For example, if a properly ratified treaty between Canada and the United States contains an entitlement for foreign companies to bid on certain projects, a provincial law prohibiting foreign companies from bidding would be unenforceable. Such circumstances are rare, because governments strive to keep laws compatible with treaties.

For example, the North American Free Trade Agreement (NAFTA) allows Canadian engineers and other professionals to obtain work visas to work legally in the United States, regardless of state statutes. Such a work visa is often referred to as a TN visa and is relatively inexpensive to obtain. But professionals should be aware that there is a critical difference between the right to work and the right to practise a profession. For example, in order to practise consulting engineering in the United States, an engineer may need to be licensed by the applicable Board of Registration in the state where the engineer intends to consult. In other words, to practise as a consulting engineer in Oregon, a Canadian engineer would need both the licence from the applicable Oregon registration body and a TN visa allowing the engineer to legally work in the United States.

Finally, tax treaties between Canada and many countries help prevent duplication of income taxes. Professionals planning to work outside of Canada should consult a tax specialist to ensure that they do not pay extra taxes. As well, those who work in certain countries with little or no income tax should also seek advice. Liability for Canadian income tax may attach to individuals, even if they are resident elsewhere for many years. Specific tax advice is essential in such situations.

1.7 Subject Areas and Principles

Most claims and disputes involving professionals deal with the principles of contract law and negligence. An understanding of the basic principles in these two areas is a prerequisite to study of related areas of law.

Contracts are voluntary agreements between two or more parties that set out the rights, responsibilities, and liabilities of the parties to each other. While contracts are often written, oral contracts are also legally enforceable, except in cases of special types of contracts.

The fundamental principles of contract law concern whether or not a contract is enforceable. Concepts such as *offer and acceptance* and *consideration* are used to determine whether a binding agreement exists, and if so, what the terms of that agreement are. Even if a binding and enforceable agreement exists, subsequent events may cause the agreement to become voidable. Chapter 6 explains these basic concepts of contract law.

Construction contracts are one type of contract of interest to architects, engineers, and geoscientists. Construction contracts follow the same rules of law as other commercial agreements. While construction contracts have some unique features, most features are applicable to contracts in other industries. Contract provisions and issues are discussed in Chapters 6 through 11.

In addition to contract law, professionals must be aware of the laws of torts. **Torts** are acts committed by one party in violation of the rights of another, which are considered sufficiently wrong as to give rise to liability. For example, the tort of **negligence** is the failure to take reasonable care for another person. Negligence constitutes the majority of tort claims. Professionals are less frequently involved in intentional torts, such as fraud, fraudulent misrepresentation, and trespass.

While contracts are a voluntary process, in that a party has a choice as to whether or not to sign a contract, tort obligations are involuntary and imposed by the law. A negligence claim is customarily asserted by a party who has been injured, either financially or physically, by the act or omission of another party. No contractual relationship between the plaintiff and defendant is necessary in a negligence claim; however, negligence claims often arise between parties with a contractual relationship. Such parties may sue simultaneously for breach of contract and negligence. The existence of a contract may provide remedies in addition to those available under the law of negligence.

To prove a negligence claim, a plaintiff must show that at the time the event occurred, the defendant could have reasonably foreseen that the plaintiff would suffer an injury or loss as a result of the defendant's failure to take due care. The plaintiff must also show that the defendant did not meet the standard of care expected of someone in his or her position. Furthermore, the plaintiff must show that the failure to meet that standard of care was the legal cause of the loss suffered by the plaintiff. All of these elements of negligence are explained in Chapter 12.

The other half of the text deals with problematic areas and issues, such as bidding, procurement, construction liens, bonds, intellectual property rights, conflicts of interest and ethical issues, and insurance. These chapters also cover more general business topics, such as business organizations, labour law, and employment law, as well as Aboriginal law and privacy law, as they relate to the architecture, engineering, and geoscience professions.

One recurring theme in this text is dispute avoidance through understanding of legal principles and recognition of common fact patterns. Despite best efforts, however, professionals cannot always avoid disputes and must instead resolve them. Chapter 14 covers dispute resolution by examining established methods of dispute resolution.

1.8 Case Citations

Case law is created by judges to create legal principles and interpret statutes. It is also used by lawyers to predict outcomes in specific circumstances. But case law is very dependent on the facts of a particular case. A slightly different set of facts can result in a very different result. In addition, the findings in one case may not be an accurate description of the entire scope of the law in a certain area. A lawyer must take into account several different cases on the same subject and determine the impact of those rulings on the facts of the new case. A lawyer also must make sure that the cases are the most current available. Thus, non-lawyers, including architects, engineers, and geoscientists, should never attempt to read a single case and then assume that the outcome will be the same in their own circumstances. An important fact may be different, or another case may impact how the facts are dealt with. Moreover, newer case law, including perhaps a case under appeal, could lead to a potentially different application of the law to specific circumstances.

The cases in this text are used to illustrate key points and are by no means exhaustive of the case law in each topic. In fact, volumes of material would be needed if this text were to include every relevant case. However, this text provides the relevant legal principles as they exist at the time of publishing.

Case citations are used to find cases in law libraries. Case citations generally look like this: *Backman v. Canada*, [2001] 1 S.C.R. 367; where "*Backman v. Canada*" is the case name, 2001 is the year of the decision, 1 is the volume number, S.C.R. is the reporting series (in this case Supreme Court Reports) and 367 is the page number. Hence, *Backman v. Canada* is found at page 367 of volume 1 of the 2001 Supreme Court Reports.[4] Recent case citations may also look like this: *Martel Building Ltd. v. Canada*, 2000 SCC 60; where 2000 is the year of the decision, "SCC" is the level of court (Supreme Court of Canada), and 60 is the number of the decision issued in that year by that court.

---

[4] Case citations may also appear in a slightly different format with the date in round brackets, such as *Walter Cabott Construction Ltd. v. The Queen* (1974), 44 D.L.R. (3d) 82 (F.C.A.). The round brackets indicate that the date is not required for locating the case in question.

CHAPTER 2

REGULATION OF THE PROFESSIONS

Overview

Most professions in Canada are self-regulated. **Self-regulation** refers to a profession's statutory authority to govern itself. The provinces have jurisdiction to regulate the professions, and each province has its own legislation governing professions.

Professional self-regulation covers two main elements: the right to title and the scope of practise. **Right to title** means that the professional regulatory authority can regulate the exclusive right for its members to use a particular title. **Scope of practise** means that the regulatory authority also has the right to regulate the exclusive right of its members to practise in a particular area or field. For example, the engineering profession has the exclusive right to the title "Professional Engineer" or "P.Eng." and the exclusive right to practise "professional engineering."[1] The professions of architecture, engineering, and, in all provinces and territories except Prince Edward Island and Yukon, geoscience, all have right to title and right to an exclusive scope of practise. The scope of practise is defined in each province's legislation. Other professions, such as interior design, only have right to title legislation.

The legislation in each province creates a regulatory body to govern each profession. These regulatory bodies typically have elected officials and staff, with the elected officials consisting mainly of members of the profession. The principal functions of the regulatory bodies are to register members, regulate the practise of its members, discipline members, and enforce the legislation against nonmembers who are practising illegally.

The provincial engineering regulatory bodies have also formed a national organization known as Engineers Canada (formerly the Canadian Council of Professional Engineers). Similarly, the provincial geoscience

---

[1] The terms engineer, engineering, professional engineer, P.Eng., consulting engineer, ingenieur, ing., ingenieur conseil, genie and ingenerie are official marks held by Engineers Canada on behalf of its constituent members.

regulatory bodies, which are part of the engineering regulatory bodies in some provinces, have formed a national organization known as the Canadian Council of Professional Geoscientists (CCPG). The Royal Architectural Institute of Canada (RAIC) plays a similar role to Engineers Canada and CCPG, but it registers individual members of provincial associations on a voluntary basis. The principal goals of these national organizations are to coordinate and foster mutual recognition among the regulatory bodies and to encourage commonality of operations.

In addition to provincial regulatory bodies and their respective national organizations, engineers, architects, and geoscientists are members of a wide variety of technical associations. These include broader groups such as the Canadian Society for Civil Engineering (CSCE), Canadian Society of Mechanical Engineering (CSME), the Institute of Electrical and Electronics Engineers (IEEE), and the Association of Consulting Engineering Companies (ACEC). They also include technical groups, such as Construction Specifications Canada (CSC), the Canadian Portland Cement Association (CPCA), and the Canadian Geotechnical Society (CGS). Membership in technical societies is voluntary but provides valuable technical training and transfer of knowledge. Many of the technical societies have their own codes of ethics, and some have copyright over specific titles.

2.1 Right to Title

Protection of the right to title of the various professions is one of the main obligations of the provincial regulatory bodies. For engineers, this means that graduates of an engineering program are not entitled to call themselves engineers or professional engineers without first obtaining the right to do so through registration. Obtaining a Bachelor of Science in Engineering is one of the elements of registration, but it does not of itself permit graduates to call themselves engineers. The same applies to architects and geoscientists.

Moreover, only registered members of a profession can use the profession's title. Everyone else is prohibited from doing so by law. For each of the three professions, a period of apprenticeship is required, during which members of the profession train university graduates. During this period, graduates can use interim titles such as "engineer-in-training" (EIT) or "member-in-training" (MIT). Misrepresenting oneself by using the title "architect" without appropriate registration can lead to enforcement proceedings, together with potential civil proceedings. The right to title is protected by the regulatory bodies through discipline and enforcement proceedings.

2.2 Scope of Practise

The architecture, engineering, and geoscience professions generally have exclusive scopes of practise. Scope of practise means that in Canada, no one can practise within the exclusive scope of practise of these professions without being licensed. There are few exceptions to this rule because legislators have deemed the ability of the professions to maintain exclusivity to be very important. Other countries generally have systems of voluntary registration with associations, such as the United Kingdom's Institute of Civil Engineering; however, professionals often cannot find employment or gain approval for drawings from public authorities without such membership. In the United States, registration is mandatory for consulting engineers only.

Each of the provinces defines exclusive scope of practise through a definition of the practise of the specific profession. These definitions can be either generic or specific. A generic definition broadly states the types of tasks that constitute the exclusive scope of practise. A specific definition identifies in detail the individual tasks that make up the exclusive scope of practise.

For example, the Engineers Canada definition of professional engineering is a generic definition:

The practise of professional engineering means any act of planning, designing, composing, evaluating, advising, reporting, directing or supervising, or managing any of the foregoing,

> that requires the application of engineering principles,
>
> and
>
> that concerns the safeguarding of life, health, property, economic interests, the public welfare or the environment.[2]

A similar definition of the profession of geoscience has been adopted by the CCPG:

> The "practise of professional geoscience" means the performing of any activity that requires application of the principles of the geological sciences, and that concerns the safeguarding of public welfare, life, health, property, or economic interests, including, but not limited to:
>
> a. investigations, interpretations, evaluations, consultations or management aimed at discovery or development of metallic or non-metallic minerals, rocks, nuclear or fossil fuels, precious stones and water resources;
>
> b. investigations, interpretations, evaluations, consultations or management relating to geoscientific properties, conditions or processes that may affect the well-being of the general public, including those pertaining to preservation of the natural environment.[3]

In each of these two cases, the definition is somewhat circular in that the key phrase is the "application of engineering [or geoscience] principles." While this wording creates some ambiguity for the courts to resolve, it reflects the difficulty of defining any profession.

The national RAIC has created a definition of the profession of architecture. However, individual provincial legislatures and regulatory bodies can choose either to adopt the national definition or to create their own. The only definitions that are binding on the public and the professions are those that have been included in the enabling legislation or regulations in each province.

Ontario's *Architects Act* provides an example of a generic, statutory definition of the practise of architecture:

> "practise of architecture" means,
>
> a) the preparation or provision of a design to govern the construction, enlargement or alteration of a building,
>
> b) evaluating, advising on or reporting on the construction, enlargement or alteration of a building, or
>
> c) a general review of the construction, enlargement or alteration of a building;[4]

Definitions also include exceptions and exclusions. The Engineers Canada definition of professional engineering also has the following exclusion, which is intended to "retain the distinction between the practise of engineering and the practise of natural science":

> Nothing in this Act shall prevent an individual who either holds a recognized honours or higher degree in one or more of the physical, chemical, life, computer, or mathematical sciences, or who possess an equivalent combination of education, training and experience, or is acting under the direct supervision and control of the individual described in the preceding paragraph from practising natural science which, for the purposes of this Act, means any

---

[2] Engineers Canada, *Guideline on the Definition of the Practise of Professional Engineering.*

[3] http://www.ccpg.ca/guidelines/definition_practise_geoscience.html

[4] Ontario *Architects Act*, R.S.O. 1990, c. A.26.

act (including management) requiring the application of scientific principles, competently performed.[5]

Statutes governing the professions typically contain other exclusions so that outside professionals regulated by their own statutes do not breach the law by practising their own profession. For example, a forester may perform work governed by a statute relating to the forestry profession, even though that same work might fall within the definition of engineering. Thus, the Engineers Canada definitions must include exceptions for related professionals.

2.3 Registration

The primary mandate of most regulatory bodies is to deal with individual members. However, some regulatory bodies also require corporations employing professionals to register; and in some cases, they require additional registration for individuals or corporations that act as consultants to the public.

Regulatory bodies register members based on experience, education, and examinations. The experience requirement for engineers is generally four years of qualifying experience. The educational component is generally a degree from an accredited university. The examination component for engineers and geoscientists is generally a non-technical examination on law and ethics, unless the candidate is from a non-accredited university program, in which case additional technical examinations may be required. The examination component for Canadian architects is a series of nine technical and non-technical examinations administered through the U.S. National Council of Architectural Registration Boards (NCARB). In the United States, the examination component for engineers is a series of examinations similar to the NCARB examinations, except that these tests are administered by the National Council of Examiners for Engineering and Surveying (NCEES). Individual provinces may have additional registration requirements depending on the experience and education of the individual candidate.

Registration is required in each province in which the individual and corporation is practising the profession. The architectural, engineering, and geoscience regulatory bodies have reciprocal agreements to facilitate multiple registrations. Some professions also have reciprocity agreements with other countries. For instance, Engineers Canada has full or partial professional-level reciprocity agreements with Mexico and the state of Texas (through NAFTA), as well as with France, Hong Kong, Australia, and Ireland. These agreements make registration in the other jurisdiction easier. Engineers Canada also has education-level agreements with the United States, Australia, South Africa, Ireland, the United Kingdom, New Zealand, and Japan, under which accredited degrees are to be generally recognized. Again, individual regulatory bodies or the reciprocity agreements themselves may have additional requirements.

2.4 The Obligations of a Professional

Professional members must practise their profession competently and in accordance with the code of ethics of the regulatory body. Chapter 3 discusses codes of ethics.

Most regulatory bodies have a mandatory requirement for continuing professional development (CPD), and some of the regulatory bodies also have practise audits to review systematic and technical competence. The goal of most CPD and practise audits is to assist individuals with following codes of ethics and maintaining competence. Some regulatory bodies assist members by providing CPD preparation courses, course information, and information on alternative methods of fulfilling requirements.

---

[5] Engineers Canada, *Guideline on the Definition of the Practise of Professional Engineering.*

2.5 Discipline and Enforcement

Legislation gives regulatory bodies the ability and responsibility to discipline members for incompetence or breaches of the codes of ethics. **Discipline** is the process of charging a member and then proceeding to a hearing to determine guilt. **Enforcement** is the process of charging a nonmember with either using the protected professional title in breach of the right to title or practising the profession in breach of the exclusive scope of practise.

Disciplinary proceedings are considered to be quasi-criminal; thus, the regulatory bodies and courts take them seriously, in part because a disciplined architect, engineer, or geoscientist may lose his or her ability to earn a livelihood. The potential penalties for discipline actions include reprimands, suspensions, fines, termination of licences, educational requirements, and mentorship requirements. Because of the nature of the potential penalties, the standards of proof and the rigours of a formal process are strict and penalties are carefully considered.

2.6 Professional Seals and Letters of Assurance

An architect, engineer, or geoscientist typically receives a *professional seal* (the seal) from the regulatory authority. Improper use of his or her seal is a disciplinary offence. Most regulatory authorities have guidelines for use of the seal. Generally, the seal is to be used on final drawings, specifications, plans, reports, or other documents prepared or checked by the professional. Sealing a document without appropriately preparing or checking the document is the source of most seal-related problems. Professionals should understand that using a professional seal means that they have personally prepared, supervised, or reviewed the documents.

Similarly, many governmental authorities require the submission of a *letter of assurance* at the end of a project. Professionals should determine before starting a project whether a letter of assurance is required. If a municipal authority requires a letter of assurance or other document that the professional may be unwilling or unable to sign and seal, then they should not take on the project. Often such letters are very broad and require the professional to certify that the building has been constructed in accordance with the plans and specifications. Professionals have to judge whether they are prepared to sign this document and, if so, what services are necessary and at what cost. Letters of assurance are often used by future owners and other parties to try to place liability on the professional.

CHAPTER QUESTIONS

1. How is the practise of architecture regulated differently from that of interior decorating?

2. Under the definition of the profession of geoscience adopted by the CCPG, which of the following does *not* fall within the bounds of professional practise?
    a) The sampling of a tailings pond
    b) The evaluation of assay samples
    c) The construction of a road into a mine
    d) None of the above

3. Which statement describes design of a portion of a building?
    a) It falls within the practise of architecture, but not engineering.
    b) It falls within the practise of engineering, but not architecture.
    c) It may be both architecture and engineering.
    d) None of the above

4. Describe what exclusive scope of practise means for a profession.

5. An architect works from an office in Saskatchewan designing a project in Yukon and will have to inspect parts of the work as they are being constructed in Ontario. Where does he or she have to be registered?

6. Describe the potential outcomes of a discipline process in which a geoscientist is found guilty of professional misconduct.

7. Which statement is true about reciprocity agreements?
   a) They mean that a professional registered in one jurisdiction does not have to register in another jurisdiction.
   b) They make it easier to register in more than one jurisdiction.
   c) They make it harder to register in another jurisdiction.

8. Is a disciplinary hearing a criminal proceeding?

9. What kind of definition is the Engineers Canada definition of professional engineering?
   a) A generic definition
   b) A specific definition
   c) Both a generic and a specific definition
   d) Neither generic nor specific

10. A graduate engineer has the following right:
    a) The right to use the title Professional Engineer
    b) The right to use the title Engineer-In-Training
    c) No right to any title, based simply on an engineering degree

ANSWERS

1. The practise of architecture is governed by statute and prohibits practise of the profession by anyone not properly registered.

2. C

3. C

4. An exclusive scope of practise means that only members of the profession can practise in certain specific areas.

5. The architect should be registered in all three jurisdictions.

6. The potential penalties for discipline actions include reprimands, suspensions, fines, termination of licences, educational requirements, and mentorship requirements.

7. B

8. No, it is a quasi-criminal proceeding.

9. A

10. C

CHAPTER 3

ETHICAL CONSIDERATIONS

Overview

Many issues of law yield shades of grey instead of black-and-white answers. For this reason, professional associations adopt ethical standards. Broadly stated, **ethics** is the theory of morality. **Moral principles** are the standard of conduct required by society or by an organization or group. The standard of conduct is not the legal standard discussed in Chapter 12; rather, it is the standard required of a profession or group of its members as set out in codes of ethics or codes of professional practice.

Professional and trade associations adopt codes of ethics to assist members in determining the correct course of action in unclear circumstances. While failure to follow these codes of ethics can result in legal liability, the primary penalty is disciplinary action against members.

Ethical issues are closely related to conflicts of interest. A **conflict of interest** is a situation in which the professional has conflicting obligations to the public, the client, the employer, the profession, or him- or herself. An ethical problem can often be solved by identifying the conflicting interests and duties and then determining which duties or interests take precedence.

Breaches of ethics are difficult to define, but a professional should recognize one when it is present. However, the vague assurance that "you'll know it when you see it" provides little assurance. Some potential breaches are easy to recognize because they are so common (*e.g.*, acting in the same matter for two competing parties is obviously a conflict of interest). Normally, when duties are owed to more than one party simultaneously, a code is needed to determine which duty governs.

3.1 The Relationship between Ethics and the Law

There is a strong connection between ethics and law. That is because laws are derived and created based on a society's collective sense of morality. It is therefore not surprising that conduct which is considered unethical or, more specifically in the case of a design professional, contrary to the applicable code of ethics will usually attract other legal consequences as well.

Legal consequences for most members of the public can be divided into two categories: criminal consequences and civil liability. In the case of a design professional, there may also be a third category of consequences, and those are consequences that may arise out of disciplinary proceedings, including temporary or permanent loss of professional status. In terms of severity, criminal consequences are the most serious, because they can result in imprisonment. Civil consequences are usually in the form of monetary damages awarded against a defendant. Disciplinary consequences for breach of a code of ethics may be in the form of a reprimand, a suspension, a fine, loss of the right to practise, or a requirement to undergo further training or review.

It is easy to think of an example of conduct that can attract criminal, civil, and disciplinary consequences simultaneously. Fraud is such an example. An engineer who defrauds a client may be charged under the Criminal Code of Canada, may be sued for damages in court by the client, and may also face a disciplinary hearing. The commonly understood rule against "double jeopardy" (or, in this case, triple jeopardy) does not apply, as that rule merely prevents a person from facing criminal charges twice for the same offence. It does not prevent criminal charges from being brought at the same time as civil and disciplinary proceedings. The fictitious engineer in this example could be sentenced to a term of imprisonment, be ordered to pay damages to the defrauded client, and lose his or her licence.

Criminal conduct committed by a design professional relating to his or her work will, in the vast majority of cases, also be considered unethical. However, the converse is not true; not all unethical behaviour is criminal, and it is fair to say that most cases in which a design professional is found guilty of unethical conduct do not involve criminal prosecution. For example, an architect who acts for two clients simultaneously where those clients have competing interests would likely be found to have breached the applicable code of ethics, but only in rare circumstances would such behaviour be found to attract criminal liability.

It is also possible that criminal behaviour and criminal conviction in a matter that does not involve professional duties, such as fraud unrelated to the professional practice, could attract disciplinary consequences. But not all such behaviour would result in disciplinary proceedings.

Many matters that attract civil liability for damages, such as negligence, may also attract disciplinary consequences. Professional engineers and geoscientists are routinely disciplined by their professional bodies for negligent conduct.

The standards of proof differ from one type of proceeding to another. Criminal charges must be proved beyond a reasonable doubt, and civil claims must be proved to a lower standard of proof, known as "balance of probabilities." Disciplinary charges for unethical behaviour must be proven to a standard that falls somewhere between those two standards. Therefore, if a design professional is found guilty of criminal fraud, that conviction would almost always be sufficient to prove civil liability and could be relied upon by a disciplinary tribunal where the charges relate to the same conduct.

Rules almost always have exceptions. Because ethics is a matter of personal morals in many cases, there are exceptions to the rule that criminal behaviour is generally considered unethical. Countries and provinces may pass unjust legislation, such as the slavery laws that existed in the United States in the past, or the internment legislation in Canada during World War II. In such cases, civil disobedience may result in criminal conviction, but is not necessarily unethical.

From a practical perspective, when a design professional is accused of having done something unethical, this accusation should alert the accused to obtain legal counsel immediately and to notify his or her liability insurer. Liability insurance will, in some cases, provide coverage for defence of the accused in disciplinary proceedings. Whether coverage exists depends on the wording of the policy and the nature of the charges. Charges that relate to intentional behaviour such as fraud are not likely to be covered, whereas negligence is more likely to be covered. Care must be taken, if possible, to avoid having the evidence in one proceeding be used in other proceedings. An accused in a criminal case has the right to remain silent and avoid self-incrimination, and if criminal charges are pending simultaneously with disciplinary proceedings, the accused should apply to suspend the disciplinary proceedings until the criminal case is concluded. The same considerations apply to civil cases. If a civil lawsuit is pending at the same time as disciplinary proceedings, the accused should apply to suspend the disciplinary proceedings temporarily.

For certain professions, codes of ethics clearly identify to whom duties are owed and in what priority. For example, lawyers have very specific rules dealing with conflict of interest between two clients, and between the duty owed to the client and the duty owed to the court. Virtually all governing bodies for architects, engineers, and geoscientists have similar codes of ethics. For individuals who are not members of professional bodies and are therefore not bound by their rules, the law sometimes provides a legal remedy for unethical behaviour.[1]

For architects, engineers, and geoscientists, duties to the following parties can usually be found in the applicable code of conduct in each province:

a) duty to the public;
b) duty to the client;
c) duty to the employer; and
d) duty to the profession.

In addition, a professional owes a duty to him- or herself. However, that duty typically ranks lower than duties owed to other parties. Many of the toughest and most problematic ethical dilemmas occur when there is conflict between two sets of duties or roles.

This chapter examines each of the above duties individually, as well as the interrelationship between the duties. The duty to the public includes the paramount duty of protecting the safety and welfare of members of the public, as well as the duty to act with fairness and integrity toward members of the public. The duty to the client includes avoiding conflict between personal gain and the client's interest, as well as maintaining confidentiality. The duty to the employer is similar to that owed to the client, particularly regarding confidentiality and noncompetition. The duty to the profession is often a formalization of basic principles of common courtesy but may include an obligation to report misconduct of other members.

3.2 Codes of Ethics

Each provincial association governing the engineering profession has a code of ethics. All are based to some degree on the national Engineers Canada Model Code of Ethics, which states its philosophical principles in the preamble:

> Professional engineers shall conduct themselves in an honorable and ethical manner. Professional engineers shall uphold the values of truth, honesty and trustworthiness and safeguard human life and welfare and the environment. In keeping with these basic tenets, Professional Engineers shall . . . .[2]

The PEO[3] Code of Ethics is structured somewhat differently but contains the same general requirements. For example, the first paragraph of that Code contains the following requirements:

---

[1] The law may impose a fiduciary relationship upon parties in order to protect one of the parties; or it may impose one of the attributes of a fiduciary relationship, such as the requirement not to disclose or make use of confidential information.

[2] Excerpt from the Engineers Canada Model Code of Ethics.

[3] Professional Engineers Ontario.

It is the duty of a practitioner to the public, to the practitioner's employer, to the practitioner's clients, to other members of the practitioner's profession, and to the practitioner to act at all times with,

i. fairness and loyalty to the practitioner's associates, employers, clients, subordinates and employees,

ii. fidelity to public needs,

iii. devotion to high ideals of personal honour and professional integrity,

iv. knowledge of developments in the area of professional engineering relevant to any services that are undertaken, and

v. competence in the performance of any professional engineering services that are undertaken.[4]

The Code of Ethics for British Columbia engineers and geoscientists mirrors the Model Code more closely. For example, the first numbered paragraph after the preamble is identical to the Model Code: "Hold paramount the safety, health and welfare of the public, the protection of the environment and the promotion of health and safety within the workplace."[5]

The codes of ethics for architects follow the same patterns. Architectural codes of ethics contain conflict-of-interest guidelines, disclosure requirements, confidentiality provisions, honesty and impartiality provisions, courtesy to fellow architects, and reporting requirements where the architect becomes aware of a violation of the code of ethics. For the most part, the codes of ethics for geoscientists are identical to those that apply to engineers.

3.3 The Duty to the Public

Professional architecture, engineering, and geoscience regulatory bodies have been established to protect the public interest; therefore, it is not surprising that the codes of ethics place the duty to the public higher than any other duty.[6] Disregard for public safety can result in negligence claims and loss of the right to practise. This duty also encompasses the duty not to undertake work outside one's area of competence.

Professionals sometimes find themselves in difficult ethical situations. Sometimes employers and clients expect them to compromise their professional judgment to save money or further the client's or employer's interest. This type of pressure can put professionals in a further conflict if loss of employment is one possible outcome. However, in such cases, the professional's decision must be clear-cut: public interest in safety must take precedence over all other duties.[7]

Yet the duty to the public encompasses more than safety. It also includes a duty to act with fairness and integrity. Professionals must act independently and impartially to all parties. For instance, members of the public need to be able to rely on professional reports when making investment decisions and must be able to assume that the professional who wrote the report believed the contents of the report.

Negligence claims often focus on the duty to the public. In such cases, the plaintiff has to prove on a balance of probabilities[8] that the defendant failed to meet the standard of care expected

---

[4] http://www.peo.on.ca

[5] http://www.apeg.bc.ca/library.actbylawscode.html

[6] Engineers Canada uses the term "paramount" for that obligation.

[7] In some jurisdictions, this pressure is explicitly recognized and addressed. For example, laws in parts of the United States require engineers to "not permit a client, employer, another person or organization to direct, control or otherwise affect the registrant's exercise of independent professional judgment in rendering professional services for the client."

[8] The balance of probabilities test means that the claim is more likely true than not. In a criminal case, the burden of proof is beyond a reasonable doubt, which is a much more difficult burden to meet.

of an average professional in the field. For this reason, the plaintiff often calls an expert witness to provide an opinion. An allegation by one professional that the work of another is unprofessional is a very serious allegation and should not be made unless the breach is clear. Therefore, expert witnesses cannot act as hired guns, saying what they are told to say. Experts are to assist the court in understanding technical matters. When experts take on the role of an advocate, they lose their independence and their credibility.[9]

3.4 The Client's Interest

The duty to the client is usually second only to the duty to the public. The following duties to the client found in the Saskatchewan Code of Ethics for engineers and geoscientists are typical of those found in other codes of ethics:

> Members and licensees shall
>
> (a) act as faithful agents of their clients or employers, maintain confidentiality and avoid conflicts of interest;
>
> (e) conduct themselves with fairness, courtesy and good faith towards clients, colleagues, employees and others . . . ;
>
> (f) present clearly to employers and clients the possible consequences if professional decisions or judgments are overruled or disregarded . . . ;
>
> (h) be aware of, and ensure that clients and employers are made aware of, societal and environmental consequences of actions or projects . . . .[10]

Architects, engineers, and geoscientists also have a duty not to accept financial compensation from any other party whose interest conflicts with an existing client or to accept any assignment where personal interest may conflict with the client's interest. For example, if a client retains an engineer to obtain zoning approval for a development that will decrease surrounding property values, and the engineer owns property in the affected area, the engineer has a conflict between the client's interest and personal interest. However, by providing full disclosure of this conflict of interest to the client, the engineer may avoid a conflict-of-interest problem.

During the course of performing duties, the architect, engineer, or geoscientist may become aware of certain confidential information.[11] For example, an architect normally knows a client's construction budget. This information must remain confidential, because a contractor bidding or working on the project could make use of it to the owner's disadvantage. An exception to the duty of confidentiality is where the duty to the public demands disclosure.

3.5 The Employer's Interest

An employee is an agent of his or her employer. One of the attributes of an agency relationship is the duty of loyalty; the employee must be loyal to the employer. Loyalty means that the employee must put the employer's interest ahead of personal interest.

In many ways, the duty to the employer is similar to the duty to the client. For this reason, the duties found in the Saskatchewan Code of Ethics, as reproduced in the above section, apply to both clients and employers.

---

[9] See Chapter 14, pp. 162–163.

[10] APEGS Code of Ethics, s. 20(2)(a).

[11] Confidential information is discussed in detail in Chapter 21.

An employee's duty to the employer requires that the employee not accept financial compensation from any other party whose interest conflicts with the employer's interest, nor accept any assignment where personal interest may conflict with the employer's interest. The most common example of such a conflict is an employee competing with the employer, either by moonlighting or by quitting to pursue an opportunity found by the employer. For example, an employee engineer who accepts engineering assignments on a personal basis, even if the work is performed on weekends or in the evening, is technically in competition with the employer. As long as it is work that the employer could reasonably perform, then the employee's acceptance of the assignment privately without the permission of the employer constitutes a breach of the employee's duty.

One remedy for such a breach is a *constructive trust.*[12] A constructive trust is a trust implied by law in order to protect the rights of one party in a relationship; it may protect the rights of the employer if an employee improperly competes with the employer. Any profit earned by an employee in competition with his or her employer is held in trust for the employer, and the employee can be forced to forfeit those profits to the employer. Of course, the employee may obtain the consent of the employer in advance, in order to avoid this problem.

As with the duty to the client, employees have a duty of confidentiality to the employer. The issue of confidentiality often arises when employment is terminated. It is not uncommon for employers to require employees to agree that the employee will not compete with the employer for a set period of time. This agreement may be requested either at the time the employee is hired or at the time of termination, particularly while a severance package is being negotiated.[13]

3.6 Duty to the Profession

The professions of architecture, engineering, and geoscience are self-governing in the sense that once legislation is in place regulating the profession, the members oversee enforcement of the profession. The members elect a board or council, which in turn establishes committees to conduct investigations and disciplinary proceedings, verify qualifications of new members, draft bylaws, and perform related functions.

One of the consequences of self-governance is that members must accept the less-than-enviable duty of policing each other. For this reason, organizations enact statutes and bylaws which require all members to report certain types of unprofessional conduct of their fellow members. Some governing bodies even require self-policing by expelling members who fail to report malpractice claims against themselves.

In Alberta, the Code of Ethics for engineers and geoscientists defines the duty to the profession in this way:

Professional engineers, geologists and geophysicists shall uphold and enhance the honour, dignity and reputation of their professions and thus the ability of the professions to serve the public interest.[14]

In British Columbia, the Code of Ethics requires that members:

Report to the Association, or other appropriate agencies, any hazardous, illegal or unethical professional decisions or practices made by engineers, geoscientists or others.[15]

---

[12] Trusts are described in Chapters 17 (p. 199) and 18 (pp. 207–208).

[13] See Chapter 21, pp. 238–239.

[14] APEGGA Code of Ethics, Rule 5.

[15] APEGBC Code of Ethics, Rule 9.

However, policing other professionals requires caution. If accusing professionals do not have access to all of the relevant facts, their criticism of the work or opinion of another member of the profession may in turn be interpreted as unprofessional conduct. While the logic of this requirement is obvious, its application can be troublesome. Consider the dilemma faced by an expert witness with limited access to documents and other evidence who is asked to express an opinion in a courtroom. In fact, this is quite common in lawsuits, because it is the job of the lawyer to introduce into court as evidence proof of the facts underlying the opinion. If experts rely on facts not in evidence, the opinion may be worthless.

An expert must provide an opinion based upon a limited set of facts, and that opinion may impugn the professionalism of one of the defendants. In order to be fair to all parties, the expert should clearly state the assumptions upon which the opinion is based and should advise the client that the opinion may change if the facts are proven to be different.

3.7 The Dual Role of the Consultant as Owner's Agent and Impartial Arbiter

Certain functions of a consulting professional, such as providing advice about selecting contractors, inspecting work for deficiencies, and negotiating with regulatory bodies, are always performed in the client's interest. For example, the architect of record for a building project is almost always required under the terms of his or her contract to perform the following additional functions, each of which can substantially affect the rights of the contractor:

- evaluate and certify requests for progress payments;
- evaluate claims for extra payment and delay;
- issue the certificate of completion; and
- act as the first arbiter of disputes between the owner and contractor.

These requirements give the architect a significant role working with the contractor. In performing these functions, the consultant must not allow any perceived or real obligation to the owner to influence his or her decision such that it creates bias or a decision that is unfair to the contractor.[16] Certainly, the consultant must not prejudice the owner's position, for example, by certifying an amount for payment that exceeds the value of the work performed. However, consultants in this position favour the owner.

Claims for extra payment can place the consultant in a conflict-of-interest position. Three interests are at work when a dispute arises over a claim for extra payment. The consultant owes a duty to the owner to protect against unfounded claims. The consultant also owes a duty to the contractor to act fairly and impartially in evaluating the claim. The third interest is the consultant's self-interest.

The consultant's self-interest exists because approving a claim for an extra may mean that there was an error or omission in the original design. In effect, the consultant is being asked to evaluate the adequacy of his or her own design. If the extra is approved in circumstances where the adequacy of the design is called into question, the owner may sue the architect to recover the extra cost. At the very least, the implicit admission of an extra lowers the consultant's reputation and may affect future work.

If a consultant honestly believes that the contractor is entitled to an extra, regardless of the cause, he or she should approve the claim. However, the consultant should be careful not to make any statement that could prejudice his or her insurer's rights. If the contractor's claim raises the possibility of a claim against the consultant, the consultant should notify his or her insurer and seek legal advice first.

---

[16] The CCDC 2 contract, clause GC 2.2.6, states explicitly that the consultant must not show favour to either party when interpreting the contract.

CHAPTER QUESTIONS

1. A contractor has entered an agreement with an owner that contains a "no damages for delay" clause, precluding recovery by the owner for any delay, howsoever caused. The owner then causes an intentional delay, which results in severe financial consequences to the contractor. The contractor asserts a claim, reasonable in all the circumstances except for the no damages clause. The architect must evaluate the claim. How should the architect deal with it?

2. A geoscientist is asked to provide assay results for a mining exploration company, which will be used in a public disclosure. The company offers to pay the geoscientist in shares of the company, rather than cash. What is the acceptable course of action for the geoscientist?
   a) He/she cannot accept payment in shares.
   b) He/she can only accept payment in shares if he/she agrees to provide an unbiased report.
   c) He/she can accept the shares under any circumstances.

3. An architect is hired as the consultant under a CCDC 2 contract. The contractor claims that it was delayed by late approval of shop drawings. The architect should do the following:
   a) Reject the claim, because it might result in the architect being sued by the owner.
   b) Evaluate the claim on its merits.
   c) Approve the claim, so that it does not appear that the architect is biased.

4. Which of the following statements is true?
   a) Duty to the employer governs over the duty to the client in all cases.
   b) Duty to public safety governs over duty to the client in all cases.
   c) Duty to one's self governs over public interest in most cases.

5. An engineer is hired as an expert witness in a lawsuit. In the course of investigating the facts, the engineer discovers a serious structural deficiency. After the lawsuit is settled, as part of the settlement, the parties enter into a confidentiality agreement. The client demands that the engineer keep confidential all facts learned during the investigation. To whom does the engineer owe duties? Which duty prevails?

6. An engineer has designed formwork for a difficult concrete pour. The engineer submits the design to her superior, who suggests that smaller structural members can be used without violating the applicable code. The engineer rechecks her calculations, but remains unconvinced. What course of action should the engineer take? Does it make any difference if the superior is not a professional engineer?

7. An engineer employee has invented a new product for use in building construction. The invention was made during the course of employment, but has not yet been disclosed to the employer or anyone else. The engineer decides to quit his job and patent the invention as his own. Is there a breach of any duty? Discuss the ethical issues involved.

8. Which statement is true for a person who is *not* an architect, engineer, or geoscientist?
   a) He/she owes no ethical duty to anyone.
   b) He/she is not bound by the codes of ethics for those professions.
   c) He/she always owes a duty to comply with the codes of ethics of those professions.

9. During the course of employment, a geoscientist working for a mining company learns about a property rich in minerals. That geoscientist wants to quit her job and buy up some of the surrounding properties. In this situation, which of the following statements is false?

   a) The geoscientist owes a duty to her employer even after leaving her job.

   b) The geoscientist can buy up properties, as long as the former employer agrees.

   c) The geoscientist owes no duty to the employer after quitting.

ANSWERS

1. The architect must evaluate the claim fairly, in accordance with the terms and conditions of the contract.

2. A

3. B

4. B

5. The engineer owes duties to the public, the profession, and the client. The duty to the client suggests that the engineer keep the problem confidential; but the engineer's duty to the public and the profession suggest that the engineer should disclose the problem. Since the duty to the public is higher than the duty to the owner, the engineer should disclose the problem.

6. Public safety issues fall under the duty to the public. The duty to public safety demands that the engineer not reduce the size of the members and must re-evaluate the design with her superior. If the superior is also an engineer, then the junior engineer must also mention her professional requirement to disclose her concerns about her superior's failure to uphold the duty to the public.

7. The engineer owes a duty of loyalty to his former employer. The invention is likely owned by the employer or held in constructive trust by the engineer for the benefit of the employer.

8. B

9. C

CHAPTER 4

PROPERTY LAW

Overview

Property is divided into two categories: real property and personal property. **Real property** refers to land and anything attached to land, such as buildings. **Personal property** is everything else.

Buildings and personal property attached to land are considered to be part of the real property, as long as the reason for attaching them was to benefit the land. For example, many mobile homes begin life on wheels, but some of them are then bolted into concrete foundations, often without removing the wheels. The concrete foundation is clearly real property; but what about the mobile home? If the attachment is only temporary, and the purpose of the attachment is to benefit the mobile home rather than the land, then the mobile home remains personal property. However, if the attachment is more permanent and intended to benefit the real property, then the bolted-down mobile home becomes real property. The diagram in Figure 4-1 shows the different types of property interests.

Personal property consists of both tangible and intangible property. **Tangible property** is property that has physical attributes. **Intangible property** is everything else.

Tangible personal property is also known as **chattels**. For example, this textbook is tangible personal property. However, the words in this text are protected by copyright, which is intangible property. Assuming that you have purchased this text, you gained the right to read and use the text, and the right to sell or otherwise use the text. However, because of the protection given by the copyright in the text, your rights are restricted in that you cannot photocopy any substantial portion of the text or sell the right to reproduce it.

One specific form of intangible property is intellectual property. **Intellectual property** is the expression of an idea in a variety of physical forms. Each form is protected in a different way: words or data on paper or in electronic form are protected by *copyright*; an invention is protected by a *patent*; a business name or logo is protected by *trademark*; and the shape of industrial objects is protected as an *industrial design*.

Understanding the distinction between real and personal property is fundamental as the law dealing with each is quite different. It is also important to understand that through its life, property can change from one

FIGURE 4-1 Types of Property

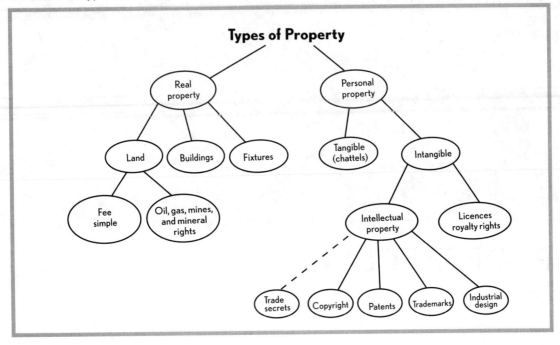

form of property to another. For example, a nail purchased from a store is a chattel, but it becomes part of real property, also known as a **fixture**, when pounded into a house. That same nail could become a chattel again if it is later removed from the house.

Property law is all about rights. One can think of the rights associated with property as being a pie, with each piece of the pie representing a different property right. Examples of these rights are use, sale, publication, reproduction, performance, exhibition, translation, and dissemination.

4.1 Real Property

Canadian law of real property is derived from English law. When Canada was formed, the government held all the rights to land. Over time, the government transferred some, but not all, of its rights to various parties. The rights that have been transferred to private parties are discussed below. Ultimately, however, the government has retained the ultimate right to all real property. In practice, the government would only assert such right through expropriation. In an expropriation, the government can take the land, but must subsequently pay fair compensation. The government did retain most of the interest in mines and minerals and then licences the right to extract such to private parties for a fee.

(a) Fee Simple Right

The greatest possible right is what most people think of as *ownership*, but which in law is known as a *right, estate,* or *interest* in *fee simple* or as a *freehold interest.* Except for governmental restrictions, such as planning requirements, environmental statutes, and matrimonial property rights, having a right in **fee simple** means that an owner can do almost anything with the property. This includes selling, leasing, occupying, or mortgaging. If the owner sell the fee simple right, the new owner holds that right.

The fee simple right can be held by more than one party at the same time. The law generally assumes that two or more parties who each have an interest in real property are **joint tenants**, meaning that they own the right concurrently, and that if one of them dies, the remaining joint tenants automatically inherit the interest of the deceased. The alternative is called **tenancy in common**, wherein each of the right holders owns a stated portion of the right, and if one of them dies, the heirs of the deceased tenant-in-common inherit his or her interest.

For example, if Joe and Mary are joint tenants, upon Joe's death, Mary inherits Joe's right and is the sole holder of the fee simple right. If Joe and Mary are tenants-in-common with each having rights to half of the fee simple interest, upon Joe's death, Joe's heirs, rather than Mary, inherit Joe's half fee simple interest.

(b) Mines, Minerals, Oil, and Gas

For the most part, the government has excluded from fee simple rights in real property the rights to minerals, oil, and gas, and air rights. While a fee simple right historically included an interest in the land from the centre of the earth into space, in current Canadian law, mineral, oil and gas, and air rights are generally not included in the fee simple right. For minerals, oil, and gas, the provincial governments hold some of the rights, and the federal government holds the rest. Governments often grant these rights through licences, for a fee to entitle the **grantee** (the party granted the right) to extract minerals, oil, and gas.

The only interaction between the fee simple holder and the holder of an interest in mines, minerals, and oil and gas rights is that the holder of the subsurface rights will need access through the surface of the fee simple holder's property to be able to extract the minerals or oil and gas from below the surface. Hence, the subsurface rights holder must first seek and obtain a **surface lease**, which provides access to the surface holder's property to permit subsurface access.

The provinces deal with various surface minerals in different ways. In many provinces, sand and gravel are not considered minerals and are owned by the fee simple holder of the surface rights.[1] For other materials, such as limestone, the determination of whether such are owned by the fee simple holder of the surface rights or by the Crown is different from province to province.

(c) Property Boundaries

All property has boundaries. Most boundaries are defined using a survey that is recorded on a detailed map or plan. Plans are kept at a land registry or land titles offices and at offices where mineral rights are recorded.

Each piece of property also has neighbouring property and is entitled to lateral support. **Lateral support** means that the land and buildings on a property will always have the same or greater support as when they were first built or purchased. Thus, a neighbour cannot undermine that support by digging in a manner that would have the effect of damaging the adjacent property. When lateral support is going to be disturbed (for instance, by digging the basement for a new house that is close to the edge of the property), it is common for the neighbouring parties to enter into an agreement to ensure support during and after construction, perhaps even including the use of anchors.

If the neighbouring property is water, or if water runs through a property, special rights are involved. **Riparian rights** provide rights to use water on or adjacent to a property. If the water is either tidal or navigable, the bed underneath the water to the average high-water mark, or, in some provinces, a specified distance above the high-water mark, is owned by either the federal of provincial government. If the water is neither tidal nor navigable, then the water and the bed underneath are owned by one rights holder, where the water is entirely on his or her property, or are shared by two

---

[1] See, for example, the Manitoba *Sand and Gravel Act*, C.C.S.M. c. S15.

owners, where the water is a boundary. In the case of tidal or navigable waterways, or where the water is a boundary, the neighbouring rights holder is entitled to use a reasonable amount of the water.

Air rights are rights that extend above the physical property to a reasonable level above the property. The principal purpose of air rights is to prevent a neighbour from erecting a building that, for instance, blocks all sunlight from reaching a neighbouring property. Where a property owner is going to infringe on the air rights of a neighbour, both parties usually enter an agreement. Air rights are also important in the creation of air space parcels. An **air space parcel** is an artificial layer of property above the land. Air space parcels are used to create different certificates of title or registry entries for different elevations on the same property. These titles would, for instance, permit one party to hold fee simple rights to the first floor of a building and a second party to hold fee simple rights to the second floor.

(d) Rights that Run with the Land

Some real property rights are said to *run with the land*, meaning that they are not just enforceable between the two parties who first agreed to them. Rather, rights that run with the land continue to affect the land despite a sale of the property to a new owner who takes the fee simple right. The new owner in fact buys the property subject to any right that runs with the land. On the other hand, interests that do not run with the land are simply contracts between two parties; they are only valid as between the parties to the contract and not binding on future owners.

(e) Registration of Real Property Rights

The western provinces use the *torrens system* for registering land rights, which is a system run by the government that creates and maintains certificates of title at land titles offices for every property. In a torrens system, the certificate of title shows the current holder of the fee simple right and lists all other rights held in a property, such as leases and mortgages. The certificate of title shows any discharges of rights, such that anyone reviewing the certificate of title can rely on the certificate of title to be up to date.

Parts of Ontario and most of the eastern provinces use a simpler registry system, in which a registry office maintains a register of only land-related rights or interests. In such a system, anyone wanting to know about other interests in a property must review the historical records to determine those interests. Unlike a torrens system, this system does not allow a person to simply search a title to an individual piece of property and determine the interests that are registered against that property. The registry is just a chronological listing of all land interests. Only rights that run with the land are registered at the registry office or on the certificate of title.

For a fee, anyone can review records maintained at a land titles office or a registry office. The records are not confidential. This means that anyone can conduct a search of a piece of property to determine whether the holder of the fee simple has **clear title**, meaning that there are no other interests in the property or charges against the property. This ability to search land titles is obviously important when someone is purchasing a piece of property to determine whether interests such as a mortgage will remain on title after the purchase. Often, conditions of purchase require that mortgage holders be paid out and interests discharged at the time when the property right is transferred.

While most of the property in Canada has been surveyed and has certificates of title or registry entries, some areas are **unpatented**, meaning that there is no certificate of title or registry entry. Property interests, such as mortgages or builders' liens, cannot be registered against unpatented lands. For example, the land that makes up most Aboriginal reserves is generally unpatented and cannot be mortgaged or liened.

Certificates of title for fee simple holders also often include the phrase "Mines and Minerals excepted," which indicates that the holder of the fee simple right does not have an interest in the mines and minerals, including oil and gas. Provincial governments maintain separate registries for mining, mineral, and oil and gas rights.

With all real property rights, new owners must register an interest as soon as they obtain it. Although subject to many exceptions that are beyond the scope of this text, the general rule is that the first to register an interest in real property has priority over subsequent registrations. For example, if a lien is registered before a mortgage, the mortgage interest holder ranks below the lien holder.

(f) Specific Real Property Rights

(i) **LEASE** A **lease** provides a party with an exclusive right to occupy all or part of the property. A lease generally runs with the land, and a new purchaser of the property takes his or her fee simple right subject to the lease. Depending on the province and the length of the lease, some leases can be registered while others cannot. This means that some leases, generally those of a short duration, may not be registered but still run with the land.

(ii) **MORTGAGE** A **mortgage** is a form of security for a loan that provides the lender with the right to be paid out of the value of the property when the property is sold, in priority to the holder of the fee simple right. Upon repayment of the loan, a mortgage is normally discharged.

(iii) **EASEMENT** An **easement** is the right to use a neighbouring piece of property, generally for the purpose of crossing it. An easement generally runs with the land and is tied to both pieces of property—the property that is being crossed and the property that is benefiting from the crossing. For example, a piece of property without access to a public road needs an easement across a neighbouring property for access to a road. Most easements are in writing and registered at the land titles or registry office; but legal easements can be created by repetitive and historical use as well.

A **right of way** is an easement that grants a right to cross land. Rights of way are used not only for private parties, but also for public utilities such as roads, power lines, and telephone lines.

(iv) **RESTRICTIVE COVENANT** A **restrictive covenant** is a set of conditions imposed on one piece of property for the benefit of a neighbouring property. A **negative restrictive covenant**, one that imposes a requirement to refrain from doing something (such as building a building), generally runs with the land; whereas a **positive restrictive covenant,** one that imposes a requirement to do something (such as building a building), generally does not run with the land. The difference between an easement and a restrictive covenant is that the conditions of a restrictive covenant can be very broad. For example, a restrictive covenant may prohibit building a building on the property, or may prohibit certain paint colours for houses on that property.

(v) **LICENCE** A **licence** is a right to use property that does not run with the land. A licence is a contract only between the holder of the fee simple right and the licensee and is not binding on future holders of the fee simple right. Licences can be used for simple access or can grant rights to harvest trees or wildlife.

(vi) **PROFIT À PRENDRE** A *profit à prendre* is a combination of an easement and a licence. A *profit à prendre* resembles an easement because it runs with the land; but it resembles a licence because it usually deals with rights such as harvesting trees or removing gravel.

(vii) **CONSTRUCTION LIENS** A construction or builders' lien is a charge or interest in land from a builder who has not been paid. The right to register a construction or builders' lien is provided by statute in each province. Construction liens are discussed in Chapter 18.

(g) Condominiums or Stratas

A condominium or strata is a legal structure for the ownership of land that permits fee simple rights for portions of the land. A condominium or strata corporation is formed by statute and owns all the bare land, the common areas (*e.g.*, the hallways), and the outer portion of the exterior wall. Individual units are created by the property developer to give each unit a fee simple title that can be sold. The owners of the individual units each have partial ownership of the corporation and have obligations to it, such as payment of monthly fees for maintenance of the corporation's property. Much like municipalities have the right to control development on individual pieces of property, the corporation can pass bylaws restricting the use of the individual units. These bylaws are often quite detailed and can include matters such as noise restrictions, the number of people who can live in a unit, and restrictions on leasing.

4.2 Chattels

Chattels are tangible personal property. Also called *goods*, chattels can be owned and the owner or owners can have many rights, including the right to use, lease, rent, or sell the chattel.

Unlike real property, chattels have no registry for ownership. Provincial government offices called personal property registries do not in fact register personal property. Personal property registries record only charges or interests in personal property, rather than property owners. Searches at a personal property registry use the name of the person or corporation. The search results do not list all the personal property owned by that person or corporation; rather, it lists the interests or charges granted by that person or corporation in their chattels or in a specific chattel. For example, interests in chattels are usually granted when a person or corporation borrows money. If a person wants to borrow money to buy a car, the lender has the borrower sign a document granting the lender an interest in personal property. This interest is recorded at the personal property registry.

As with real property, an interest in personal property should be registered at the personal property registry as soon as it is obtained. The general rule is that the first to register an interest in personal property has priority over subsequent registrations.

The sale and purchase of chattels is governed by legislation in each province known as a sale of goods statute. A sale of goods statute is a complicated piece of legislation that traces its history back to English law before Confederation. These statutes normally create implied terms in contracts involving the sale of chattels, such as implied warranties of fitness for purpose, warranties of merchantability or quality, and terms relating to the passing of title to the goods. Many cases have provided more precise meanings of these statutes in different circumstances. However, detailing sale of goods legislation and its related case law are beyond the scope of this text.

4.3 Intellectual Property

As with other forms of property, intellectual property is a set of rights, often referred to as a *bundle* of rights in the same sense as real property rights bundles.

Intellectual property law does not protect ideas, but rather protects the expression of ideas. For example, the concept of the world's best mousetrap cannot be protected. Only when the inventor is producing or has documented the process of creating the mousetrap can he or she seek protection through a patent or copyright. The only way that ideas themselves can be protected is through the concepts of confidential information and trade secrets discussed in Chapter 21.

Each form of protection for intellectual property discussed in this chapter is defined by federal legislation. Additionally, international treaties Canada has entered into provide international rights for intellectual property, although parties seeking to protect intellectual property should specifically review those rights and not presume that protection based on international treaties exists.

(a) Copyright

(i) **DEFINITION** Under the *Copyright Act*,[2] **copyright** protects the expression of words and data in original literary, musical, dramatic, and artistic works, and must be in written form, or performed, recorded, or communicated in a form such as radio or television. Literary works include computer programs as well as printed literature, while dramatic works include movies, radio, television, and live performances. As stated above, the ideas behind these works are not protected, meaning that anyone may reproduce the same concepts using different words without breaching copyright.

(ii) **REQUIREMENTS TO GAIN PROTECTION** To access copyright protection, an author must ensure that the work is original and permanent, using the author's own work or skill. It must also be published,[3] meaning that copies are provided to the public. The work must also fall within the categories listed above.

Copyright protection does not require registration. The copyright register in Ottawa is used mainly for works such as books, movie scripts, and music. All an author has to do to gain international copyright is to place in the work the © symbol or the word copyright followed by the author's name and the year of publication (*e.g.*, © DEF Company Ltd. 2005, Calgary, Alberta).

(iii) **PRINCIPAL RIGHTS** The principal economic rights provided by copyright protection are publication, reproduction, performance, exhibition, translation, dissemination, and authorization of such uses. It is assumed under the *Copyright Act* that the original author is the owner of the copyright,[4] but this assumption may not be true for commissioned engravings, photographs, or portraits,[5] or work created during employment.[6] Contracts involving copyrightable material should always specify who owns what rights and under what conditions. For example, in a design contract, the client may want to own the copyright, in which case the author would have to assign it to the client or be satisfied with a licence to use the work. Other important rights include the right to convert one copyrighted work to another form, such as converting a book into a movie.

An author also has moral rights in a work. **Moral rights** protect the artistic integrity of the work, such as the manner in which a painting is displayed. This right cannot be assigned to someone else, but it can be waived. Moral rights include the right to prohibit distortions, mutilations, and modifications, and the right to prevent others from claiming authorship.

Copyright protection is not unlimited. The only rights protected are those listed in the Act. There are a number of defences to copyright infringement, which are beyond the scope of this text.

(iv) **TERM OF PROTECTION** Copyright is protected for a term of 50 years after the end of the calendar year during which the author or last living author dies.

(v) **PLANS AND SPECIFICATIONS** A professional generally owns copyright in plans he or she prepared for a client. However, this ownership is usually specified by contract. If an employee

---

[2] R.S. 1985, c. C-42.

[3] See *Copyright Act, s.* 2.2.

[4] See previous note at s. 13(1).

[5] See previous note at s. 13(2).

[6] See previous note at s. 13(3).

creates plans during the course of employment, the employer generally owns the copyright. It is prudent to deal with copyright and ownership issues in employment or work contracts.

(b) Patents

(i) DEFINITION Patents under the *Patent Act*[7] protect inventions, which are defined as "any new and useful art, process, machine, manufacture or composition of matter, or any new and useful improvement in any art, process, machine, manufacture or composition of matter."[8]

(ii) REQUIREMENTS TO GAIN PROTECTION In order to qualify for patent protection, an invention must be new, such that it is not generally known to the public; it must be useful, such that it must have a commercial application and actually do what it claims to do; and it must be inventive. The process for registering a patent with the patent office in Ottawa is much more stringent, lengthy, and costly than the copyright process. Given these factors, some companies decide to keep their otherwise patentable inventions as trade secrets.

In Canada, if the applicant discloses an invention to the public before filing a patent application, the applicant has one year to file the application or forfeits the ability to patent it. If a third party discloses an invention to the public before the applicant files a patent application, the ability to file a patent is automatically lost.

(iii) PRINCIPAL RIGHTS The principal rights protected are the exclusive right to make, sell, and use the invention. In return, the invention must be sufficiently described to permit others to make the invention after the term of protection expires. This means that an individual or organization must take significant measures to protect a patentable invention from public disclosure prior to filing the patent application.

As between an employee and employer, the presumption is that the employer holds all rights to an invention developed in the course of employment. This may extend to an invention created on an employee's own time as long as it relates to the employment. If the parties intend for an employee to have such rights, this should be clearly dealt with in the employment contract.

(iv) TERM OF PROTECTION The term of protection for any new patent is 20 years from the date that the application is filed.

(c) Trademark

(i) DEFINITION Under the *Trade-Marks Act*,[9] **trademarks** protect marks used to distinguish goods or services. They are most commonly used to protect the name associated with a product, a service, or a product and a service, but they can also cover designs.

(ii) REQUIREMENTS TO GAIN PROTECTION A trademark may be registered if a company or individual has been using it[10] or making it known in Canada. They can also register their trademark through international treaties. Trademarks must be distinctive, although multiple trademarks can be issued to different applicants for the same trademark as long as the applicants can show that there is no likelihood of confusion between the two trademarks (for example, the products or services are used in different markets).

---

[7] R.S.C. 1985, c. P-4.

[8] See previous note at s. 2.

[9] R.S.C. 1985, c. T-13.

[10] See previous note at s. 4.

(iii) PRINCIPAL RIGHTS A registered trademark provides the owner with the exclusive right to use the trademark in association with specific goods and/or services.

(iv) TERM OF PROTECTION The term of protection is 15 years, but it can be renewed indefinitely. However, owners must use trademarks and actively protect them by monitoring and prosecuting infringements. Failure to use or protect a trademark results in loss of the trademark.

(d) Industrial Design

(i) DEFINITION The *Industrial Design Act*[11] protects the shape, configuration, and general look of mass-produced items. Examples of items protected by the Act include furniture, toys, household items, and vehicles.

(ii) REQUIREMENTS TO GAIN PROTECTION Designs must be novel and original but cannot be purely utilitarian. The shape, configuration, or general look must have an aesthetic purpose.

(iii) PRINCIPAL RIGHTS A registered industrial design provides the owner with the exclusive right to use the design.

(iv) TERM OF PROTECTION The term of protection is 10 years from the date of registration.

(e) Integrated Circuits

(i) DEFINITION The *Integrated Circuits Topography Act*[12] protects the design of integrated circuits. An integrated circuit is a series of layers of semi-conductors, metal, insulators, and other materials, and topography refers to the configuration of these layers.

(ii) REQUIREMENTS TO GAIN PROTECTION Originality is the key element to integrated circuit protection. The topography cannot simply be a reproduction of another circuit design.

(iii) PRINCIPAL RIGHTS The owner of the rights has exclusive right to reproduce and manufacture the protected integrated circuit topography.

(iv) TERM OF PROTECTION The term of protection is 10 years from the date that the application is filed.

(f) Infringement and Remedies

Infringing any of the types of intellectual property occurs when someone, without consent from the owner of the intellectual property, does something (*i.e.*, copying a copyrighted document) that only the owner (or a licencee) has the right to do. An owner's remedies could include an injunction preventing further infringement, an order to obtain all copies of the infringing material, and damages. Damages for infringing some intellectual property rights could be calculated based on the loss of profit of the owner or, in some cases, the profits earned by the infringing party. There are also potentially serious penalties under the various intellectual property statutes.

---

[11] R.S.C. 1985, c. I-9.

[12] S.C. 1990, c. 37.

CHAPTER QUESTIONS

1. Who owns an unpublished design drawing?
   a) The originator (*i.e.*, design engineer)
   b) The party who makes the copy of the drawing
   c) The client who hired the engineer to do the design
   d) None of the above.

2. What is the difference between real property and personal property?

3. Why is it important to register property at land registries, land titles offices, and personal property registries as quickly as possible?

4. Which of the following is not an example of intangible property?
   a) Copyright
   b) *Profit à prendre*
   c) Licence
   d) Patent
   e) None of the above

5. If a land owner wants to excavate the property, and such excavation requires rock anchors to be installed into the neighbour's property, it is necessary to obtain permission for this reason:
   a) Failure to do so would be a trespass.
   b) Without permission, there would be insufficient lateral support.
   c) It would violate the neighbour's air rights.

6. The owner in fee simple of a property has this or these rights:
   a) To sell it
   b) To mortgage it
   c) To mine it for gravel
   d) All of the above
   e) Only A and B

7. If a holder of a fee simple interest in real property fails to properly support her neighbour's property, is she potentially liable? Why or why not?

8. Give an example of property that begins as a chattel and then becomes real property.

9. Is it necessary to register at the Copyright Office to gain copyright protection? Describe the requirements to obtain copyright protection.

10. What is the purpose of a patent?
    a) To protect the idea that an inventor came up with
    b) To protect the expression of an idea
    c) To prevent a person from copying a drawing
    d) None of the above

ANSWERS

1. A

2. Real property is land, or something affixed to land for the purpose of benefiting the land. Personal property is all other types of property.

3. The time for registration is important because it often determines the rights of parties.

4. E

5. A

6. D

7. Yes. All property owners owe a duty to properly support their neighbour's property.

8. A piece of drywall, once installed on a property, becomes part of the real property.

9. No. In order to gain copyright protection, the work must be published with the © symbol together with the author's name and the year of publication.

10. B

CHAPTER 5

BUSINESS ORGANIZATIONS

Overview

There are many different methods of organizing a business. Each method has inherent strengths and weaknesses.

The three principal forms of business organization in Canada are sole proprietorships, partnerships, and corporations. **Sole proprietorship** is the legal term describing an individual carrying on business. The individual then gains all of the profits and is responsible for all debts and obligations of the business. A **partnership** is a group of individuals or corporations that pool their resources to run a business. The partners share all of the profits and are jointly responsible for all debts and obligations of the business. In contrast, a **corporation**, or **company**, is a separate legal "person" created pursuant to federal or provincial statutes. A corporation is owned by shareholders and has directors and officers who control the business. The corporation gains all of the profits and is responsible for all debts and obligations of the business.

A **joint venture** is a fourth option. A joint venture is an organization where two or more joint venturers act together but retain their separate legal status.

5.1 Sole Proprietorships

A sole proprietor is an individual carrying on business either in his or her own name or under a business name. The sole proprietor obtains all of the profits and takes all of the losses of the business. In fact, there is no difference between the individual and the sole proprietorship. All of the individual's assets are at risk for the liabilities of the business.

A sole proprietor can register a **business name**, which is simply a trade name under which the individual carries on business. The business name does not change the status of the business, and the sole proprietor remains liable for all of the business obligations. Provincial legislation generally requires a business name to be registered. For example, J. Smith could register as Apple Engineering or JS Engineering or any other name, provided, however, that a business name does not violate trademarks and will not be confused with other corporate or business names in the province of registration. As well, statutes may place restrictions on use of profession-related words, such as "engineering" or "architecture."

Sole proprietors must also follow all other legislative requirements for a business, including obtaining a business licence. If sole proprietors hire people, they are bound by mandatory labour legislation.

The principal disadvantage of operating as a sole proprietor is liability. The individual is completely and fully liable for all debts and obligations of the business. A second potentially serious disadvantage is that at certain income levels, individuals are taxed at a rate higher than corporations. The advantages of operating as a sole proprietor are that the set-up costs are lower than those of a corporation, there are no corporate reporting and filing requirements, and, at low income levels, there may be tax advantages. Compared with partnerships, sole proprietorships have the advantage of not exposing an individual to liability for mistakes made by partners.

5.2 Partnerships

Partnerships are governed by the *Partnership Act* in each province. It is a good business practice for partners to enter into a partnership agreement that governs the way the partnership conducts business. The partnership agreement generally sets rules including authority to sign contracts, the division of profits, procedures for business decisions, and procedures for dissolving the partnership.

Subject to the terms of the partnership agreement, each **partner** has authority to enter into contracts and carry on business in the name of the partnership. Even where the partnership agreement limits the authority of the partners, parties doing business with the partnership are unaware of the terms of the partnership agreement and therefore assume that every partner has unlimited authority and liability. This joint authority is very important, because each partner is jointly liable for all of the debts and obligations of the partnership, and such liability is not limited to that partner's proportionate share. For example, if Smith of Smith & Jones buys a new computer system for the business, the computer store can sue both Smith and Jones for the unpaid amounts for the computer system. However, if the computer system was purchased for Smith's personal use and not as part of the business, the computer store can only pursue Smith.

(a) Forming a Partnership

The circumstances of a partnership determine whether it is a legal partnership. If two or more parties sign a partnership agreement, they are partners, although other indicators of partnership still have to be present.[1] However, even without a partnership agreement, they may still be considered partners under the terms of a *Partnership Act*. If two or more parties jointly participate on a continuing basis in the management of a business, declare themselves to be partners, and/or make a joint monetary contribution to start a business, many *Partnership Acts* consider them legal partners. For example, if three musicians put on a concert, they are unlikely to be considered a partnership; but if they put on a series of concerts, the law may consider them partners.

Most provinces require partnerships to register. Like sole proprietors, partnerships must follow all other legal requirements, including obtaining a business licence. Partnerships that hire people are also bound by mandatory labour legislation.

Partnership Acts have a number of assumed rules in common. But these rules can often be

---

[1] *Backman v. Canada*, [2001] 1 S.C.R. 367.

modified by a partnership agreement. For example, *Partnership Acts* generally assume that each of the partners will gain an equal share of the profits; but the partnership agreement could replace that assumed provision with an unequal split of profits. Prospective partners should review their province's *Partnership Act* before agreeing to form a partnership, especially if they plan not to have a partnership agreement.

Like individuals, companies can become partners. Partnerships can also consist of individuals and companies. For example, architectural partnerships often comprise personal corporations, each corporation being owned by either an architect or an architect's spouse.

Professionals considering a partnership need to consider the advantages and disadvantages relative to other business forms. One of the advantages of a partnership over a corporation is the lack of incorporation costs and ongoing filing fees. However, partners face potentially greater liability than shareholders, and therefore liability is seen as a disadvantage of partnerships. One advantage over sole proprietorships is the sharing of expenses, risk, and expertise. However, sharing of risk is a double-edged sword, and can be both an advantage and a disadvantage. Finally, taxes for partnerships can be complicated. Prospective partners should consult a tax accountant or a tax lawyer about the tax consequences of using a partnership.

(b) Fiduciary Duty

A fiduciary duty[2] is a relationship of special trust in a partnership. Each partner owes the other partners a fiduciary duty. For example, fiduciary duty forbids partners from operating separate competing businesses and from taking profits of the business solely for themselves. In addition, they must always act in the best interest of the partnership and declare conflicts of interest.[3] Moreover, a partnership agreement cannot negate or alter fiduciary duties.

(c) Limited Partnership

Limited partnerships have at least one general partner and one limited partner. A **general partner**, who also runs the business, has unlimited liability. **Limited partners** are liable only for their cash contribution to the business[4] and, in some provinces, for the profits that they derive from the business. Limited partnerships are available in every province and provide some of the benefits of a corporation to the partnership structure.

Limited partners cannot actively participate in the business, or they lose their limited liability status. Active participation includes being involved in controlling the business or providing services to the business.

(d) Limited Liability Partnerships

In some provinces, professionals can form limited liability partnerships. A **limited liability partnership** is a partnership in which the liability of a partner is limited only to the liability of that partner. If Joe, Hanna, and Erin are limited liability partners and Joe is negligent, only Joe's assets and the assets of the partnership as a whole are at risk; Hanna's assets and Erin's assets are not at risk. Limited liability partnerships must carry a prescribed level of liability insurance.

---

[2] See Chapter 12, p. 141.

[3] See Chapter 3.

[4] For example, see in Ontario the *Limited Partnership Act*, R.S.O. 1990, c. L.16, s. 9.

5.3 Corporations

A corporation is a legal entity that permits large numbers of individuals to invest in a common business venture, while limiting the liability of the business to the new legal person, the corporation. A corporation must use one of the words Corporation, Corp., Incorporated, Inc., Limited, or Ltd. as part of its name in order to provide notice to parties doing business with it that liability for corporate debts and obligations is limited to corporate assets and does not extend to the shareholders, officers, directors, or employees.

Corporations can be created pursuant to the federal *Canada Business Corporations Act*[5] or provincial statutes. In either case, a corporation must be registered in every province in which it does business.

Provincial statutes governing the practices of engineering, architecture, and geoscience limit the rights of corporations to practise a profession, and require corporations employing professionals to register or have a professional licence to practise where those corporations offer professional services to the public. These requirements are in addition to the registration or licensing requirements for individuals employed by those corporations. With respect to the professions of engineering and geoscience, a corporation may be required to obtain a Certificate of Authorization (in Alberta, the document is referred to as a permit) from the governing association that entitles it to practise, and such certificates are issued only where the corporation meets specified criteria, including having registered or licenced individual members on staff. With respect to the practice of architecture, the governing legislation in many provinces contain similar requirements, including the requirement that the majority ownership of the corporation be held by licenced architects. Care should be taken to examine the requirements before establishing any professional corporation in a particular province, as the requirements vary from province to province.

(a) Separate Legal Entity

A corporation is considered a separate legal person. For example, Bill Smith may own all of the shares of Smith's Painting Ltd., but that does not mean that Bill *is* Smith's Painting Ltd. For this reason, Bill and Smith's Painting Ltd. must keep separate records and separate bank accounts and must act as separate persons. If Bill treats his corporation as himself, then Smith's Painting Ltd. risks losing its limited liability status, and Bill could become personally liable for the company's debts and obligations.

However, even though legally Bill and Smith's Painting Ltd. are separate legal persons, some liabilities are shared, even if Bill maintains appropriate separation. For example, if Smith's Painting Ltd. is a small company, its bank and many of its larger suppliers will probably require Bill to sign a personal guarantee for the debts owed by Smith's Painting Ltd. before they advance funds or supplies because they suspect that Bill's Painting Ltd. does not have significant assets.

In addition, some statutes ignore the limited liability status of a corporation and place personal financial responsibility on the directors and officers. These exclusions to what is known as the corporate veil (the limited liability shield of the corporation) are discussed below.

There are circumstances under which a court will treat both the corporation and its principal shareholder(s) as one person, and, in effect, ignore the fact that they are separate legal entities. This is known in law as "piercing the corporate veil." These circumstances are usually limited to acts of fraud. Where a company is either set up with the intention of perpetrating a fraud or used as a vehicle to perpetrate a fraud, a court may pierce the corporate veil in order to hold the individual

---

[5] R.S.C. 1985, c. C-44.

who directed the fraud personally responsible. Cases in which the court has pierced the corporate veil typically involve closely held companies, for example, companies having one or a very small number of individuals as shareholders and directors.

(b) Corporate Organization and Control

Corporations are owned by **shareholders** in proportion to their shareholdings. Individuals, partnerships, and corporations can be shareholders. Shareholders vote to elect directors and authorize fundamental changes to the corporation.

Directors are individuals who direct the business of the corporation. The directors elect **officers**, often from amongst themselves, such as the president, vice-president, and corporate secretary, who provide a closer operational direction for the business. Officers and directors must be individuals and cannot be corporations.

Legal documents create the corporation and define the types and number of shares to be issued and the name of the corporation. These documents are called *articles of incorporation, memoranda of association,* or *letters patent,* depending on the jurisdiction where the corporation is first registered. Bylaws and articles of association created when the corporation is formed list more detailed rules relating to the authority of officers and directors and the holding of meetings.

(c) Capacity

Like a sole proprietor or partner, a corporation can enter into contracts. Much like a partnership, wherein each partner has authority to enter into contracts, a corporation's employees, officers, and directors have authority to enter into contracts on behalf of the corporation. But this authority is usually limited and defined. While most day-to-day contracts, such as the purchase of office supplies, can be authorized by employees, the corporate rules may restrict individual authority for contracts above a specific monetary value or of a certain type. For example, a corporation's bylaws or articles of association may require the signatures of two signing officers and the issuance of a resolution of directors for the execution of contracts in excess of $50 000. When conducting business with a corporation, if there is any question about signing authority, a party should obtain a corporate resolution authorizing a transaction and indicating the appropriate signing parties.

(d) Debt and Equity

A corporation's value is based on its debt and equity. **Debt** is an obligation to pay or render something, generally money, to someone else. **Equity** is the residual value of a property or business after deducting mortgage and liability costs. *Debt financing* usually comes from a bank, while *equity investment* usually comes from shareholders. For example, shareholders in a start-up corporation often contribute to a corporation's equity through shareholders' contributions. A corporation also creates equity through the earning of profits.

(e) Public and Private Corporations

A **private corporation**, often called a **closely held corporation**, is a corporation in which all of the shares of the corporation are held by a small group of shareholders. The articles of incorporation for a private corporation have significant restrictions on the use of shares: for example, shareholders cannot offer them to the public for sale. One advantage of a private corporation is that documents such as records of profits and losses do not have to be made public. However, the names and addresses of directors, officers, and shareholders must always be available for public inspection.

In contrast, a **public corporation** is one in which shares are publicly traded, generally on a stock exchange. Public corporations operate under stringent legislative rules, including public access to and filing of accounting records.

(f) Officers, Directors, and Shareholders

A corporation's officers, directors, and shareholders must obey specific rules governing their actions. If a professional is considering becoming an officer or director, he or she should review the specific obligations under federal and provincial legislation before accepting the appointment. In addition, officers and directors owe fiduciary duties to the corporation, as discussed below. Historically, lawyers were often eager to serve as directors for large clients because doing so can reinforce the close connection that often exists between corporate counsel and his or her client. But because of the increasing risk of lawsuits and other liabilities, the role of officer and director in both for-profit and not-for-profit corporations is becoming less attractive. For examples, in sports leagues, many people have started refusing to accept roles as officer or director. Other professionals, such as architects, engineers, and geoscientists, should exercise similar caution before accepting an appointment as an officer or director.

(i) **FIDUCIARY DUTY** Officers and directors owe a fiduciary duty[6] to the corporation. For example, officers and directors cannot operate separate competing businesses or take profits of the business solely for themselves. In addition, they must always act in the best interest of the corporation, be loyal to the corporation, act honestly and in good faith, and declare all conflicts of interest. These obligations are absolute; but they are only owed to the corporation, not to its shareholders. As a result, only the corporation can sue officers and directors for breaches of these duties. However, forcing a corporation, which is controlled by the officers and directors, to sue those same officers and directors can be difficult. Thus, in order to provide a remedy for oppressed minority shareholders in a corporation reluctant to sue its officers and directors, some provinces now permit *derivative actions*, which are court actions by the oppressed minority shareholders against a corporation's officers and directors.

(ii) **CONFLICTS OF INTEREST** Conflicts of interest are discussed in detail in Chapter 3. In general, for corporations, a conflict of interest occurs when the interests of the corporation conflict or could conflict with the interests of the officer or director. For example, if a person with a business selling office supplies is also director of a corporation about to enter into a long-term contract for the purchase of office supplies, then that person has a conflict of interest. He or she must declare the conflict of interest in the purchase discussions and avoid participating in any way in the discussion or decision about the contract.

Conflicts of interest can also occur when the personal interest of the officer or director is in conflict with his or her duty to the corporation. For example, a director should avoid accepting personal gifts from parties doing business with the corporation. Officers and directors must ensure that they do not accept gifts from parties that are capable of influencing their decisions.

(iii) **GOVERNMENTAL LIABILITIES** Several statutes and regulations place specific obligations and liabilities on officers and directors. For example, labour standards legislation generally places an obligation for unpaid employees' wages on officers and directors if the corporation goes bankrupt and fails to pay such wages. Other legislation, such as the environmental legislation discussed in Chapter 23, places due diligence obligations on officers and directors.

---

[6] See Chapter 12, p. 141.

Under professional and other statutes, such as the *Aeronautics Act*,[7] the *Defence Production Act*,[8] and the *Export and Import Permits Act*,[9] directors and officers of a corporation that has committed an offence can be found individually guilty of the same offence as the corporation. Directors and officers also have specific duties under the *Bank Act*,[10] the *Bankruptcy Act*,[11] and the *Consumer Products Warranties Act*.[12] In addition, officers and directors can also be held liable for unpaid wages under provincial labour standards acts; unpaid GST under the *Excise Tax Act*;[13] failure to remit Canada Pension Plan deductions; failure to maintain and remit income tax under the *Income Tax Act*;[14] breaches of occupational health and safety legislation; and breaches of trust under construction lien statutes.

As outlined in Chapter 23, the principal defence for officers and directors against liability is due diligence. **Due diligence** means that officers and directors took all active, reasonable steps to ensure that the corporation was not in breach of the statute in question. For some statutory obligations, it must be proven that the officer or director participated in the offence; for others, being passive or ignorant does not constitute a defence. Thus, officers and directors must take their roles seriously and fulfill all corporate and statutory obligations.

Officers and directors can take steps to protect themselves. Obtaining a corporate indemnity to pay for costs associated with the potential liability of officers and directors is a good first step; however, indemnities are simply contractual promises, and a bankrupt company will be unable to fulfill a promise to pay under an indemnity. Liability insurance is another option; however, some statutes impose not only financial liability but also criminal penalties against officers and directors. Neither an indemnity nor officers and directors liability insurance prevents an officer or director from being sentenced to prison.

Thus, professionals should not become an officer or director in a publicly traded corporation without considering the risks and rewards. They should be aware that becoming an officer or director is a serious undertaking.

(iv) **INSIDER TRADING** **Unlawful insider trading** occurs when investors use or are provided with privileged, non-public information to trade on securities or commodities markets in contravention of the law. **Insider trading** may include the purchase or sale of shares prior to the disclosure of a corporate news release; or it may involve the purchase or sale of shares on the basis of information that will never be disclosed to shareholders.

Not all insider trading is unlawful. Shareholders with a connection to the corporation may trade shares of the corporation, as long as they do not violate the rules of the stock exchange or the legislation, which generally means that they cannot make use of information that is unavailable to the public. Engineers and geoscientists are often insiders in the sense that they own shares in corporations for which they possess confidential information.

Unlawful insider trading can also occur if an investor gains non-public information about another company. A person is an insider if he or she has any kind of privileged, non-public information about a company. Insider trading also applies to the commodities markets.

---

[7] R.S.C. 1985, c. A-2.

[8] R.S.C. 1985, c. D-1.

[9] R.S.C. 1985, c. E-19.

[10] S.C. 1991, c. 46.

[11] R.S.C. 1985, c. B-3.

[12] R.S.S. 1978, c. C-30.

[13] R.S.C. 1985, c. E-15.

[14] R.S.C. 1985, c. 1.

For example, if a geoscientist with shares in the gold mining company she works for learns that the company's largest gold strike is 25 percent of the size previously reported and therefore sells her shares prior to a news release announcing this fact, she is guilty of insider trading.

The liabilities for insider trading also apply to providing privileged, non-public information to others. This practice is commonly referred to as *stock tipping*. Both civil liabilities and criminal penalties are available for insider trading and stock tipping. These offences and liabilities potentially apply to anyone who has participated in the insider information exchange.

5.4 Advantages and Disadvantages of Organizational Structures

Table 5-1 on the following page summarizes some of the advantages and disadvantages of the three forms of business organization.

CHAPTER QUESTIONS

1. Which of the following is not a form of corporate organization?
   a) Partnership
   b) Corporation
   c) Sole proprietorship
   d) Construction association

2. Officers and directors must always act with due diligence. Define due diligence, and provide an example.

3. Define the roles of officers, directors, and shareholders.

4. Describe two steps that corporate officers and directors can take to reduce their risk of liability.

5. Ted works for a high tech company and owns shares in the company. A few days after the company makes a public announcement about the failure of its key technology, Ted sells his shares. Is Ted guilty of insider trading? Why or why not?

6. Which statement is true about a limited partnership?
   a) It is essentially the same as a corporation.
   b) It must have at least one limited partner but does not require a general partner.
   c) It must have at least one general partner but does not require a limited partner.
   d) None of the above

7. A corporation may *not* use one of the following as part of its name to designate that it is incorporated:
   a) Corporation, or Corp.
   b) Company, or Co.
   c) Limited, or Ltd.
   d) Incorporated, or Inc.

8. If one partner in a partnership is negligent, can the other partners be liable as well?

TABLE 5-1 Advantages and Disadvantages of Organizational Structures

SOLE PROPRIETORSHIP

Advantages

- Relatively low start-up costs
- Greatest freedom from regulation
- Owner in direct control of decision making
- Minimal working capital required
- Tax advantages to owner
- All profits to owner

Disadvantages

- Unlimited liability
- Lack of continuity in business organization in absence of owner
- Difficulty raising capital

PARTNERSHIP

Advantages

- Ease of formation
- Relatively low start-up costs
- Additional sources of investment capital
- Possible tax advantages
- Limited regulation
- Broader management base

Disadvantages

- Unlimited liability
- Lack of continuity
- Divided authority
- Difficulty raising additional capital
- Difficulty finding suitable partners
- Possible development of conflict between partners

CORPORATION

Advantages

- Limited liability
- Specialized management
- Ownership is transferable
- Continuous existence
- Separate legal entity
- Possible tax advantage
- Easier to raise capital

Disadvantages

- Closely regulated
- Most expensive form to organize
- Charter restrictions
- Extensive record keeping necessary
- Possible development of conflict between shareholders and executives

Source: Modified from "Advantages and Disadvantages of Each Form of Business Organization." Reprinted with permission from the Government of Saskatchewan, Saskatchewan Industry and Resources, Business and Co-operative Services. http://www.cbsc.org/servlet/ContentServer?cid=1081945275353&pagename=CBSC_ON%2Fdisplay&lang=en&c=Guid eFactSheet.

9. Directors may be personally liable for which of the following?
   a) Unpaid wages of the corporation
   b) Breach of trust committed by the corporation under a construction lien statute
   c) Unpaid rent for the corporation's offices
   d) Unpaid taxes of the corporation
   e) All of the above
   f) All of the above, except C

10. Conflict of interest rules can apply to which of the following?
    a) Officers and directors
    b) Partners
    c) Sole proprietor
    d) A and B only
    e) All of the above

ANSWERS

1. D

2. Due diligence means that all reasonable steps have been taken to satisfy statutory obligations. An example of due diligence is the creation of policies and procedures relating to protection of the environment. By creating these policies in response to environmental legislation, an officer or director is acting with due diligence. If an employee breaks these policies, or physical circumstances suddenly change to render the policies ineffective, the officer will not be held individually responsible because he or she had developed policies in due diligence.

3. Directors are individuals who direct the business of the corporation. Officers, such as the president, vice-president, and corporate secretary, provide a closer operational direction for the business. Shareholders own the business and do not necessarily have a role in the business.

4. They can obtain an indemnity from the company they will be working for, and they can purchase directors' and officers' liability insurance.

5. Not guilty. The information was made public prior to the shares being sold.

6. D

7. B

8. If the negligence relates to the business of the partnership, the answer is yes.

9. F

10. D

CHAPTER 6

CONTRACTS

Overview

A **contract** is an enforceable voluntary agreement between two or more parties. The terms *contract* and **agreement** are interchangeable. An **enforceable contract** is one that a court upholds. Contracts take many forms: they may be written, oral, or partly written and partly oral. Contracts consist of express terms and implied terms. **Express terms** are words, phrases, or conditions that have been discussed and agreed to by the parties. **Implied terms** are those terms that have never been discussed or agreed to between the parties but which are taken for granted. While the law requires certain contracts to be in writing in order to be upheld by the courts, such as contracts relating to purchase and sale of land, contracts for construction projects, software development, or consulting services need not be written. However, written contracts are always much easier to prove.

From a theoretical perspective, the purpose of a contract is to set out the rights, responsibilities, and liabilities of the parties involved. But to the parties involved in a contract, the real purpose is to allocate risk and obligations between the parties. A party to the contract is said to be **privy** to the contract; the parties to the contract are *in privity*. Only parties to the contract can enforce a contract. Even if the contract had been created for their benefit, parties not privy to the contract may not enforce the contract.[1]

This chapter explains the basic principles of contract law. Most principles relate to whether a contract is enforceable. **Enforceability** refers to the likelihood that a court would uphold a contract or a portion of it in the event of a dispute. Some agreements are unenforceable due to a flaw in the contract or in its formation; and some become unenforceable due to events that occur subsequent to the creation of the contract.

Two key principles determine whether an enforceable contract exists: contract formation consisting of an *offer* and an *acceptance*; and *consideration*. In addition, the parties must have legal capacity to contract and an intention to create legal obligations, and the contract must have a lawful purpose.

---

[1] In Canada, intended third party beneficiaries *cannot* enforce a contract, except in rare circumstances. Until recently they could not enforce them at all. But the case of *London Drugs Ltd. v. Kuehne & Nagle Int'l Ltd.*, [1992] 3 S.C.R. 299, now allows employees to enforce exclusion clauses in contracts entered into by their employer. However, in the United States, intended third party beneficiaries of a contract may enforce the contract. This is one of the few significant differences in contract law between Canada and the United States.

Contracts may become voidable due to events such as *duress, frustration, impossibility, mistakes, misrepresentation,* and *unconscionability.* A **voidable contract** can be terminated or ended by a party that is not in breach of the contract; but that party may also choose to continue with the contract. Parties sometimes raise these events as defences to claims that arise during or following the performance of a contract.

6.1 Contract Formation, Offer, and Acceptance

A contract is formed between two or more parties when there is an offer capable of being accepted and an acceptance of that offer. An **offer** is a proposal by an offeror to an offeree, containing the essential terms of a proposed contract. An **acceptance** is an unequivocal agreement to an offer. Disputes sometimes arise regarding whether a statement constitutes an offer, *i.e.,* whether it is capable of acceptance. A vague statement is not capable of acceptance. In addition, if an offer does not define essential terms, such as the contracting parties, the price, the time for performance, and the scope of the contract, it probably does not qualify as an offer. A request for offers is called an **invitation to treat** and is not an offer because it cannot be accepted. However, once in a bidding process, an invitation for bids may be considered an offer.[2] The law relating to bidding, the bid process, and its impact on offer and acceptance principles, is discussed in Chapter 10.

Many contracts are formed through negotiations. Parties send offers and counter-offers back and forth until the offer or one of the counter-offers is accepted. However, if an offeree rejects an offer or counter-offer, he or she cannot later accept it unless the offeror revives it. In other words, a counter-offer has the same effect as a rejection: it terminates the previous offer. A **rejection** is an express or implied refusal to accept an offer; the offeree must communicate the rejection to the offeror to make the rejection effective.

Furthermore, obligations of good faith are important to contracts. In *Martel Building Ltd. v. Canada,*[3] the Supreme Court of Canada established that once a contract exists, both parties have an implied good faith obligation in the performance of the contract. The parties also have an obligation of good faith during precontractual negotiations if they explicitly agree to negotiate in good faith, or if there is a statutory requirement to do so, as in collective bargaining. Either party can walk away from negotiations with impunity if they have not formed a contract or explicitly agreed to negotiate in good faith, since there is no implied obligation of good faith in precontractual negotiations.

Generally, a party can revoke an offer at any time before it has been accepted. In order for a revocation to be effective, the offeror must communicate it to the offeree. If an offer does not give a time limit for acceptance, then it is assumed to be open for a period of time commercially reasonable under the circumstances.

Once an offer has been accepted, a contract has been formed. Only in rare circumstances is the formality of signing or executing the contract necessary. In many circumstances, the commencement of performance by both parties of an unexecuted contract provides proof of an enforceable contract. In such circumstances, neither party can later deny the existence of a contract.

The offeror is entitled to specify the mode of acceptance, such as fax transmission, email, registered mail, or verbal agreement. An offeror who uses regular mail may be presumed to be willing

---

[2] On this point, Canadian law and American law differ. In Canada, an invitation for bids is legally an offer, whereas in the United States, it is not.

[3] 2000, SCC 60

to use the same method for acceptance. However, when regular mail is used, acceptance is deemed to have been communicated when the communication is placed into the mail system. This is known as the **Postal Acceptance Rule**.

Acceptance may also be communicated by conduct. For example, if a potential client offers to pay a consultant to perform a site inspection, the consultant accepts the contract by starting to perform the inspection.

However, one important exception to the general offer and acceptance rule exists in the procurement context. **Procurement** is the purchase of goods and/or services. Large scale procurement often occurs through a bid process.[4] A **bid** or **tender** is an offer made in compliance with a fixed set of contract terms in a competitive process. An offer in the form of a bid is considered fixed for a set period of time before the bid is accepted. Parties soliciting bids need time to evaluate them and may insist that they remain irrevocable until the end of the evaluation period, usually 30 to 60 days. Similarly, bidders relying on suppliers and subcontractors in order to put a bid together require that the supplier and subcontractor prices be irrevocable for the same period.

6.2 Consideration

For an offer to be considered irrevocable, it must contain an enforceable promise not to revoke.[5] What makes the promise enforceable is *consideration*. **Consideration** is an old legal concept meaning that something of value, however small, has been given or promised by each party to the contract.[6] In order for a contract to be enforceable, each party to the contract must receive consideration (see Figure 6-1). The primary consideration given by the client in a consulting contract is the promise to pay the consultant's fees and expenses. The primary consideration given by the consultant is the promise to perform the design work.

FIGURE 6-1 Contractual Consideration

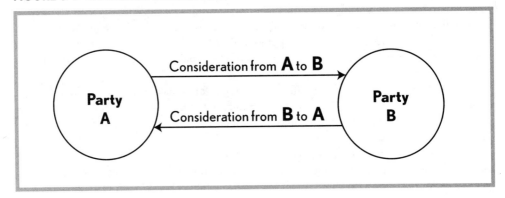

---

[4] See Chapter 10.

[5] An alternative way to enforce an irrevocable offer is to prove that the promisee relied upon it and has suffered a detriment as a result. This is known as *promissory estoppel*. For example, a general contractor who relies on a subcontractor's price would be able to force the subcontractor to hold that price. Otherwise, the general contractor would be stuck with that price component in his bid, without anyone to do the work at that price.

[6] One common law definition of consideration is "either some right, interest, profit, or benefit accruing to the one party, or some forbearance, detriment, loss, or responsibility, given, suffered, or undertaken by the other." *Currie v. Misa* (1875), L.R. 10 Ex. 153, at p. 162.

The consideration given by each party does not have to be of equal value. One party may benefit more than the other. Courts do not usually consider who got the better deal, unless it is so one-sided as to be unconscionable. As long as an agreement was reached, courts generally try to enforce it. A promise without consideration is a gift, and courts do not enforce a promise to give a gift.

Consideration does not have to be a promise to work or to pay money: it can also be a promise by one party to give up a legal right or to perform an obligation. For example, a contractor may agree to give up his or her right to sue the owner for scheduling delays in exchange for the owner's granting an extension of time to complete the contract. This kind of arrangement normally constitutes an enforceable agreement.

Contracts can be amended during performance due to changed circumstances, as described later in this chapter. However, each amendment must be supported by new consideration to be enforceable. Problems with consideration arise for contracts where one party demands changes to the contract midstream without also adding further consideration.

Courts place a great deal of emphasis on consideration. In one case,[7] a contractor was contracted by a builder to deliver and install windows by a certain date. As the date approached, the contractor refused to install the windows on time unless the owner agreed to increase the contract price. The owner reluctantly agreed in order to avoid a delay to the final completion date. Once the work was completed, the owner refused to pay the additional money, arguing that there had been no consideration. The Court agreed that the window contractor had already been under an obligation to perform the work by the given date, and thus did not offer anything new or relinquish any rights in exchange for the promise by the owner for more money. The contractor's duty to perform was found to be past consideration, not new consideration. Had the contractor promised to accelerate the schedule by one week in exchange for the price increase, the amendment to the contract would not have failed for lack of consideration.

Case Study 6.1

Smith v. Dawson (1923), 53 O.L.R. 615 (Ont. C.A.)

Facts

The plaintiffs built a house for the defendant for a fixed price. When the house was close to completion, it was severely damaged by fire. The defendant was insured; the plaintiffs were not. After the fire, an agreement was reached whereby the plaintiffs would complete construction in consideration for the additional payment of the insurance proceeds.

Issue

Two of the judges in this case, Riddell J. and Middleton J., commented on the issues of the case. Riddell J. stated:

> The situation then seems quite clear: the plaintiffs, learning that the defendant had received some insurance money on the house, objected to go on without some kind of assurance that

---

[7] *Modular Windows of Canada v. Command Construction* (1984), 11 C.L.R. 131 (Ont. H.C.).

they were to get the insurance money; the defendant demurred, as she had lost considerably by the destruction of her furniture, but finally said, "All right, go ahead and do the work." If this constitutes a contract at all, it was that she would give them the insurance money she had received, if they would go ahead and do the work they were already under a legal obligation to do.

In some [other jurisdictions] a doctrine has been laid down that (at least in building contracts) the contractor has the option either to complete his contract or to abandon it and pay damages. These Courts have accordingly held that the abandonment by the contractor of his option to abandon is sufficient consideration for a promise to pay an extra amount.

But such a course is to allow a contractor to take advantage of his own wrong, and other Courts reprobate it. This is not and never was the law in Ontario, and is not and never was the law in England. It has long been textbook law that "not the promise or the actual performance of something which the promisee is legally bound to perform" is a consideration for a promise.

I am of the opinion that the promise (if there was one) to pay for the work to be done was not binding for want of consideration, and would allow the appeal.[8]

Middleton J. stated:

Unless the legal situation is kept clearly in mind, the case seems to present some aspect of hardship. The plaintiffs undertook to build the house for the contract price and to hand it over complete to the defendant. In the absence of any provision to the contrary in the contract, the destruction of the building by fire would not afford any excuse for non-performance of the contract.

When the work was going on, the material and labour which went into the building became the defendant's property subject to any lien in the plaintiff's favour; so she had an insurable interest in the property, and she effected an insurance for her own protection.

The builders had an insurable interest, not only because of their lien, but also because the destruction of the property by fire would injure them, as under the building contract they would be bound to replace. They did not insure, preferring to carry the risk themselves. There was no obligation on the part of the owner to insure for the benefit of the contractors, and the contractors have no equitable or other claim on the money received by the owner as a result of her prudence and expenditure.

As I understand the evidence, there was no more than a demand by the owner to the contractor to complete his contract. If there was more, it did not amount to a new contract, as there was no consideration.

In its essence the defence is an attempt to shift the loss resulting from the fire—legally a loss falling on the contractors—to the shoulders of the owner, who, fortunately for her, is not liable. . . .[9]

Questions

1. According to the contractor, what was his consideration for the additional payment?

---

[8] *Smith v. Dawson* (1923), 53 O.L.R. 615 (Ont. C.A.) at para. 26.

[9] See previous note at paras. 27 to 29.

2. What is meant by "abandonment by the contractor of his option to abandon"?

3. Would the result have been different if the contractor had abandoned the contract, been sued by the owner, and then entered into a settlement with the owner in which the contractor would receive the insurance proceeds? Should the result depend upon whether there is an actual abandonment?

Analysis

1. The contractor's argument was that he had given up something of value, namely his right to abandon the project. In other words, he argued that his forbearance in not exercising that option was consideration for the increased payment. It is questionable whether the contractor had a "right" to abandon, although he certainly had an option to abandon and suffer the consequences, which would include damages for breach.

2. That is the description of the "consideration" which the contractor relied upon to support his claim.

3. A different result is likely. Once actual abandonment occurs, the owner must sue the contractor to recover damages. At this point, there is a dispute between the parties, and a settlement agreement to resolve a *bona fide* dispute is usually enforceable. Most courts enforce settlement agreements, regardless of who had the stronger case.

For many years, it has been accepted as well-established law that consideration was essential not only to make a contract enforceable, but also to make contract amendments enforceable. In other words, amendments to a contract, which are typically referred to as "change orders" in construction contracts, have been subject to the same requirements for enforceability. This long-established rule is no longer as absolute as it used to be.

As described and illustrated in the case of *Smith v. Dawson* above, there must be fresh consideration to make a promise enforceable. However, in the case of *Greater Fredericton Airport Authority Inc. v. NAV Canada*,[10] one involving a dispute between a local airport and the federal aviation authority over which party was required to pay for the installation of a navigational beacon, the New Brunswick Court of Appeal held that it should not be bound by the rigid application of the consideration requirement in all circumstances. It should also be noted that the New Brunswick Court of Appeal, in that case, went on to modify the law as it applies to economic duress. It is possible that this development in the law of consideration in New Brunswick will become part of the law in other parts of Canada.

There is one well-recognized exception to the requirement that consideration flow to and from all parties to a contract. If one party is, for example, promising to give a gift, so that it is not receiving consideration from any other party, the party giving the gift may make that promise in a written document, and then apply a seal to that document. The seal is meant to act as consideration and is recognized by our courts as such. If the party is a corporation, the corporate seal is used. Individuals do not generally have personal seals anymore, so nominal consideration, such as one dollar, may be used.

6.3 Agreements to Agree

An agreement to make an agreement in the future is not an enforceable contract. Parties often sign a **letter of intent**, or a term sheet, which is a document to set out the basic terms of a business arrangement

---

[10] 2008 NBCA 28, para. 19.

in advance of agreeing to all of the detailed contractual terms. The parties then commence to negotiate the detailed terms. But if such negotiations fail, there is no binding contractual obligation.

However, such agreements create three problems. Firstly, not all parties may be clear that the preliminary agreement is unenforceable. Secondly, and more likely, both parties may commence performance without creating a formal contract. Thirdly, if a letter of intent is worded in such a way as to suggest that it is accepting an offer or a bid, without creating additional conditions that must be met, then this letter would be more properly considered an acceptance rather than a letter of intent, although many of the terms of the contract would be unclear and may have to be inferred by a court.

6.4 Voiding a Contract

Disagreements over contracts often result because one party realizes that the contract is not in his or her best interest. These disagreements often occur before the performance of the contract even begins. However, courts void a contract only in rare circumstances. Such circumstances include *mistake, misrepresentation, duress, unconscionability, frustration,* and *impossibility.*

(a) Mistake

A **mistake** is a misunderstanding with respect to a term of a contract. To make a contract voidable, a mistake must have all of these three features:

1. It must be significant and not trivial. In other words, it must be a *material* mistake.
2. It must be a **mutual mistake**, which is a mistake made by both parties.
3. It must have existed at the time the agreement was made. A mistake made after the agreement was entered into is insufficient.

For example, in a contract for $1 million, a trivial mistake would be $50 whereas a significant mistake would be $300 000. An example of a mutual mistake is a situation in which both parties assumed that an environmental permit would be unnecessary but is in fact necessary. In the most common examples of mutual mistake, both parties are unaware that the subject matter of a sale of goods no longer existed at the time of the sale contract.

Where the mistake is not a mutual mistake, there is normally no remedy available; however, courts may consider providing a remedy if one party was either aware or should reasonably have been aware of a mistake made by the other party.

(b) Misrepresentation

A **misrepresentation** is an untrue factual statement that is made by one party and induces the other party to enter into the contract. Misrepresentations fall into three categories: innocent, negligent, and fraudulent. Both innocent and negligent misrepresentation involves a statement made by one party to another that is false or misleading but involves no intent to deceive or mislead. However, a negligent misrepresentation is made without due care or skill. A fraudulent misrepresentation occurs when the party making the statement is aware that the statement is false. A party who has entered into a contract in reliance on a misrepresentation may be able to rescind the contract; but rescinding the contract may be more difficult if there was no fraudulent intent. Damages are generally available for fraudulent or negligent misrepresentations, but they are much more difficult to obtain for innocent misrepresentations.[11]

---

[11] For further discussion of misrepresentation, see Chapter 12, pp. 139-140.

In some construction cases, allegations of fraudulent misrepresentation have been made in the context of subsurface conditions. Clauses in construction agreements often require the contractor to completely investigate the site before submitting a bid in order to be aware of site conditions. Often owners commission a geotechnical investigation for bidders, together with a disclaimer that the owner makes no representations regarding the accuracy of the investigation or the nature of subsurface conditions. In some cases, the bid documents require the contractor to conduct his or her own soils investigation. Given the realities of putting a bid together, the constraints of time and cost often make it unreasonable to expect any contractor to undertake an independent study of any real value.

Where the contract provides for an increase in price due to unforeseen subsurface conditions, the appropriate remedy is to authorize an extra to the contract.[12] Difficulty arises if the contract contains no such mechanism, and the owner had information in advance that contradicts the bid assumptions. In many jurisdictions, the owner has an obligation to make available to bidders all relevant information that is not reasonably available to the bidder; and failure to do so may constitute a fraudulent or negligent misrepresentation.

(c) Duress

Duress is improper pressure, threats, or coercion used to induce a party to enter into a contract. Construction cases involving duress usually focus on modifications to the contract made during performance. A good example is the *Modular Windows*[13] case, where a contractor demanded that the owner amend the agreement or else face a long delay. In order for duress to exist, the party claiming duress must prove that he or she did not have any real choice but to comply with the other party's demands.

There is a fine distinction between legitimate business pressure and illegitimate pressure that amounts to duress. Courts examine whether there were practical alternatives available to the party claiming duress. One of the most commonly cited definitions for economic duress is taken from the Privy Council decision in *Pao On v. Lau Yiu*,[14] and quoted with approval by the Ontario Court of Appeal in *Ronald Elwyn Lister Ltd. v. Dunlop Canada Ltd.*,[15] as well as in many other cases:

> Duress, whatever form it takes, is a coercion of the will so as to vitiate consent. Their Lordships agree with the observation of Kerr J. in *The Siboen and The Sibotre*, [1976] 1 Lloyd's Rep. 293 at 336, that in a contractual situation commercial pressure is not enough. There must be present some factor which could in law be regarded as a coercion of his will so as to vitiate his consent . . . In determining whether there was a coercion of will such that there was no true consent, it is material to enquire whether the person alleged to have been coerced did or did not protest; whether, at the time he was allegedly coerced into making the contract, he did or did not have an alternative course open to him such as an adequate legal remedy; whether he was independently advised; and whether after entering the contract he took steps to avoid it.[16]

(d) Unconscionability

An **unconscionable contract** is one that is so unfair, oppressive, or one-sided that it would be offensive for the court to enforce it. Courts will not void a contract or portion of a contract simply

[12] The CCDC 2 Agreement GC 6.4 contemplates such an increase.
[13] *Modular Windows of Canada v. Command Construction* (1984), 11 C.L.R. 131 (Ont. H.C.).
[14] [1979] 3 All E.R. 65 (P.C.).
[15] (1979), 27 O.R. (2d) 168.
[16] *Pao On v. Lau Yiu*, [1979] 3 All E.R. 65 at 78 (P.C.). Also *Ronald Elwyn Lister Ltd. v. Dunlop Canada Ltd.* (1979), 27 O.R. (2d) 168.

because it favours one party over another. In order for a contract to be unconscionable, inequality of bargaining power at the time the contract was formed must be extreme. Commercial contracts are rarely held to be unconscionable.

A clause or contract that is particularly harsh toward one party may be evidence of an inequality of bargaining power. A court faced with such a contract or clause can deal with it in one of two ways; it can refuse to enforce it, or it can interpret the contract or clause as favourably to the oppressed party as possible. An example of such a clause is the *no-damages-for-delay* clause, which some owners try to insert into construction agreements. These clauses prevent the contractor from suing the owner for any delay. Courts are reluctant to find such clauses unenforceable but will often narrowly interpret them in order to avoid a harsh result for the contractor.

(e) Frustration

Frustration, or *impossibility*, occurs when an unforeseen event occurs that makes the performance of the contract either impossible or of no value. In such cases, a party may be able to claim that he or she does not have to perform the contract. For the courts to agree, the risk of the event must *not* be a risk that was originally allocated to one of the parties.

Most standard form contracts include clauses governing foreseeable risks. For example, typical contracts contain a ***force majeure* clause**, which provides for relief due to events listed in the clause that the parties agree are beyond their control. The list of *force majeure* events used to be limited to acts of God but now often include such events as labour disputes, fires, delays by transport companies, and unavoidable casualties. Frustration and impossibility are thus difficult to prove under modern contracts, because most clear *force majeure* clauses foresee and allocate most risks and eventualities.

6.5 Amendment of Contracts

Parties sometimes need to amend contracts. An **amendment** is a change or alteration to a contract. Many contracts contain provisions that permit amendments made in writing. While some courts have decided that oral amendments are enforceable even with such clauses in the contract, parties would be imprudent to ignore the contractual provision for written amendments and act on verbal changes.

A contract amendment is in some ways a mini-contract. In other words, the same requirements that make a contract enforceable, such as consideration, offer and acceptance, legal purpose and legal capacity, also make an amendment enforceable. Similarly, a contract amendment may be voidable as a result of mistake, misrepresentation, duress, or any of the other grounds discussed above that apply to contracts generally.

Most contracts stipulate that amendments need to be made in writing, prior to any performance of the amendments. In practice, however, changes to the contract are often discussed orally and executed before the paperwork is complete, much less signed by all parties. Disputes may develop if one party later decides that it does not want to pay for certain changes and tries to rely on a clause that requires that changes only be made in writing. In these disputes, the courts will have to decide whether that party waived his or her right to have changes agreed in writing.

6.6 Waiver and Estoppel

Even though a party has rights under a contract, those rights may not all be enforceable in all circumstances. A **waiver** of rights occurs when by words or conduct, a party ceases to enforce certain of his or her contractual rights. When one party waives rights, those rights are subject to **estoppel**: the other party in the contract starts relying on the representation or action of the first party, rather than on the contract. For example, if a lease requires rental payments by the first day of each month, but the landlord consistently accepts payments on the third of the month, then the landlord is estopped from requiring payment on the first. In order to again enforce the strict terms of the lease, the landlord must provide reasonable written notice that he or she intends to enforce the original lease.

This raises the question of how and when a waiver can occur. Where by word or conduct, a party to a contract causes another party to reasonably believe that certain rights will not be enforced, and where subsequent enforcement of those rights would be unfair, Canadian courts generally rule that a waiver has occurred. The exceptions and other legal requirements to this principle are beyond the scope of this text.[17]

A general rule may be gleaned from the many cases that deal with waiver: if a party wants to rely upon the strict requirements of a contract and be able to enforce its rights, it cannot do so selectively or inconsistently. In one case, *Kei-Ron Holdings Ltd. v. Coquihalla Motor Inn Ltd.*,[18] the parties were operating under the 1982 version of the CCDC 2 standard form construction contract. That contract required all changes to be approved in writing before they were performed; but the parties made changes based on oral agreement. The Court found that a waiver had occurred as a result of the parties' practice because both had stated that they had been aware of the contractual terms relating to changes, and both chose to ignore them.[19] In another case, *Banister Pipeline Construction Co. v. TransCanada Pipelines Ltd.*,[20] the contract contained procedures for prior approval of extra work; but the parties had not followed those procedures. Again, the Court found that a waiver had occurred. Finally, in *Keen Industries Ltd. v. Hegge Construction Ltd.*,[21] the Court again concluded that the notice provisions had been waived, merely because the paperwork authorizing one change had been done after the change had been performed. The Court found that "performance of the contract was somewhat casual on both sides in the sense that strict formality was not observed by either side."[22] Further, the Court found that, where an owner (or a contractor, in the *Keen* case) orally directs a subcontractor to perform extra work, there is an implied promise to pay. Hence, this action establishes the waiver.[23]

6.7 Quasi-contract

Courts may decide that contracts are inappropriate for determining compensation as a result of fundamentally changed conditions, mutual mistake, or abandonment of a contract by one party. Yet if the other party has received the benefit of the work, the law says that such party should pay something for it. The principle of *quantum meruit* applies when one party should compensate the

---

[17] Waiver and estoppel are known in law as "equitable" remedies, meaning that they originated in the Courts of Equity in England. Equity in this context is effectively the same as fairness. Courts do not grant equitable remedies unless certain requirements are met. For example, the plaintiff must come to court with "clean hands"; in other words, equitable remedies will not be granted to a party who has acted improperly in the relationship.

[18] (1996), 29 C.L.R. (2d) 9 (B.C.S.C.).

[19] See previous note.

[20] [2003] A.J. No. 1008, 2003 ABQB 599.

[21] [1993] 8B.C.J. No. 1448 (S.C.).

[22] See previous note.

[23] See also *Redheugh Construction Ltd. v. Coyne Contracting Ltd.* (1996), 26 B.C.L.R. (3d) 3, 29 C.L.R. (2d) 39 (C.A.), which cites Goldsmith and Heintzman's *Goldsmith on Canadian Building Contracts*, 4th ed. (Toronto: Carswell, 1995) Sections 4-17 and 4-18: "[R]equirement for a written authorization for extra work may be waived by an owner (or presumably a contractor in the case where the other party is a sub-contractor), and that such a precondition may also be waived by the owner's conduct or acquiescence."

other party that performed the work. ***Quantum meruit*** simply means "the amount it is worth." The party that performed the work must receive pay that approximates the value of the work, including a reasonable profit. But for *quantum meruit* to apply, there must be no valid contract in place that applies to the work.

In the construction industry, this principle applies most frequently to changes in circumstances. If the change is one that is covered by the change order provisions of the contract, then *quantum meruit* does not apply. However, if the changes are such that they fundamentally alter the nature of the contract and are therefore beyond the scope of the change order provisions, *quantum meruit* may be the best way to deal with compensation.

Most construction contracts contain provisions for assessing the monetary value of changes. They frequently require the owner and contractor to agree on a price before they approve a change. But changes often occur quickly in the construction business. Thus, the paperwork to formalize the change, known as a **change order**,[24] may follow the performance of the change by weeks or even months. Therefore, most standard form agreements allow the owner to issue a **change directive**,[25] which is an order that the contractor proceed with a change to the work, to be paid for on a cost-plus basis.[26] In essence, even though the contract set a fixed price for the entire project, the change is performed as though it were under a separate cost-plus agreement. This result is similar to that of *quantum meruit* cases.

Quantum meruit changes also occur for **constructive changes**—changes to the contracted work that the owner refuses to acknowledge. For example, a developer might refuse a contractor access to a portion of the work site needed to perform the contract requirements, instead forcing the contractor to work at greater expense from a less convenient area. If the owner refuses to acknowledge that this refusal constitutes a change to the contract, the courts will consider this refusal a constructive change and require the owner to pay for it on a *quantum meruit* basis.

CHAPTER QUESTIONS

1. An engineer has not been paid the last instalment for his consulting work on Project A. The client says, "If you agree to give up your claim for the remaining fees, I will consider hiring you as the consultant for Project B." The engineer agrees. But Project B is ultimately awarded to another engineer. Is the agreement to forego the fees enforceable? Explain.

2. A contractor meets a prospective client at a party. The client says, "I'm looking for someone to build a 50 000 ft^2 warehouse with cast-in-place concrete walls on my property on the north side of the city." The contractor replies, "I can build those for $50 per square foot." The client sends a fixed-price contract to the contractor in the amount of $2.5 million. The contractor refuses to sign it. But is there an enforceable agreement already in place? Explain.

---

[24] See Chapter 9, pp. 71–74.

[25] See Chapter 9, pp. 73-74.

[26] CCDC 2 clauses GC 6.1, 6.2, and 6.3. In a cost-plus agreement, the owner agrees to pay the contractor for costs plus a fixed amount, in contrast to a fixed-price contract, where the contractor gets a fixed price regardless of the costs.

3. A contract is said to be a "legally enforceable promise." However, a court will refuse to enforce a contract if the following is true:
   a) It is an obviously bad business deal for one of the parties.
   b) The purpose of the contract is illegal.
   c) Neither A nor B
   d) Both A and B

4. "Consideration" is best defined as the following:
   a) One party doing a favour for the other party.
   b) Money paid by one party to another party
   c) Something of value that is exchanged or given by a party
   d) The thought process that a party goes through in deciding to enter the contract

5. In which of the following ways may contracts be formed?
   a) Entirely oral
   b) Partly oral and partly written
   c) Entirely written
   d) All of the above

6. If two parties reach an agreement orally, but make a mistake in documenting their agreement in writing, then what happens?
   a) No contract exists because a mistake was made.
   b) A contract exists as set out in the written document.
   c) A contract exists according to the oral agreement because the parties did not intend to change the contract when they put it in writing.

ANSWERS

1. The agreement to forego fees is the consideration flowing to the client. From that perspective, the agreement would be enforceable. However, the consideration flowing to the consultant is presumably that the client will "consider using" the consultant. This promise is likely too vague and uncertain to constitute consideration, and the agreement to forego fees is likely unenforceable for that reason.

2. The elements of an enforceable contract include the following:
   - Offer and acceptance
   - Enforceability
   - Legal intent
   - Capacity to contract

Legal intention to contract is an issue here. This problem illustrates the principle that a contract does not have to be in writing in order for it to be an enforceable agreement (although in this case there is no enforceable agreement because there is no offer and acceptance). Professionals find that clients seek advice in informal settings and must take care not to provide casual advice without proper deliberation. The fact that this hypothetical problem takes place in an informal setting should not distract the professional from the real issue, which is contract formation. Did the engineer in fact make an oral contract? On any objective standard, a third party is extremely unlikely to conclude that he did. The contractor's reply was: "I can build those for $50 per square foot." This is not an offer to do so, but rather a statement of capabilities. On this basis alone, a court would likely decide that there has been no offer, and therefore there can be no acceptance.

3. B

4. C

5. D

6. C

CHAPTER 7

BREACH OF CONTRACT

Overview

Contractual disputes often occur as a result of events that cause financial loss to one of the parties, such as an accident, a downturn in the real estate market, or deficiencies in the performance of the contract. Sometimes disputes are precipitated not by events but by the realization by one of the parties that the deal they agreed to will not turn out as planned. Negotiation resolves many of these disputes. Those that cannot be so resolved are turned over to an impartial tribunal, such as a judge, jury, or arbitrator.[1]

When a contract dispute is turned over to a court, the court must determine whether the disputed issue constitutes a risk or obligation that one of the parties assumed at the outset. If it does, then the party who assumed the risk should bear the loss. For example, if the cost of materials increases after the contract starts, and the contract is a fixed-price agreement, then the courts would conclude that the contractor had accepted that risk. However, if the contract is a cost-plus agreement, then the owner has accepted the risks.

However, the triggering event for the majority of contract disputes is the failure by one party to perform his or her obligations. This failure is called **breach of contract**. Breaches and remedies for breach of contract, the most common of which is the payment of money called damages by the party in breach, are discussed on pp. 59-61.

7.1 Breach

Parties to contracts sometimes fail to fully perform all their obligations. This failure constitutes a breach of the contract if any of these conditions apply:

- the inability of one party to perform the contract
- inadvertence

---

[1] Dispute resolution is discussed in Chapter 14.

- disagreement as to the requirements of the contract
- profit (or lack thereof).

(a) Inability

Inability refers to a situation in which one party would like to perform its obligations but is unable to do so for reasons such as lack of financial resources.

(b) Inadvertence

Inadvertence means that one of the parties to the contract has unintentionally failed to carry out one of its promises or obligations.

(c) Disagreement

Disagreement means that two of the parties to the contract interpret the meaning of the contract differently. For example, a disagreement can occur in a contract between a business owner and a software developer. After the contract has been signed, the owner demands that additional features be incorporated into the software. The software developer refuses on the basis that the work is outside the scope of the contract. The owner then threatens the software developer, promising that if the software developer refuses to perform, the owner will not pay for the product. The software developer is faced with a dilemma: perform the work and risk not getting paid extra for it; or refuse to perform and hope that a court finds the work to be outside the scope of the contract.

Such a disagreement occurred in the *Peter Kiewit v. Eakins* case.[2] The contractor chose to perform the work that it viewed as being outside its scope and sued for extra payment. The Supreme Court of Canada held that by performing the work, the contractor lost the right to recover the extra payment. The contractor could have preserved his right to that payment only by refusing to do the extra work until an agreement had been reached for payment.

At the time, many observers considered this result to be unduly harsh for the contractor. As a result, many provinces enacted statutes entitling one party to perform the disputed work "under protest."[3] **Performing under protest** means that the protesting party gives the other party written notice that it will be performing the work on the understanding that it is doing so without prejudice to its right to argue that it is entitled to additional payment for that work. This principle applies to all commercial contracts. One example of the performance under protest provision is as follows:

> If a dispute arises between the parties to a contract respecting the obligations of a party under the contract, the party whose obligations are disputed may elect to perform the contract in accordance with the requirements of the other party, and the electing party is then entitled to compensation from the requiring party for any
>
> a) service performed,
> b) property supplied or transferred,
> c) liability assumed, and
> d) money paid

---

[2] *Peter Kiewit Sons' Co. of Canada Ltd. v. Eakins Construction Ltd.*, [1960] S.C.R. 361.

[3] For example, in British Columbia, the performance under protest provisions are included in s. 62 of the *Law and Equity Act*, R.S.B.C. 1996, c. 253.

by the electing party in the course of that performance beyond that which the contract required the electing party to do.[4]

Statutes and contracts that provide for performance under protest typically require that the protesting party give notice of protest.

(d) Lack of Profit

Breach of contract may sometimes occur because one party finds that it is more costly to perform than to face the legal consequences of non-performance. For example, a party may be losing money on a contract and may calculate that it would cost less to let the company go bankrupt than to inject more capital into the company to complete the contract.

7.2 Damages

When a breach of contract occurs, the court may grant one of a number of remedies to the innocent party. A court may order one party to pay damages to the other party. **Damages** refers to the compensation the court awards to an injured party, payable by the party or parties that caused the injury. A court may also order one party to perform specific acts, known as **specific performance**. Courts rarely order specific performance for architectual, construction, engineering, and geoscience contracts. Specific performance is typically used where the seller of a house refuses to convey title to the property. A court may also prohibit a party from doing something, known as an **injunction**; or it may make a **declaratory order**, in which the court determines the rights of the parties but does not require either of the parties to do anything (such as paying damages).

By far, the most common remedy for all breaches of contract is damages. The amount awarded is supposed to put the innocent party in the same position as the final outcome of the contract if no breach had occurred. For a party whose contract to perform work or provide services has been wrongfully terminated, the damages would include lost anticipated profit. For an owner whose opening date has been delayed, the damages may include lost rents or financing costs or both.

Recovery of damages is subject to three limitations. The first limitation is mitigation. **Mitigation** means that the party who has suffered an injury or loss must take reasonable steps to reduce, or mitigate, the injury or loss. For example, if a client suffers a delay caused by a software engineer's failure to deliver part of a portion of the software on time, and as a result, the sale of the software falls through, the client must try to sell the product to another buyer and then recover from the software engineer only the difference in sale price and other additional costs incurred.[5]

The second limitation is that the damages cannot be speculative. For example, a party claiming loss of anticipated profits must be able to prove that it is more likely than not that a profit would have been earned and must show how much that profit would have been.

The third limitation is remoteness. **Remoteness** is the lack of a connection between a wrong and an injury or loss. If a loss was not reasonably foreseeable at the time the contract was made, it is considered too remote to be claimable. That the delay of a completion date might cause a party to lose revenue is likely reasonably foreseeable; but that such a delay would bankrupt that party may not be reasonably foreseeable. The party's bankruptcy is an example of a loss that might be considered too remote to be recoverable. For this reason, serious consequences of a possible breach that are not ordinarily foreseeable should be brought to the attention of the other contracting party

---

[4] *Law and Equity Act*, R.S.B.C. 1996, c. 253, s. 62(2).

[5] If the new sale price is higher, there may be no loss suffered.

before entering the contract. Failure to advise the other party of unusual risks might preclude recovery of damages for those risks on grounds of remoteness.

In many contracts, such as those for the supply of goods, suppliers often insist on a consequential damages exclusion or limitation clause. **Consequential damages** is a phrase which has different meanings in different contracts and different contexts but is generally understood to mean indirect losses, such as loss of business. For example, the supplier of a seal for a marine engine produces a seal for $5; but the failure of the seal may cost the boat owner $500 000 in repairs and $1 million in lost business during repairs. The boat owner can choose to either have a seal for $5 and accept the risk of the seal failing, or he can expect the seal to cost $100 000, which is what the seal manufacturer would have to charge to account for the risk of failure. To avoid this cost-*vs*-risk problem, suppliers generally require buyers to accept the risk by agreeing to a provision in the contract that excludes damages for consequential loss. Courts carefully interpret such clauses and construe any ambiguity against the party that drafted the clause. For this reason, these clauses should be drafted by a lawyer.

Consequential damages clauses are less common in building construction or design contracts than in supplier sales contracts. In contrast, liquidated damages clauses are more common.[6] **Liquidated damages** are genuine estimates of loss written into the original contract, before any breach has occurred. The parties agree beforehand that if a certain type of breach occurs, such as a delay, damages will be fixed at a certain amount, such as $1000 per day. In some cases, a party may set liquidated damages unreasonably high in order to coerce a party into performing. However, courts may consider such a clause to be a penalty, rather than liquidated damages. A **penalty** is a sum of money included in a contract as punishment for breach of the contract, rather than as compensation for the breach. Courts generally do not enforce penalty clauses. Professionals should accurately estimate the amount of damages that would be payable in the event of a breach before signing the contract and keep a record of this calculation. A **bonus clause** is one that entitles one party to additional payment if its performance exceeds what has been promised, such as finishing prior to the contractual completion date. A bonus clause is not required in the same contract as a liquidated damages clause to make the liquidated damages clause enforceable.

Finally, in cases involving correction of deficiencies, sometimes a court awards the actual cost of correcting the deficiencies. Courts make such awards when the repairs have been carried out prior to trial, and the repair costs are reasonable. Sometimes, however, courts choose to calculate damages based on diminution (or reduction) in value of the property, rather than the cost of the repairs. The rationale for such a decision is outlined in this recent Ontario case:

> The standard measure of damages in most building contracts cases is the cost of performance—the cost of rectifying, repairing and/or replacing the unsatisfactory work. (Hitch and Snyder, *supra*, at 2-5.) It is not uncommon, in the construction context, for the cost of performance or remediation to exceed the diminution in value, or even to exceed the total value of the property itself, given the high costs associated with both demolition of unsatisfactory projects or products, and the installation of replacement ones. (See Hitch and Snyder, *supra*, at 2-8.) If the injured party is awarded damages for the diminution in value, this might result in an award for nominal damages only.
>
> The relationship between the cost of performance and the benefit of performance (which can

---

[6] A liquidated damages provision may be used in conjunction with a consequential damages provision.

be considered analogous or logically similar to the diminution in value) is, therefore, central to the analysis of the proper measure of damages in building contracts cases. . . .

Following the logic that courts are reluctant to award a windfall to an injured party, and that courts are concerned instead with accurately measuring the loss itself, the likelihood of the injured party actually performing the restoration, once they are awarded the cost of performance, will be an important factor. (See Waddams, *supra*, at [paragraph] 1.2420; see also Hitch and Snyder, *supra*, at 2-14.) Clearly, where the injured party has already carried out some or all of the work, courts will be more likely to award the full cost of performance. However, where the injured party appears to have no "genuine interest" in carrying out the repairs or replacement or restoration, but simply intends to pocket the money as a windfall, courts will either reduce the award for cost of performance, or choose to award the diminution in value instead.[7]

7.3 Contract Termination

Contracts must eventually end. The most common way to end a contract is to complete the performance. After each party has performed all of his or her obligations under the contract, and after all warranty periods have expired, the contract comes to an end.

Other events may cause a contract to terminate. As discussed earlier in this chapter (see pp. 57–59), these events include an event that frustrates the contract or makes the contract impossible for the parties to perform. The parties may also mutually agree to terminate their agreement before it has been completely performed. A mutual agreement to terminate a contract should set out the terms of termination, including final payments, if any, by one party to another.

In some cases, one party might decide to terminate the contract without the consent of the other. This unilateral termination will either be provoked by a breach of contract, or else will itself be a breach of contract.

(a) Fundamental and Simple Breaches

Breaches of contract can be divided into two types: fundamental breaches and simple breaches. A **fundamental breach** is a breach that goes to the root of the contract and deprives the innocent party of all or substantially all of the benefit of the contract. For example, failure to pay monies that are due and owing is a fundamental breach. If one party commits a fundamental breach, the other party has two options: to continue to perform and sue for damages, or to declare the contract to be terminated and sue for damages.

A **simple breach** is a breach that does not entitle the innocent party to treat the contract as ended or permit the innocent party to stop performing their part of the contract. The only remedy is to sue for damages. Courts decide whether a particular breach amounts to a fundamental breach. If the innocent party treats a simple breach as a fundamental breach and proceeds to terminate the contract, the innocent party will have itself committed a fundamental breach and will be liable for damages. Deciding to unilaterally terminate a contract is a very serious step that may have potentially disastrous consequences. One should never proceed with this type of termination without careful deliberation and legal advice.

---

[7] *Safe Step Building Treatments Inc. v. 1382680 Ontario Inc.* (2004), 37 C.L.R. (3d) 281 (Ont. S.C.).

(b) Repudiation and Anticipatory Breach

In **repudiation** of a contract, one party by words or conduct lets the other party know that it does not intend to perform its obligations. Repudiation without justification constitutes a fundamental breach. In **anticipatory breach**, one party communicates to the other party before the time for performance of an obligation that he or she intends to breach that obligation of the contract. For example, if payment were due on the last day of the month, and an owner indicated in advance of such date that he or she would not be making payment on the due date, the owner would have committed an anticipatory breach. Anticipatory breach may also occur if parties find themselves in a position in which they cannot perform. If the anticipatory breach is fundamental in nature, rather than simple, then the anticipatory breach would also be considered to be a repudiation.

(c) Termination Clauses

Most contracts contain termination provisions. A **termination clause** lists the acts by either party that entitle the other party to terminate the contract.[8] Most provisions require that the party about to terminate give notice to allow the party in breach to remedy the breach promptly. The terminating party must strictly comply with all limitation periods and notice requirements, or else the termination might be considered wrongful, and the terminating party will then itself be in breach.

Some contracts also contain a "termination for convenience" clause that may be used by the owner or, in some cases, by either party to terminate the contract without cause or reason. In some cases, such clauses will require that the non-terminating party be paid a termination sum or lost profit. Termination for convenience clauses can be useful in that they permit the contractual relationship to end without allegations of breach, but such clauses should be carefully considered in light of the circumstances and the compensation that is available.

CHAPTER QUESTIONS

1. A geoscientist's client has failed to make a significant progress payment when it is due. The geoscientist delivers a letter that states: "Unless we receive payment by noon tomorrow, we will consider your failure to pay a fundamental breach and the contract at an end." The owner replies: "We will pay you tomorrow at 4:00 p.m. We hope that is acceptable." The geoscientist replies that it is unacceptable. At 4:00 p.m., the client presents payment, but the geoscientist has abandoned the site and sues for fundamental breach. Discuss the legal issues, and who should prevail in the lawsuit.

2. If a party to a contract breaches the contract, the innocent party may obtain damages. What is the purpose in awarding damages?
    a) To punish the breaching party
    b) To deter further breaches by the breaching party or other parties
    c) To place the innocent party in the same position he would have been in had the guilty party properly performed the contract
    d) To place the innocent party in the same position he would have been in had the contract never been entered into

---

[8] The CCDC 2 contract contains provision for termination by the contractor and by the owner for cause. Those provisions are in clause GC 7.1 and 7.2.

3. A contractor has entered into an agreement with an owner with payment to be based on hours worked at specific labour rates. While performing the work, the contractor's level of efficiency is lower than expected. The owner refuses to pay the full amount of the contractor's costs. What legal issues are involved? What other facts are needed in order to resolve the dispute?

4. Which of the following event(s) entitles a party to be released from its obligations under an existing contract?
   a) Fundamental breach of contract
   b) Destruction of the contract documents
   c) The parties cannot agree on an amendment
   d) Party forgetting to sign the contract documents, prepared after work has started

ANSWERS

1. Failure to pay under a contract is almost always a fundamental breach. Given that the payment was "significant," the client likely committed a fundamental breach of the contract. The geoscientist then waived the failure by not immediately terminating. This created a new condition whereby payment was required by noon on a day set by the geoscientist. The question then is whether the failure to pay by noon was a significant breach. Given that the payment was made four hours later, unless there were specific circumstances such as the last flight for a week leaving from a remote location was leaving at 2:00 p.m., a court will likely find that the four-hour delay was not a fundamental breach.

2. C

3. The principal issue is whether productivity is a term of the contract. If the owner has agreed to pay on an hourly basis, with no reference to productivity, the owner will likely have to pay on that basis. However, there may be either an express or implied term that the contractor will perform at a level of competence and productivity that is standard in the industry. If the contractor's performance falls below such level, the owner may be entitled to pay a lesser amount.

4. A

CHAPTER 8

INTERPRETING AND DRAFTING CONTRACTS

8.1 Interpretation

When a court interprets a contract, it ascertains the meaning of the words used by the contracting parties and determines the legal effect of those words. Interpretation of contracts involves the application of many legal principles:

- the overriding principle of the objective intention of the parties;
- the canons of contract construction; and
- the introduction of additional evidence.

The role of the court in interpreting a contract is not to rewrite the contract for the parties nor to impose a contract where the parties failed to form a valid one themselves. If the literal meaning of the contract is clear, then there is no room left for interpretation. However, where the contract contains clauses that are ambiguous or uncertain, the court must analyze and interpret the contract to determine its legal meaning. In some cases, the court may also allow the introduction of additional evidence to assist in interpretation.

(a) The Intention of the Parties

The principal and overriding objective in contract interpretation is to ascertain the reasonable mutual intentions of the parties regarding the legal obligations they assumed under the written contract. During contract interpretation, the court is said to apply an *objective standard*, which requires determining what hypothetical reasonable parties would have intended. The objective standard contrasts with the *subjective* standard, which involves determining what the actual parties themselves intended. In a contract dispute, the court is interested in the objective intentions of the parties. In this manner, the parties' agreement should be understood in the way in which the contract language would appear to the ordinary reasonable person looking at it from the outside.

However, determining the objective intention of the parties does not confine the interpreter to looking strictly at the contract itself. An interpreter may use the *canons of contract construction* and the *contractual context* as guidelines.

(b) Canons of Contract Construction

The canons of contract construction are not rules of law; they are simply principles employed by the courts to assist in determining the objective intention of the parties.

(i) PLAIN AND ORDINARY MEANING The *golden rule* of contract interpretation is that the words of a contract should be construed in their grammatical and ordinary sense and should be assigned their plain and ordinary meaning, except to the extent that interpretation or modification is necessary in order to avoid absurdity, inconsistency, or repugnancy. Therefore, in the absence of ambiguity, the natural or literal meaning of the words set out in the contract should be adopted.

(ii) SPECIAL MEANING Where words have a customary or special meaning, as often occurs with technical documents in construction contracts, the special meaning takes the place of the ordinary meaning for the purpose of construing the contract. Use of a dictionary is appropriate for determining the special meaning of words, as is use of contractual headings and marginal notations. In some cases, expert evidence is required.

(iii) READING THE CONTRACT AS A WHOLE Courts consider each contract as a whole. As such, each clause should be considered not in isolation but in the context of the whole contractual document. If there is a conflict between two parts of a document such that both clauses cannot be enforced, the interpretation that reflects the true intentions of the parties is adopted.

(iv) GIVING EFFECT TO ALL PARTS OF A CONTRACT In interpreting a contract, a court will give effect to all parts wherever possible. In other words, no part of the contract should be treated as inoperative, surplus, or meaningless.

(v) RESTRICTION BY EXPRESS PROVISIONS Standard form contracts often have added to the standard form special or supplemental conditions. In these circumstances, courts assign greater importance to the special conditions; and in the case of conflict between the general conditions and the special conditions, the special conditions prevail.

(vi) COMMERCIAL PURPOSE Courts interpret commercial agreements to reflect the most commercially reasonable result and the most commonly accepted business principles, unless the language clearly precludes this approach. The courts do not imply terms unless they are necessary to give effect to the intention of the parties.

(vii) CONTEXT The courts determine word meanings by their context. They do not interpret broad, general words to be as broad or general as they appear in isolation, but instead assign meaning based on the context in which they are used.

(viii) CONTRA PROFERENTEM *Contra proferentem* is a Latin phrase meaning that where a contract is ambiguous and there are two alternate interpretations, the courts choose the interpretation that favours the party that did not write the contract.

(c) Introduction of Additional Evidence

In contract interpretation, the primary evidence available is the contract document itself. Unless there is evidence of fraud or mistake, the parties are bound by the terms of the written contract that they have signed. However, sometimes other extrinsic evidence may be relevant to contract interpretation.

Courts have held that where a contract is entirely written and its written language is clear and unambiguous, extrinsic evidence is not admissible to add to, vary, or contradict the written words. This rule is known as the **parol evidence rule**. However, the common law provides for some exceptions to the parol evidence rule in instances where injustice might otherwise result. For example, the parol evidence rule does not preclude the contract interpreter from considering evidence of the factual background known to the parties at or before the date of the contract (sometimes referred to as the *factual matrix*). As well, Canadian courts have adopted the view that the subsequent conduct of the parties can be a useful guide to interpreting a written contract, in circumstances where the contractual language is ambiguous.

Case Study 8.1

Chinook Aggregates v. Abbotsford, [1989] B.C.J. No. 2045 (B.C.C.A.)

Facts

Abbotsford's tender documents contained the words, "the lowest or any bid will not necessarily be accepted." After receiving the bids, the owner wanted to award a contract based on award criteria that were not in the tender documents. They defended their right to do so based on the wording of the above-quoted clause. Chinook Aggregates, which was the lowest bid, sued for breach of contract.

Question

1. Is the owner entitled to rely on the literal meaning in order to accept any bid in its sole discretion?

Analysis

1. According to the Court, the tender documents did not provide the owner with unfettered discretion. The clause meant only that the owner could choose not to award a contract to any of the bidders, or could refuse to award to the low bidder for a legitimate reason.

8.2 Drafting Contracts

Three principal purposes govern the drafting of a contract:

- reflecting the business arrangement between the parties
- providing guidance to the parties in resolving future disputes by accurately defining and allocating risks and obligations, and dealing with foreseeable issues
- providing proper protections and incentives for each party.

Good drafting is a matter of fulfilling these needs, using clear language, and making the assumption that there will be problems throughout the term of the contract. Too many contracting parties assume that the project will proceed perfectly. However, if that were the case, formal contracts would not be necessary.

The following is a list of contract types from best to worst:

- The contract is negotiated on a detailed level between the parties, and the written contract accurately represents their mutual intention.
- One party drafts a written contract, which the other party accepts.
- The contract is a short written document, including a purchase order.
- The contract is oral, but is confirmed by letter, hopefully acknowledged by the other party.
- The contract is entirely oral.

In terms of content, careful attention must be paid to the fundamental elements of the contract. Some of these include defining the parties, the scope of each party's obligations and risks, including the cost, and the required time for performance. Parties often focus on ancillary contract terms and fail to address the fundamental obligations. In order to draft an accurate contract, parties must properly define and allocate the fundamental elements.

Moreover, parties must ensure that all elements of the contract are suited to the business arrangement. Standard form contracts, which are discussed in Chapter 11 (see p. 114), are often good starting points for drafting a contract but are rarely good end points. Proper contract drafting requires that the contract suit the business deal between the parties.

In addition, the language used in the contract must be clear, concise, and unambiguous. Only defined terms should be capitalized. Indices[1] should be used for longer agreements. Statements of risk or obligation allocation should be stated clearly. Each portion of the contract, including the general conditions and detailed specifications, must be consistent with the other portions. Only necessary documents should form part of the contract.

Lawyers are generally not competent at drafting detailed technical specifications; however, many technically proficient professionals believe themselves to be competent to draft contracts. For this reason, parties should seek legal input before agreeing to a contract. Correcting a contract before it is concluded is always easier than trying to undo a finished contract.

CHAPTER QUESTIONS

1. If there is a conflict between the literal interpretation of a clause and the intention of the parties, which governs?

2. Give an example of a situation in which expert evidence may be required in order to interpret a clause.

ANSWERS

1. The intent of the parties governs.

2. Where there is a customary or special meaning, as often occurs with technical documents in construction contracts, an expert may be needed to help interpret the contract.

---

[1] *Indices* in this context refers to numerical organization of the document.

CHAPTER 9

SELECTED CONTRACT ISSUES

Overview

Contract issues are at the heart of many disputes. Certain issues occur relatively rarely, such as the incapacity of the parties to contract or impossibility of performance. Other issues occur frequently. It is these frequent issues—agency, indemnities, change orders, subcontracts, unforeseen conditions, specifications, and contract administration—that are the subject of this chapter.

9.1 Agency and Authority

An **agent** is a person authorized to act on behalf of another party known as the **principal**. The agency relationship is essential to virtually all projects. There are three parties to an agency relationship: the principal, the agent, and the third party. For example, in a construction context, the owner is the principal, the architect or engineer is the agent, and the contractor is the third party.

The principal can provide authority to the agent in one of two ways: express authority or apparent authority. **Express authority**, sometimes called *actual authority*, is often created by a contract (which may or may not be written) between the principal and the agent. For example, an agreement for design services between the owner and architect should define the limits of the architect's authority to bind, or make agreements on behalf of, the owner. Any change authorized by the architect that fits within the express authority is binding on the owner, even if the owner does not like the change.

Apparent authority, sometimes called *ostensible authority* or *implied authority*, is created by representations (*e.g.*, statements) made by the principal to the third party (see Figure 9-1). The contract between the owner and the contractor almost always contains such representations. The CCDC 2 standard form construction contract contains the following language: "The Consultant will have authority to act on behalf of the Owner only to the extent provided in the Contract Documents, unless otherwise modified by written agreement. . . . "[1]

---

[1] CCDC 2, clause GC 2.1.1.

FIGURE 9-1 Actual and Apparent Authority

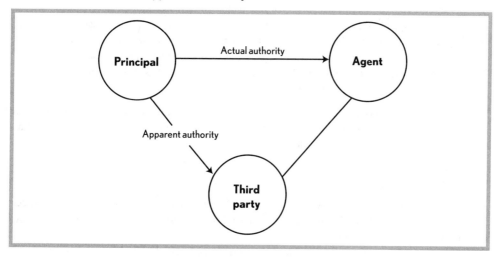

Even in the absence of any agreement between the principal and the agent, a representation by the principal to the third party that the agent has authority creates apparent authority in the agent. In fact, implied authority may be created simply by the principal's appointing an agent. If the principal knows that the agent is exceeding his or her authority but does nothing to correct the situation, that may expand the scope of apparent authority.

As illustrated by the above extract from the CCDC 2 document, not all grants of authority are absolute. An agent may be authorized to do only certain acts or may be authorized to enter into only one transaction on behalf of the principal. In addition, the grant of authority may place monetary limits on the agent's authority, such as the right to approve change orders with a value of less than $10 000. The principal must make these limits known to the third party; otherwise, the principal is bound if the agent exceeds the limits of actual authority but is within the apparent authority.

If *either* actual authority or apparent authority exists, the principal is bound by the contract or contract amendment negotiated by the agent. Both types of authority don't have to be present.

An agent owes a duty to act in the principal's best interest. However, not every act performed by a consultant is performed as an agent. The consultant must act impartially in some cases, such as when providing reports and reporting findings. The consultant must always be aware whether the act being performed requires impartiality or must be done in the best interest of the client.

Note that a statement by the agent that he or she has authority is not sufficient to create apparent authority. The following statement from the case of *Pacific National Leasing v. Marina Travel Agency Ltd.* contains a concise explanation of this principle:

I have further reference to *Bowstead and Reynolds on Agency, supra*, at page 366:

Where a person, by words or conduct, represents or permits it to be represented that another person has authority to act on his behalf, he is bound by the acts of that other person with respect to anyone dealing with him as an agent on the faith of any such representation, to the same extent as if such other person had the authority that he was represented to have, even though he had no such actual authority.

Further at page 370 the learned author states:

Therefore the representation must be made by the principal . . . It is usually said that a representation by the agent himself that he has authority cannot create apparent authority, unless the principal can be regarded as having in some way instigated or permitted it, or put the agent in a position where he appears to be authorized to make it. 'All ostensible' authority involves a representation by the principal as to the extent of the agent's authority. No representation by the agent as to the extent of his authority can amount to a 'holding out' by the principal.[2]

Thus, if a person holds himself or herself out as an agent and has no authority to act as an agent, the purported principal is not responsible; but the agent will be personally bound by whatever contracts he or she has entered into.[3]

9.2 Indemnities

An **indemnity** is an agreement by one party to bear the financial loss of another party for a specified event. This is sometimes known as a "hold harmless" agreement. The most common type of indemnity is an insurance policy.

Many contracts contain an indemnity provision. A contractual indemnity is a means of allocating risk. When one party (the indemnitor) indemnifies another (the indemnitee) against a specified risk, if that risk materializes, then the indemnitor bears the loss. The CCDC 2 contract contains the following indemnity clause:

> The Contractor shall indemnify and hold harmless the Owner and the Consultant, their agents and employees from and against claims, demands, losses, costs, damages, actions, suits or proceedings (hereinafter called "claims"), by third parties that arise out of, or are attributable to, the Contractor's performance of the Contract, provided such claims are:
>
> 1) attributable to bodily injury, sickness, disease, or death, or to injury to or destruction of tangible property, and
>
> 2) caused by negligent acts or omissions of the Contractor or anyone for whose acts the Contractor may be liable, and
>
> 3) made in writing within a period of six years from the date of Substantial Performance . . . or within such shorter period as may be prescribed by any limitation statute of the province. . . .[4]

The effect of this provision is to make the contractor liable for any loss that the owner may suffer as a result of a claim by a third party (someone who is not a party to the contract), which is attributable to the contractor's negligence (or the negligence of a party for whose acts the contractor is liable), and which results in property damage or personal injury (provided, of course, that the claim is made within the six-year limitation period described in the clause).

Under the CCDC 2 indemnity provision, the contractor indemnifies the owner for the contractor's negligence. The contractor is essentially saying, "I'll pay you if I am negligent and it causes you to incur a loss." In Canada, most indemnities simply reflect what would otherwise be assumed: that each party bears the risk of its own failures, whether to the other contracting party or to third parties. Some indemnity provisions (more commonly found in the United States) make the indemnitor

---

[2] *Pacific National Leasing v. Marina Travel Agency Ltd.*, [1996] O.J. No. 2827 (Ont. H.C.) at para. 23.

[3] One exception to this rule is that an "undisclosed principal," upon learning of the contract, may choose to ratify it, in which case the principal (and not the agent) will be bound to the contract.

[4] CCDC 2, clause GC 12.1.1.

liable for the negligence of the indemnitee. This would be akin to the contractor saying, "I'll pay you if *you* are negligent and incur a loss." In essence, it makes the indemnitor an insurer. However, some American courts refuse to enforce such provisions on the ground that they violate public policy, in that they may promote a disregard for safety. Hence, contracts made in the United States require close scrutiny, especially for indemnity provisions.

The indemnity clauses in many standard form construction contracts do not expressly exclude liability for design errors.[5] If the owner is sued by a third party for a loss caused by a combination of design and construction error, an indemnity that is not limited to contractor negligence may expand the potential liability of the contractor. Some jurisdictions have adopted *comparative negligence statutes* that reduce the liability of each party to the percentage of fault attributable to them; but a wide indemnity could negate that effect. The common law rule is that in the absence of comparative negligence statutes, if two defendants contribute to the same loss, both parties are each liable for the full amount. Some jurisdictions (especially in the United States) have enacted negligence statutes that eliminate joint and several liability entirely. These provisions are critically important in determining liability and should be examined carefully during negotiations.

An indemnity provision may create liability where none existed; but unless it contains an express statement that eliminates liability, it will not take away liability that otherwise exists. Many standard form contracts contain language that reinforces this principle.[6]

Many indemnity provisions also include legal costs as one of the losses indemnified. Inclusion of legal costs would mean that the owner would be reimbursed for the reasonable defence costs associated with such a claim.

In the absence of an indemnity clause, the contractor would still likely be liable to the owner and to third parties for the contractor's own negligence. As well, with or without the clause, the contractor may also be liable at common law for the negligence of its employees, under the doctrine of vicarious liability. Vicarious liability is a rule of law which makes one party liable for the acts of another. For example, an employer may be vicariously liable for the negligence of his or her employees, and a partner may be vicariously liable for some of the acts of his or her partners.[7]

9.3 Change Orders

Construction agreements almost always contain a provision for changes to the scope of work. Changes that result in an increase in the contract price are called *extras*, and those that result in a decrease are called *credits*. Unlike other types of commercial agreements, construction agreements frequently have hundreds of change orders issued during a large and complex project; and even on small and simple jobs, one can anticipate that some changes will be made. For example, the owner may want to enhance or change the design; unforeseen conditions may become apparent which necessitate additional work; one of the parties may suggest cost-saving measures; design error may necessitate changes; or the municipal authorities may demand changes to suit their interpretation of the building code. The cause of the change generally determines which party is responsible for paying for it.

---

[5] The CCDC 2 does not contain an express exclusion. The AIA A201 indemnity, on the other hand, is "only to the extent caused" by contractor (or subcontractor) error.

[6] The CCDC 2 provision is clause GC 1.3.1, which states that "Except as expressly provided in the Contract Documents, the duties and obligations imposed by the Contract Documents and the rights and remedies available thereunder shall be in addition to and not a limitation of any duties, obligations, rights and remedies otherwise imposed or available by law."

[7] Vicarious liability is discussed further in Chapter 13, p. 151.

Of course, the owner is responsible for requested enhancements or changes to design. In addition, the owner is usually responsible for changes caused by hidden conditions.[8]

(a) Extras for Design Negligence

Sometimes, the designer has a different interpretation of the building code than the officials who enforce it. Failure by the designer to anticipate the official interpretation does not necessarily constitute negligence. If the designer's interpretation was reasonable, and reasonable inquiries were made, that should exclude liability for negligence. However, in this type of situation, the contractor is generally entitled to an extra for any increase in cost flowing from the designer's modifications.

Participants in the construction industry understand that no design is perfect, and small changes due to design adjustments should be expected. This is one reason owners often build a contingency into the budget. Contingencies usually range from 5 percent to 10 percent of the contract amount, but the precise amount reflects the complexity, size and duration of the job. In theory, the owner is entitled to recover from the designer extra costs due to design negligence; but in practice, if the cost is within the acceptable and expected range, the owner does not generally attempt to recover from the designer.

(b) Impact Costs

When owners approve quoted prices for changes to the contract, they usually do so expecting that the price already includes all costs associated with the change. If the contractor completes the job and then declares that the costs are much higher, the owner can justifiably refuse to pay for the extra costs.

For example, in an often-cited case known as *Doyle Construction v. Carling O'Keefe*,[9] the contractor performed a number of changes to the contract, for which firm prices had been agreed. At the conclusion of the project, the contractor submitted a claim for "impact costs," which the contractor described as "those costs that arise from the inefficiency created by delays, interference, and changes of the sequence of the work." The owner argued that adding impact costs without giving notice at the time the change order was priced deprived the owner of "the opportunity of analyzing the reasons and exploring methods of cost reduction." The Court agreed, expressly adopting the statement of Mr. Justice Mahoney in a former case, *Walter Cabott Construction Ltd. v. The Queen*:[10]

> Again, it seems to me that when a person engages a contractor and when an extra price is agreed to in respect of a particular item, that person has a right to assume that the contractor has taken into account all of his costs, direct and indirect, flowing from the change in circumstances that led to the renegotiation and that he will not later be presented with a bill for additional compensation.[11]

In most standard form agreements, change order clauses require the contractor and owner to agree on a fixed price for the change. This practice provides some certainty to the owner, who must ultimately decide whether the change should proceed. Yet many contractors submit their change order quotations with the following qualification (or words to this effect):

---

[8] This risk can be negotiated, and often is. Some contracts place the risk on the contractor. Methods of assigning risk for hidden conditions are described on pp. 76–78.

[9] *Doyle Construction Co. v. Carling O'Keefe Breweries of Canada Ltd.* (1988), 27 BCLR (2d) 89 (C.A.).

[10] (1974), 44 D.L.R. (3d) 82 at p. 90 (Fed. Ct. T.D.). Note: the trial decision was successfully appealed; but in the authors' opinion, the extract from the trial judgment still represents good law.

[11] *Doyle Construction Co. v. Carling O'Keefe Breweries of Canada Ltd.* (1988), 27 B.C.L.R. (2d) 89 (C.A.) at para. 96.

The price quoted does not include any allowance for delay or impact associated with this change. The Contractor expressly reserves the right to claim at a later date for all costs associated with delay and impact resulting from this change.

If the owner accepts the quotation subject to this qualification, the contractor may not be barred from asserting an impact claim at a later date. The owner would also not likely obtain any certainty from the price, because the impact claim might be left open for later substantiation. For this reason, owners often include language in the request for quotation or in the contract itself which requires the contractor to include in the price all costs associated with the change, including impact costs, and identify *before the change is approved* how much of a delay or impact is being claimed.

One reason owners and contractors take these positions is the cumulative impact of many small changes. One small change may not cause a significant delay; but 100 small changes could amount to 10 percent (or more) of the contract price and result in a lengthy delay to completion. Contractors therefore try to protect themselves from this impact by reserving the right to claim for it at a later date.

(c) Timing and Pricing of Changes and Performance Under Protest

Standard form contracts may require the contractor to obtain approval for the change before proceeding with the work or else proceed at their risk. The CCDC 2 contract states that "The Contractor shall not perform a change in the Work without a Change Order or a Change Directive."[12] Owners often insist on a price from the contractor before allowing the change to proceed; but in some cases, the urgency of the change does not allow the owner this luxury.

Documentation of change orders frequently cannot keep up with construction activity. Contractors are under pressure to perform work that has been requested by the owner, often with the promise that the documentation will follow. Disputes develop when the documentation does not follow, and there is subsequent disagreement over what was said or promised.

Almost all construction contracts contain a provision allowing the owner or consultant to order work to proceed in the absence of a change order, with the cost and schedule issues to be resolved later.[13] These provisions are necessary in three situations: where time does not permit proper pricing of the work before the change must be implemented; where no agreement is reached on price; and where there is fundamental disagreement on whether the work is truly a change or is part of the original scope of work (see Figure 9-2 on page 74).

As discussed in Chapter 7 (see p. 158), the *Kiewit* case[14] involved a contractor who performed extra work as directed by the owner and then was denied the right to compensation on the basis that the act of performance was deemed an admission that the work was part of the contractor's responsibility. Subsequent performance under protest legislation in each province has made such problems easier to resolve.[15]

---

[12] Clause GC 6.1.2. A *change directive* is an order to proceed with the work when the owner and contractor cannot agree on the price or method of pricing the change.

[13] CCDC 2 clause 6.3 allows the owner to issue a "construction change directive" in the absence of complete agreement on the terms of a change order.

[14] *Peter Kiewit Sons' Co. of Canada Ltd. v. Eakins Construction Ltd.,* (1960) S.C.R. 361.

[15] See Chapter 7, pp. 58–59.

FIGURE 9-2 Change Orders and Change Directives

	Approval	**Timing**	**Pricing**
Change Order	Must be signed by owner	Cost must be agreed prior to commencement of work	Typically lump sum
Change Directives	May be issued by consultant	Cost may be determined at a later date	Typically cost-plus

9.4 Subcontract Issues[16]

Building construction has become a highly specialized industry, to the degree that the general contractor rarely has the expertise to perform all aspects of the work. Instead, *general contractors* (also called *prime contractors*) hire specialized *subcontractors* (also known as *trade contractors* or *subtrades*) to do most of the work. Electrical work, mechanical work, elevator building, drywalling, painting, millwork, glazing, roofing, and other work tend to be subcontracted. Often, the prime contractor performs his or her own carpentry and concrete work, although in some cases a prime contractor subcontracts everything except supervision and general overhead.[17]

Subcontracts are governed by the same contractual rules that govern prime contracts. However, because of the subcontractor's lower position in the contractual chain, subcontracts are subject to these additional issues:

- Subcontract formation through the bidding process
- Pay-if-paid clauses
- Incorporation by reference
- One-tier bonds
- Project delivery.

(a) Subcontract Formation Through the Bidding Process

The most common questions about subcontract formation involve whether there is an obligation on the part of a prime contractor to use a subtrade that was named in the prime contractor's bid; under what circumstances a subtrade is allowed to walk away from his or her bid; and what liability a subtrade or prime contractor faces when there is a breach of the "bidding contract." These issues are discussed in Chapter 10.

---

[16] Much of the content of this section was presented by Brian Samuels at a British Columbia Continuing Legal Education (CLE) seminar in November 2004, on the subject of Construction Law. It is reproduced with the kind permission of the Continuing Legal Education Society of British Columbia.

[17] *General conditions* refers to overhead items including (but not limited to) crainage, trailers, supervision, scheduling, bonding, and set-up costs.

(b) Pay-If-Paid Clauses

A **pay-if-paid clause** is essentially a clause that shifts the risk of non-payment by the owner from the general contractor to the subcontractor. It makes payment to the subcontractor contingent on the general contractor being paid by the owner. In short, the question is whether the prime contractor or the subcontractors should bear the risk of non-payment by the owner.

Many cases make a distinction between pay-if-paid clauses and pay-when-paid clauses. The former is a *condition precedent to payment,* a condition that must be satisfied before the legal obligation (in this case, payment) is created. In contrast, a **pay-when-paid clause** applies only to the timing of payment, not to whether payment must eventually be made. In these situations, courts do not forgive payment by the general contractor to the subcontractor if the owner never pays.

In the absence of a pay-if-paid clause, the only defences generally available to a non-paying prime contractor are failure by the subcontractor to satisfy conditions precedent, such as failure to perform the work properly, and failure by the architect to certify the work performed. Whether the owner pays the prime contractor is not relevant.

However, courts have sometimes been reluctant to enforce as conditions precedent clauses that shift the risk of non-payment from the contractor to the subcontractor because there are fundamental differences between such clauses and other conditions precedent. First, the triggering event for a pay-if-paid clause does not involve a breach or misconduct on the part of the subcontractor. In addition, non-payment by the owner is an event beyond the control of either the prime contractor or the subcontractor. Canadian courts sometimes try to soften the harsh effect of pay-if-paid clauses by interpreting them narrowly. The trial-level decision in *Arnoldin Construction & Forms Ltd. v. Alta Surety Co.* provides an excellent discussion of the differences between pay-if-paid and pay-when-paid clauses, the conflicting case authority in both Canada and the United States, and the policy considerations behind these cases.[18]

Some courts have interpreted these clauses as pay-*when*-paid clauses, meaning that the clauses do not shift the risk of non-payment; rather, they affect only the timing of the payment. Under such an interpretation, if owners do not pay the prime contractor on time, prime contractors have only a reasonable amount of time to seek to enforce payment from the owners. However, if prime contractors are ultimately unable to obtain such payment, they must still pay subcontractors, regardless of their inability to obtain those funds from owners.

In those jurisdictions where courts have enforced pay-if-paid clauses, the clauses usually include wording that explicitly creates a condition precedent. For example, the clause might say "payment by the owner to the Prime Contractor is a condition precedent to any obligation on the part of the Prime Contractor to pay the Subcontractor." For greater certainty, if the parties truly want to make such a clause enforceable, they should include words that discuss the assumption of risk (*e.g.,* " . . . and the Subcontractor accepts the financial risk of insolvency by the Owner . . . ").

---

[18] *Arnoldin Construction & Forms Ltd. v. Alta Surety Co.* (1995), 19 C.L.R. (2d) 1 (N.S.C.A.), reversing (1994), 13 C.L.R. (2d) 307 (N.S.S.C.).

(c) Incorporation by Reference

Virtually all standard form subcontracts and most custom-designed subcontracts contain a clause that incorporates into the subcontract the terms of the prime contract, insofar as those terms are applicable. For example, a standard form known as the BCCA 200,[19] once widely used in British Columbia but now obsolete, contained the following provision: "And, Whereas the Sub-Contractor has agreed with the Prime Contractor to be bound by the provisions of the Prime Contract, where applicable . . ."[20]

The danger to the subcontractor inherent in such a clause is that the subcontractor may not have reviewed the prime contract. The bidding process does not always allow sufficient time for a subcontractor to examine documents that are not specifically related to his or her trade. In addition, the prime contract may not be attached to the bidding documents and may only be available for review at the consultant's offices. However, if the documents are available, there is no legal excuse for failure to review. Furthermore, the prime contract may contain very onerous provisions, such as a *no-damages-for-delay clause*. No-damages-for-delay clauses prevent the prime contractor from recovering damages from the owner on account of delay, regardless of the cause. Such clauses, like pay-if-paid clauses, are not always enforced as written, because of their harsh and sometimes unfair result (see Chapter 19, p. 220). A prudent subcontractor should always request a copy of the prime contract and carefully review it before submitting a price.

For some projects, the prime contract may not have been negotiated at the time the subcontractor submits its bid price; yet the bid will be based on clauses that require the subcontractor to agree to a standard form contract which contains terms incorporated by reference. In these cases, the subcontractor may be able to successfully argue that his or her agreement to abide by the terms of the prime contract contemplated only a standard form prime contract.

(d) One-Tier Payment Bonds

In larger projects, owners often require that the prime contractor provide a *labour and material payment bond* in an amount equal to 50 percent of the prime contract price. This bond provides an additional remedy for an unpaid subcontractor.[21] Under a standard form labour and material payment bond (known in Canada as a CCDC 222), only a "claimant" is entitled to claim against the bond; and a claimant is defined as "one having a direct contract with the Principal for labour, material, or both . . . " This is known as a one-tier bond, because it covers only the first tier of subcontractors, suppliers, and workers under the principal.[22] Thus, if a subcontractor about to bid on a project is concerned about the possibility of non-payment, he or she should make inquiries whether there will be a bond in place and whether the subcontractor is included within the definition of "claimant" under that bond.

9.5 Unforeseen Conditions

The phrase "unforeseen conditions" is actually a misnomer, since the phrase usually refers to conditions known to possibly exist but that are not expected to occur. The standard CCDC 2 "unforeseen

---

[19] Published by the British Columbia Construction Association.

[20] There is a further provision in the BCCA 200 document that in the event there is a discrepancy between the terms of the prime contract and the subcontract, the subcontract prevails.

[21] Bonds are explained in greater detail in Chapter 17.

[22] In the United States, labour and material payment bonds often cover not only subcontractors, but sub-subcontractors as well. These are known as *two-tier bonds*.

conditions" clause[23] refers not only to conditions that were unforeseen but also to those not reasonably foreseeable. In other words, it is not sufficient for a contractor trying to prove entitlement to additional compensation to show that he or she did not anticipate the conditions. If the reason for not anticipating the conditions is misinterpretation of the contract documents or technical report or failure to carry out a required investigation, then the claim will likely be unsuccessful. However, such claims are still referred to as "unforeseen conditions" or "hidden conditions" in the industry.

A claim for unforeseen conditions is usually made by the contractor in the form of a request for a change order. However, on rare occasion, where conditions are found to be easier than anticipated, the owner may issue a change order seeking a credit. This owner-initiated class of claims is most prevalent in two situations:

- on renovation projects, wherein existing structures are often not accurately documented by drawings, and detailed investigation not may be possible until demolition is performed; and
- on projects involving subsurface (soils) conditions.

Owners may also attempt to shift the risk of unforeseen conditions to the contractor during the bidding process. The instructions to bidders require the contractor to carry out his or her own investigations and agree to accept this risk. For example, in the case of *Begro Construction Ltd. v. St. Mary River Irrigation District*,[24] the following clause was included in the contract:

> Site conditions. The Tenderer must examine the site of the work before submitting his Tender, either personally or through a representative and satisfy himself as to the nature and location of the work, local conditions, soil structure and topography at the site of the work, the nature and quality of the materials to be used, the equipment and facilities needed preliminary to and during the prosecution of the work, the means of access to the site, on site accommodation, all necessary information as to risks, contingencies and circumstances as may affect his Tender, and all other matters which can in any way affect the work under the Contract. The Tenderer is fully responsible for obtaining all information required for the preparation of his Tender and for the execution of the work. The Tenderer is not entitled to rely on any data or information included in the Tender Documents as to site or subsurface conditions or test results indicating the suitability or quantity or otherwise of site or subsurface materials for backfilling or other uses in carrying out the construction of the work. If the Tenderer requires additional time to conduct his own investigations or is of the opinion either that the site or subsurface conditions or that site or subsurface materials differ materially from that indicated by data or information included in the Tender Documents he shall promptly request such additional time or notify the Engineer in writing of his opinion before the time of Tender submission. The Engineer will either request the owner to extend the time for submission of Tenders to enable Tenderers to carry out further investigation or issue an addendum modifying the Tender Documents or both as the circumstances may permit.[25]

However, there are two inherent flaws in this type of approach. First, where there is little available information on the hidden conditions, such as in a superficial geotechnical report with

---

[23] CCDC 2 clause GC 6.4.1 authorizes extra payment if conditions differ materially from those indicated in the contract, or if "physical conditions of a nature ... differ materially from those ordinarily found to exist and generally recognized as inherent in construction activities of the character provided for in the Contract Documents."

[24] (1995), 15 C.L.R. (2d) 150 (Alta. Q.B.).

[25] See previous note at para. 161.

very few boreholes, prudent contractors include a contingency in their price to cover this risk. This contingency generally exceeds what the owner would pay if it retained the risk because the contractor must adequately protect itself from the potential risk. Second, courts sometimes recognize that contractors cannot reasonably do detailed investigations during the very short time available in the bidding process and therefore place the risk with the owner. The case of *Edgeworth Construction Ltd. v. N.D. Lea & Associates*[26] discussed the same issue in the context of engineering drawings rather than geotechnical investigation. In *Edgeworth*, the question was whether the contractor was entitled to rely on the engineering drawings available during the bid process, or whether the contractor should have conducted his own independent engineering review. The Court found that it was unreasonable to expect a bidder to undertake a review of that degree and detail:

> In the typically short period allowed for the filing of tenders—in this case about two weeks—the contractor would be obliged, at the very least, to conduct a thorough professional review of the accuracy of the engineering design and information, work which in this case took over two years. The task would be difficult, if not impossible. Moreover, each tendering contractor would be obliged to hire its own engineers and repeat a process already undertaken by the owner. The result would be that the engineering for the job would be done not just once, by the engineers hired by the owner, but a number of times. This duplication of effort would doubtless be reflected in higher bid prices, and ultimately, a greater cost to the public which ultimately bears the cost. . . .[27]

This reasoning also applies to geotechnical reports. Each bidder cannot realistically hire his or her own geotechnical expert.

On occasion, an owner will withhold pertinent information from bidders because the disclosure of that information will cause bidders to inflate prices. For example, if a report shows rock below grade, any prudent bidder reviewing that report will include a contingency for removal of the rock. The owner is not entitled to withhold such information. In the *Begro* case, the Court chastised the owner for withholding such information:

> If a party is induced to enter into a contract by not providing to the party all of the information that is available to the owner and the consulting engineers, it is not a complete answer to say to the contractor, "if you had used due diligence you would have found out that this information was available". Common sense dictates that this information in the hands of the engineers and owners was of vital importance to the contractor and all reports should have been made available to each contractor who submitted bids on the Chin project.[28]

9.6 Specifications and Drawings

Specifications detail the work requirements, including the quality and standard of materials and workmanship. They generally define the scope of the work to be undertaken in the contract and are read together with the drawings. Well-written specifications are clear, concise, and complementary, forming an integral part of the contract documents. Poorly drafted specifications include those that rely heavily on "industry standards," fail to complement the contract documents, and contain contradictions.

---

[26] [1993] 3 S.C.R. 206.

[27] See previous note at p. 220.

[28] *Begro Construction Ltd. v. St. Mary River Irrigation District* (1995), 15 C.L.R. (2d) 150 (Alta. Q.B.) at para. 166.

While lawyers often help write the "front-end" portion of agreements, they rarely look at detailed specifications for fear that they will be held responsible for technical details. The front-end portion of the contract should detail the allocation of risk and obligation, while the specifications should detail scope of work.

When a specification becomes the subject of litigation, the object of discussion is often one paragraph among hundreds of pages of text or one drawing detail among dozens of details. The court carefully considers each clause and each drawing as part of the whole contract. Here are some examples of potential challenges involved in interpreting specifications:

1. Inconsistency of terms and language
2. Repetition of material, often containing different information
3. Incorrect or inconsistent drawing notations
4. Lack of coordination between the drawings, between the specifications, and between the drawings and specifications
5. Use of drawings and specifications without appropriate amendment (this is even more prevalent with the advent of CAD and word processors)
6. Incorrect use of legal jargon
7. Descriptions containing not only the required result (which is acceptable) but also mandating a methodology (which should be solely for the contractor to determine)
8. Using drawings and notes, with no written specifications.

Specifications can be either performance specifications or standard specifications. A **performance specification** sets out the operating parameters that must be met by the final product, whereas a **standard specification** describes in detail all of the individual components of the final product. For example, a performance specification for elevators may specify the number of cars, the operating speed of the cars, and other performance criteria. Such performance specifications should be used in circumstances where the trade contractor has highly specialized design expertise and where the owner prefers to rely on that trade contractor to both design and construct the end product. Certainly, there are advantages and disadvantages to performance specifications; but mixing the two types of specification by imposing detailed components and methodology in a performance specification is not recommended.

9.7 Contract Administration

(a) Authority

As discussed in on pp. 68–70, an architect or engineer administering a contract is likely the owner's agent. There may be limits to the scope of that agency either in the contract between the owner and the contract administrator or in the contract to be administered (*e.g.,* the construction contract). The contract administrator must clearly understand his or her role in administering the contract and the limits on his or her authority. Some contracts clearly state that the contract administrator is not the agent of the owner, while at the same time, they give the contract administrator the authority to make decisions without consulting the owner. Parties entering a contract should avoid this type of confusion and take care to ensure that the role is properly described.

While contract administrators are generally engaged to protect the interests of the clients, there are also times when they must be impartial (see p. 69). For example, when the contract administrator is designated in the contract as the first interpreter of the contract, that person must do so impartially. Contracts should very clearly state the circumstances under which the contract administrator must be impartial. Contract administrators should also be wary of clauses that make them decision-makers in the event of disputes and should ensure that they understand their obligations regarding such a responsibility.

Once a contract administrator understands the scope and limitation of his or her authority, he or she must take care not to exceed that authority. For example, in a construction context, a contract administrator must enforce the requirements of the contract but should not dictate the contractor's means and methods for arriving at the result. If the contract administrator imposes such requirements or gives advice, and the contractor suffers a loss as a result, the contract administrator may be liable for the contractor's damages.

Contract administrators must follow the contractual procedures to the letter of the contract. As discussed in Chapter 6, failure to strictly enforce the contractual terms may lead to a waiver or estoppel and effectively alter or amend the parties' contractual rights. Similarly, approving change orders (see pp. 71–73) without the owner's authorization may lead to liability for the contract administrator.

(b) Timeliness

The contract administrator is often required to review shop drawings, conduct field reviews, provide certain advice, and make decisions. He or she must fulfill all of his or her roles in a timely fashion; failure to do so may result in delays to contract completion and impact the ability of the parties to perform the contract. For example, if an owner must make payment within 30 days of receiving an invoice, and the contract administrator has failed to review and process the invoice after 15 days, that delay may cause the owner to be in breach of his or her obligation to pay within 30 days. Thus, an administrator's failure to perform in a timely fashion can lead to delay and impact claims, as discussed in Chapter 19.

(c) Field Reviews

A contract administrator must also be aware that the contract documents may not spell out every detail of his or her obligations. For example, while the contract may call for reasonable reviews, it may not prescribe the number or frequency of these reviews. In litigation, a court will imply a reasonable number and frequency on the basis of what a reasonable contract administrator would do in the circumstances. Such a ruling is effectively a *tort standard*. Even if the contract does indicate a number and frequency of reviews required, a court may redefine that contractual element if it believes the contract was insufficient. Therefore, contract administrators must have a clear understanding of the level of reviews required. Moreover, if the owner will not agree to pay for a reasonable level, then the contract administrator must either decline the engagement or perform the sufficient reviews at his or her own expense.

Another reason to ensure clear understanding of the review process is the risk of differences between owner expectations and contract administrator intentions. Owners may expect the contract administrator to be a guarantor of the work product, while the contract administrator may believe that he or she is only supposed to be doing periodic reviews to determine general conformity to the contract documents. This discontinuity is often the source of considerable tension and litigation between the parties.

Once a problem is identified, a contract administrator should first make all of the parties aware of the problem and then follow up to ensure that it is corrected. Throughout this entire process, accurate and complete documentation is the best defence to a claim relating to a review.

Coast Hotels Ltd. v. Bruskiewich[29] illustrates this issue. The contract in question involved the retrofit of the domestic hot-water piping system in a hotel. The owner retained an engineer to design the hot-water system and a plumbing contractor to install it. Eighteen months after completion, the system began to leak; and the owner commenced an action against the contractor and the engineer for breach of contract and negligence.

The engineer's proposal had made provision for 10 site visits, based on the contractor's reputation in the industry, the engineer's previous experiences working with the contractor, and an anticipated "team" approach with participation by the contractor, the engineer, and the owner. The engineer testified that had this been a tendered job, it would have required 15 to 20 field reviews to account for uncertainty about the contractor's work.

Instead, the engineer conducted five site visits during the design phase, six during construction of the first phase of the project, and two during the second phase. While conducting the site reviews, the engineer was typically accompanied by representatives of the owner and the contractor. The reviews normally consisted of a walk-through of the project and included reviews of samples of the work being done. None of the engineer's observations created any cause for concern. However, the Court concluded that of the 13 site visits, only four had taken place while work was being done on the project. The balance had taken place at the beginning, before significant work was done, and at the end, after the walls had been closed in.

The Court concluded that the problems with the system did not lie with the design but rather with inadequate supervision, substandard materials, and poor workmanship. As a result, the Court apportioned liability 80 percent against the contractor and 20 percent against the engineer. With respect to the liability of the engineer, the Court acknowledged that it is not the engineer's responsibility to review all of the work done with respect to the project. However, it held that

> . . . there was also the duty of care not to supervise but to inspect the work performed by the contractor. The contract called for review of the existing site conditions and the owner's requirements, and included carrying out general contract administration for mechanical work, including site instruction and changes, and "carry out periodic site field reviews and provide formal written field reports."[30]

Thus, while a contract administrator must determine how many field reviews are necessary on a given project, the court will review the timing of the field reviews to determine whether they were adequate to perform the functions required of the field reviews. In this case, the Court concluded that while the number of field reviews was adequate, the timing was not.

In determining whether the field reviews themselves were adequate, the Court considered the following factors:

1. Alterations made to the plumbing meant there were no current as-built drawings.
2. The plumbing was affected by "aggressive" water, which had resulted in corrosive pitting in the old hot-water system.

The Court concluded that these factors should have resulted in increased vigilance during the site reviews. Instead, the engineer placed undue reliance on the employees of the owner and the contractor. The Court concluded that the engineer failed to meet the duty of care required by the circumstances:

---

[29] [2002] B.C.S.C. 1499 (S.C.).

[30] See previous note.

While the number of field reviews may sound adequate, were they effective and were they timed to perform the functions of such reviews? Were they performed to achieve their objective, that is, review the work and give effective assurances? In my view, they were not.[31]

It is important to note that the Court did not consider the budget available for field reviews. In general, professionals must conduct sufficient reviews to meet their professional standards regardless of the number of reviews paid for by the client and embodied in the contract. In this case, the actual number of visits did exceed the contracted amount but was still inadequate. From the Court's comments, it is also clear that reviews need to occur while work is actually proceeding.

CHAPTER QUESTIONS

1. An architect asks a contractor to add some concrete foundation walls at substantial additional cost. The work is urgent and must be started before a price can be agreed upon. Following performance, the contractor submits a price containing impact costs. The architect rejects the price but is prepared to agree if the impact costs are deleted. The architect threatens to delay the approval process unless the contractor agrees. Reluctantly, the contractor agrees. After the project is complete, the contractor claims for those impact costs. Is the contractor entitled to such costs?

2. An agreement between an owner and contractor states that the engineer can approve change orders with a value of less than $10 000. However, the owner instructs the engineer not to approve changes with value greater than $5000. The engineer approves a change, without the owner's permission, with a value of $8000. The owner is informed after the fact and refuses to pay for the change. Is the owner bound by the change order? Why or why not?

3. A contractor indemnifies an owner "for any and all losses, including pure economic loss, attributable in whole or in part to the negligence of the contractor." The contract is subject to the law in a jurisdiction that has not abolished joint and several liability. Due to a combination of design error and construction error, the owner suffers a $100 000 loss. The architect is 50 percent at fault. What is the maximum possible liability of the contractor? Why?

4. An owner developing a warehouse project hires a general contractor, who in turn subcontracts the concrete work to a subcontractor. The subcontractor hires a sub-subcontractor for some of the work. Can the sub-subcontractor claim against a standard form Canadian payment bond taken out by the owner? Explain.

5. Which of the following must inspectors do?
   a) Ensure absolutely that all work complies with the contract documents
   b) Refrain from interpreting the contract
   c) Ensure that they act within their authority

6. Explain why a performance specification should not describe the detailed components of the work.

7. Under which circumstances is a subcontractor bound by the terms of the prime contract?
   a) Only if he or she has read that prime contract before bidding

---

[31] See previous note at para. 76.

b) Whether or not he or she has read it before bidding if the subcontract clearly states that the subcontractor is bound by the prime contract

c) Only if the prime contract is reasonable

8. To what extent are pay-if-paid clauses enforceable?

a) Under all circumstances

b) Under no circumstance

c) In some cases

9. Explain the difference between actual authority and apparent authority.

10. The number of inspections is always directly proportional to the amount of the fee. True or False? Explain.

ANSWERS

1. This is a difficult scenario. On one hand, the contractor represented to the architect that the impact claim was being withdrawn, and the architect relied on such withdrawal in issuing the change. On the other hand, the contractor was under duress (although likely not enough to void the agreement) when the withdrawal was made. On balance, if the contractor had wanted to maintain the claim, a dispute should have been raised at the time. It is now too late for the contractor to make the claim, despite the duress.

2. The owner may be bound by the change order if the owner did not make the contractor aware of the monetary limit of the engineer's authority. If the owner did make the contractor aware that the engineer had limited authority, then the owner would not be bound by the change order.

3. The contractor could potentially be liable for the entire $100 000 loss because of the language of the indemnity.

4. A standard form Canadian payment bond is a one-tier bond, meaning that only subcontractors, and not the subcontractor's subcontractors and suppliers, can claim against the bond.

5. C

6. A performance specification is intended to define the intended outcome of the work, rather than the details of what is to go into the work. Therefore, the details should not be included.

7. B

8. C

9. Actual authority means a party has authority by contract, whereas apparent authority means a party has been held out by the principal as having authority.

10. The consultant's contractual obligation is limited by what is in the consulting contract. However, the consultant's tort obligation is based solely on what a reasonable professional consultant would have inspected in the circumstances. While the amount of payment for those services may be a factor in the court's determination of the appropriate number of reviews, it is certainly not determinative.

CHAPTER 10

GETTING TO A CONTRACT

Overview

The goal of every contract is the same: work done on time, on budget, and of highest quality. Achieving two out of three of those objectives is relatively easy; but achieving all three is very difficult. A buyer has many options for choosing project participants, but each option has inherent compromises and traps.

Methods for contracting the purchase of goods and services differ between jurisdictions, although many legal elements are common. For example, one popular method is the **call for tenders**, a request from a buyer of goods or services, often called the owner, to a group of potential sellers of those goods or services, often called **bidders**, to submit a bid for delivering a defined set of goods or services. While Canadian and American calls for tenders look very similar, Canadian calls for tenders strictly follow Canadian tendering rules, whereas the American practice is more relaxed, using the tendering process simply to narrow the choice to two or three potential bidders and then negotiate from there.

While many of the examples in this chapter are from the construction industry, the principles apply equally to the contracting for any good or service. The term "project" used in this chapter describes any procurement rather than just construction projects. Hence, engaging a forestry engineer to provide silvaculture services can be a project in the same sense as the building of a bridge.

Project delivery systems for projects are constantly evolving. In the past, design-bid-build, design-build, and construction management were the most common methods of project delivery for construction. Today, owners, consultants, and contractors can choose from a variety of new and innovative delivery models for major capital projects. These different contract delivery forms are discussed in Chapter 11.

In addition to choosing a delivery system, a buyer must also choose a method of payment. In the past, construction owners sometimes sought a "fixed price"; but experience has shown that achieving an end price that resembles the "fixed" price is often elusive. Many buyers now use cost-plus or cost-plus with incentives for cost reductions. Public buyers have also used public-private partnerships (P3s, privately funded initiatives, or PFIs). In some jurisdictions, alliance agreements have also been used.

Each process, delivery system, and payment method has distinct advantages and disadvantages. However, project participants are often unaware of the fundamental assumptions and compromises inherent in the process or delivery system in which they have become involved. For a project to be successful, all parties must fully understand the fundamentals of the form of agreement upon which the project is founded.

This chapter examines the processes used to arrive at agreements and the main project delivery methods. The relationship between these concepts is shown in Table 10-1.

TABLE 10-1 Contract Delivery Processes, Methods, and Forms

Choosing Whom to Contract With (the Process)	Choosing How to Contract (the Delivery Method)	Choosing How to Pay (see Chapter 11)
Request for Qualifications	Design-Bid-Build	Fixed Price or Stipulated Sum
Tender	Design-Build	Cost-Plus
Request for Quotation	Design-Build-Operate	Unit Price
Request for Standing Offers	Design-Build-Operate-Finance	Alliance
Request for Proposals	Construction Management	Public-Private Partnership
Pre-Qualification		
Hybrid Methods		

10.1 Delivering a Project

There are several different commonly used methods for delivering a project from inception to completion and beyond. But with so many different options, parties are often confused which to choose. Such confusion can lead to increased pricing and fewer willing participants in the process.

All projects involve the same steps: conceptualizing the project, designing it, constructing it, inspecting it, financing it, and finally putting it into service. Each project delivery process has its own method of executing these steps. The differences between the most popular delivery processes are shown in Table 10-2.

TABLE 10-2 Responsibilities in Each Project Delivery Method

Project Delivery	Design-Bid-Build	Construction Management	Design-Build	Design-Build-Operate	Design-Build-Operate-Finance
Conceptualize	Owner/Consultant	Owner/Consultant	Owner/Consultant	Owner/Consultant	Owner/Consultant
Design	Owner/Consultant	Owner/Consultant	Contractor	Contractor	Contractor
Build	Contractor	Contractor	Contractor	Contractor	Contractor
Inspect	Owner/Consultant	Owner/Consultant	Contractor	Contractor	Contractor
Pay	Owner	Owner	Owner	Owner	Contractor/Owner
Use	Owner	Owner	Owner	Contractor	Contractor

For example, the most common construction project delivery method used in North America is design-bid-build. In the **design-bid-build** process, the project owner and his or her consultant conceptualize and design the project in its entirety; the owner then hires a contractor to build it, generally through a tendering process; the consultant inspects the construction; and the owner finances the project and puts it into service. A variation on that theme is the **design-build** process, wherein an owner, with or without a consultant, conceptualizes the project; hires a contractor to design it, build it, and inspect it, with or without independent secondary inspection; and then finances the project and puts it into service.

10.2 Transfer of Risk and Obligation

The goal of each buyer of a good or service is to create the optimum level of risk and obligation for each party. While most parties focus on risk, the obligations of each party are also important and can be altered by the method of delivery. Control over the design is one example of an obligation that more than one party may want.

Figure 10-1 shows a list of the most common risks and obligations in a project delivery method. The contract will assign each of these risks and responsibilities to certain parties:

FIGURE 10-1 Risks and Obligations in Project Delivery

- Design control
- Time requirements
- Warranty
- Changes in legislation or common law (*i.e.*, an increase in tax)
- Site conditions (*e.g.*, subsurface conditions)
- Revenue loss from the project not functioning properly
- Cost increases
- Ability to perform
- Authorities and permits (*e.g.*, an environmental permit)
- Environmental impacts (*e.g.*, uncovered contaminated soil)
- Limitations on liability
- Termination rights
- Responsibility for insurance
- Weather (*e.g.*, damage caused by a hurricane)
- Insurance
- Transfer of ownership (*e.g.*, when does risk of loss transfer from seller to buyer)
- Liability for third party claims

Each of the "variables" in Figure 10-1 can be wholly or partially transferred from one party to another in the contract between the parties, although parties must recognize that there is generally a cost associated with each transfer. Risk allocation should be carefully considered (See Figure 10-2). In some instances, the parties can also choose to transfer risks and obligations assigned to them to third parties, such as suppliers, insurers, and bonding companies.

FIGURE 10-2 Tips for Transferring Risk and Obligation

- Rather than a complete transfer of risk, consider a capped contingency for certain risks (*i.e.*, the buyer pays for the first $100 000 in respect of a certain risk, and the seller bears the balance of the risk) or the reverse (*i.e.*, the seller pays the first portion of the risk and the buyer retains the balance).

- Examine risk and obligation allocation as an optimization exercise. Placing risks or obligations on the wrong party will compromise the project from a cost perspective both in terms of adding unwarranted costs and driving away potential participants.

10.3 The Processes for Choosing Project Participants

Buyers usually choose a consultant before other project participants. Traditionally, a buyer chooses a consultant through references and developed relationships.[1] Through the consultant, the buyer then develops tender documents and chooses a seller based on submitted bids. This chosen seller then contracts suppliers to complete the project. Buyers use several methods to select these project participants.

The first decision a buyer has to make is whether a formal tender process suits the needs of the project. For government institutions, this decision may be regulated by law. For example, because of trade agreements, most significant Canadian government purchases must be conducted as competitive open procurements. Other organizations need to decide whether a formal tender process is more advantageous than negotiating a contract with one party.

In determining the best approach, a buyer has to consider the time, cost, and risk associated with each process. These considerations include the need for thorough identification of the goods or services being purchased, preparation of detailed documentation, response to questions from bidders during the bid process, and formal evaluation of bids and award of contract. At the other extreme is a time-tested method for arriving at an agreement—negotiation. In between, there are a myriad of alternatives that can be defined by the buyer.

A buyer's policies may require use of a formal tendering processes. For those who need to maintain neutrality in the selection of a supplier, open tender may be the best choice because it theoretically ensures that the successful bidder is selected through a formal evaluation of published and transparent criteria. This reduces the likelihood of allegations of favouritism or bias and reduces the risk of disputes or complaints.

Buyers need to understand that once a formal bid process has been started, a series of legal rights and obligations come into effect; and in some cases, obligations under national or international trade agreement become relevant. In Canada, the courts have generally held that tendering replaces negotiation, such that negotiating with bidders as part of the tendering process is illegal, unless the inclusion of a negotiation step is very clearly explained in the call for tenders.

---

[1] Consulting contracts, also called professional service agreements, are discussed in Chapter 11.

Buyers should be aware that a court looks behind the name of a document to determine its fundamental purpose. Simply calling a request a call for tenders does not make it a tender; conversely, calling a tender a request for qualifications or request for expressions of interest does not exempt the document from the strict tendering rules if it indeed is a tender. One significant difference between a call for tenders and a request for proposals is the level of detail available in the **scope of work**, which is the definition of the goods and services to be supplied pursuant to a contract. A call for tenders may have 500 000 pages of highly detailed specifications to build a building, whereas a request for proposals may have generic information about potential advertising services. While true requests for proposals do not need to follow the strict tendering protocols, good faith and fairness may compel an owner to follow the terms and conditions of a tender process anyway.

(a) Types of Delivery Systems

(i) **REQUEST FOR QUALIFICATIONS** A **request for qualifications** (RFQ) is a process used to choose a project participant based solely on the qualifications of the respondents, rather than on other criteria, such as price. Requests for qualifications are often used for selecting professionals. They can also be used for selecting design-builders, participants in an alliance agreement,[2] or participants in public-private partnerships.[3] Given the increased levels of trust and the longer-term nature of the relationship between the buyer and the seller in alliance agreements and public-private partnerships, many buyers have found that bottom-line price is neither the sole nor necessarily the best selection criteria.

A request for qualifications typically involves only a generic project description (*e.g.*, a new computer software system) together with a framework outline of the services required (*e.g.*, a software engineer to design and implement).

(ii) **INVITATION TO TENDER OR TENDERING** An **invitation to tenders,** also known as a call for tenders or tendering, is commonly used when numerous potential sellers can supply goods or services. The invitation or call defines the procurement requirements to allow the owner to evaluate bids against mandatory criteria. Tendering is normally used when buyers are looking only for the lowest price response, which will be accepted without negotiation.

Tendering is a very formal process. In a tendering process, the buyer generally has a clearly defined scope of work and requests prices from either a selected list of potential sellers or more generally from the industry at large. Tendering sometimes includes a preliminary process of prequalifying or shortlisting potential sellers based solely on qualifications, rather than price, so that only sellers with a realistic prospect of being successful submit a full tender.[4]

(iii) **REQUEST FOR QUOTATION** A **request for quotation** is an informal process through which separate requests are made with no formal closing time or tender closing conditions. A buyer uses a request for quotation when the objective is to find the lowest price response. It is also useful when procurement requirements have not been adequately defined, due to constraints of time or costs; thus, further negotiation may be necessary.

---

[2] See Chapter 11, pp. 119–120.

[3] See Chapter 11, p. 120.

[4] See Section vi p. 89.

(iv) REQUEST FOR STANDING OFFERS A **request for standing offers** is a process through which the buyer prearranges prices, terms, and conditions with sellers for frequently ordered goods or services. The goods or services are not purchased until the buyer issues a "call-up" or requisition, which can take place any time during a specified time period. The contract is to supply goods or services at set or predetermined rates if and when they are needed.

(v) REQUEST FOR PROPOSALS (RFPS) A **request for proposals** (RFP) is a process through which a buyer invites potential sellers to propose a solution to a particular problem; but the details in these potential solutions are less well developed than those in a call for tenders. The success of a proposal submitted in response to a RFP is normally based on the quality of the proposed solution, not only on price.

An RFP has more defined project parameters than a request for qualifications. For an RFP, the buyer creates a more detailed framework for the services needed (*e.g.*, design-build services to construct a new water treatment plant with specified criteria). Each potential seller that submits a detailed proposal responding to the RFP is called a **proponent**. While an RFP is a more formal process than a request for qualifications, the buyer expects to negotiate a final contract with the best proponent or one of the top proponents, rather than proceeding through a formal tender.

(vi) LETTER OF INTEREST OR PRE-QUALIFICATION **Pre-qualification** is a screening process through which potential sellers pre-qualify, which limits the ultimate call for tenders or request for proposals to a small number of qualified sellers. Sellers submit a *letter of interest* about a future request for proposal or invitation to tender, so that the buyer can identify the seller's level of expertise. A pre-qualification process transforms the first step in the bidding process (determining if a seller is qualified) into a precondition for bidding. One benefit is that underqualified sellers are spared the time and effort of bidding.

Generally, if the bidders pre-qualify, the buyer will have less flexibility to reject a subsequent low bid on the grounds of the seller's ability or their financial or commercial circumstances, unless the seller's circumstances have changed adversely since pre-qualification, or unless the buyer can demonstrate a *bona fide* concern about information not disclosed by the seller. However, buyers may still be able to take into account differences between the sellers' qualifications if they have indicated that such would be considered in awarding the contract.

(vii) HYBRID METHODS Buyers can create their own unique selection process. For example, a tendering process could include negotiations at certain stages. Except for government contracts, the process itself does not have to be fair, although a patently unfair process may allow a court to step in and provide compensation. However, once buyers have established and started a process, they must follow it to completion.

Each process has distinct features with different legal consequences that both buyer and seller must recognize and understand in order to make informed choices.

(b) Procedural Issues

(i) FAIRNESS AND GOOD FAITH The term "fairness" is overused and under-defined as it relates to procurement. To each party, fairness has a different meaning. In Canada, the courts have held, subject to some exceptions, that there is no obligation of fairness in precontractual negotiations between two negotiating parties.[5] On the other hand, the courts have also required that any form of open procurement process must have an element of fairness and good faith. These decisions are often misunderstood and taken well beyond the meaning the courts intended.

Does the process defined in the procurement documents have to be fair, or does only the application of that process have to be fair? In Canada, the answer is that the process itself does not need to be fair or equal for all parties; but once defined, the process must be applied with fairness and good faith. Indeed, the strengths and weaknesses of individual sellers mean that no system could be perceived as "fair" by all parties. The inclusion or exclusion of particular criteria within the process will be less or more advantageous to each seller (See Figure 10-3). Including a criteria that favours sellers with whom the buyer has a favourable history would arguably be unfair to sellers with an unfavourable history or no history whatsoever; but not taking the historical relationship into account would be unfair to other sellers, who have worked long and hard to establish the relationship, and to the buyer, who wants security. Subjective criteria as well as objective criteria are needed to make a good business decision.

FIGURE 10-3 Tips for Defining a Tendering Process

> - Define the simplest approach that meets the needs of the project. Defining an elaborate, complex, and expensive process for a small procurement will limit the number of interested parties.
>
> - Use criteria that emphasize value rather than simply relying on the lowest bid approach. As pointed out by the Supreme Court of Canada, "The purpose of the [tender] system is to provide competition, and thereby to reduce costs, although it by no means follows that the lowest tender will necessarily result in the cheapest job." The Supreme Court recommends that buyers take a more nuanced view of "cost" by including in the criteria for awarding the contract factors other than price. This approach may be impacted by the fairness considerations discussed in this chapter, since this approach relies in part on subjective criteria.

Source: *M.J.B. Enterprises Ltd. v. Defence Construction (1951) Ltd.,* [1999] 1 S.C.R. 619.

Given the impossibility of creating a set of criteria that is fair to all, and recognizing that even if this were possible, it might not result in the best procurement process for the buyer, Canadian courts have simply required adherence to the process once established and have not to date permitted inquiry into the fairness of the system itself.

(ii) OPENNESS OR TRANSPARENCY Some sellers and buyers believe that the reasons and results of a public procurement process must also be made public. Certainly, some buyers choose to make available some or all of the reasons and results; but there is no legal requirement to do so. However, public buyers usually reveal the results publicly.

---

[5] *Martel Building Ltd. v. Canada,* 2000 SCC 60.

Does transparency require that everyone should have access to the reasons and results of a procurement process? Many government and some non-government buyers reduce their objective and subjective criteria to a weighted list and then publish it, in order to create the perception of transparency. The difficulty with this kind of "openness" is that the weighting must be created before the buyers really understand their projects. This practice can create a negative atmosphere: for example, a seller who receives fewer points on the "ability" criteria than the competition may decide to correct the perceived injustice through litigation.

For these reasons, it is risky to share the details of the application of the process. Instead, buyers should simply make the decision they believe to be most advantageous and enter an agreement with the successful party.

Figure 10-4 suggests ways to implement a complex procurement process.

FIGURE 10-4 Tips for Implementing a Complex Procurement Process

- There appears to be a trend toward more complex, multi-layered procurement processes to promote fairness and transparency. Such processes emphasize the risks associated with litigation from bidders disappointed with specific elements of the procurement process; but they do not fully insulate the buyer from such litigation. Buyers that are not bound to follow open procurement processes should consider negotiation or a combination of an RFQ followed by negotiation with a short list of potential sellers.

- The costs and risks associated with complex procurement processes create disincentives to potential sellers. Buyers should consider paying potential sellers all or some of the costs associated with bidding, at least after they have reached a certain stage in the process. (e.g., The three short-listed bidders will each receive $50 000 as part payment for their tender costs.)

- Buyers should define the process and the role of the process evaluator such that both can use subjective criteria. While subjective criteria make it more difficult to "prove" fairness and transparency, using only objective criteria reduces value and likely creates unfairness to some potential sellers.

(iii) **THE COST OF FAIRNESS AND TRANSPARENCY** In an effort to be "fair and transparent," some buyers, particularly public buyers, implement selection processes that are very complex. These selection processes include fairness monitors, fairness commissioners, ministerial review panels, and engaged fairness consultants. The problem with many of these attempts is that they sometimes fail to define the parameters for success.

A *fairness monitor* is a monitoring committee established with the express purpose of demonstrating fairness in applying the procedures for the procurement process; ensuring that the transaction process adheres to the principles of fairness, openness, transparency, and integrity; and assessing whether the design and execution of the process was fair from a third party perspective. Thus, the role of the fairness monitor is not to measure degree of compliance with the process; rather, it is to check that procedures have been followed.

Professional evaluators are members of an Evaluation Society. The United Kingdom, Canada, and the United States each have chapters of the Evaluation Society. The stated goal of the Evaluation Society is to promote and improve the theory, practice, understanding, and utilization of evaluation and its contribution to public knowledge and to promote cross-sector and cross-disciplinary dialogue and debate.

Membership in the society is voluntary and has no qualification requirements, other than the payment of dues. However, each evaluation society has developed guidelines for good practice. These guidelines define principles for establishing a proper evaluation process.

In addition to providing useful guidelines, professional evaluators can provide useful consultation to a buyer or can supply a formal report which can improve the buyer's future selection processes.

In addition, some public bodies have appointed individuals or groups to review complaints about procurement fairness and intervene as necessary. This intervention generally deals only with whether the government or agency followed the required process, although it can in some cases also deal with whether the process itself was appropriate. The *fairness commissioner* is similar to a fairness monitor, except that he or she only becomes active when problems arise. The fairness commissioner process can be useful in assisting applicants and public buyers to achieve fair and transparent processes.

Finally, many public bodies have instituted formal *appeal or contestability procedures* for dealing with procurement disputes. These panels have varying powers, but all attempt to provide an alternative to litigation for the resolution of procurement disputes. The agency overseeing the appeal process must ensure that the individuals hearing the appeal are independent and appropriately skilled and that the appeal process is well defined.

10.4 Tendering

The bidding process is a mechanism used to facilitate contract formation. The terms "bid" and "tender" can be used interchangeably, as can the terms "**bidding**" and "**tendering**," and the terms "buyer" and "owner" and "seller," "bidder," and "contractor." The law of contract formation (offer and acceptance) explained in Chapter 6 applies generally to all contracts; however, when the bidding process is used, the courts have implied terms and interpreted clauses in very specific ways. The bid process has also resulted in some unusual rules for subcontract formation.

The bidding process requires every party involved to be familiar with the rules of the game. Failure to play by the rules may result in liability for damages. Large amounts of money can be won or lost by bidders and buyers before the project really starts, simply as a consequence of breaking the rules of bidding. For example, a buyer who improperly awards a contract to a bidder based on a non-compliant bid may in some circumstances be held liable to the lowest compliant bidder and may be ordered to pay damages equal to the lowest compliant bidder's anticipated profit on the project.

The law relating to mistake is relevant as well, with respect to the right of the bidder to withdraw a bid and the right of the recipient of a bid to enforce a bid bond if the bid contains a mistake. However, courts are reluctant to allow bidders to withdraw, even where a mistake has been made, except under well-defined circumstances.

For those watching the bidding process for the first time, the frantic activity and secrecy at the closing of a bidding process may seem surprising and unnecessary. Surely there must be a more orderly and civilized way of choosing a seller and its suppliers for a project. Yet to those involved in the process, the reasons for such behaviour are self-evident. Bid depositories (see p. 101) have been developed in order to deal with some of the problems inherent in the bidding process, including bid shopping.

(a) Contract Formation in the Bidding Process

Historically, the law relating to offer and acceptance is derived from the English common law. That law drew and continues to draw a distinction between an *offer* and an *invitation to treat*. An offer can be accepted in order to create a contract, whereas an invitation to treat cannot. An invitation to treat is considered a request for someone else to make an offer. In other words, it is an invitation that is not made with the intention that it shall become binding upon acceptance.

The difficulty with this distinction is whether an invitation for tenders is considered at law to be an offer or an invitation to treat. Until the Supreme Court of Canada decision in *Ontario v. Ron Engineering*[6] in 1981, an invitation for tenders was considered merely an invitation to treat. That assumption was overturned in *Ron Engineering*. As a result, any analysis of current Canadian tendering law begins with *Ron Engineering*.

The starting point in analyzing *Ron Engineering* is to understand that when a seller submits a bid in response to an invitation, that bid may be considered an "acceptance" of the invitation; therefore, a "bidding contract" has been created. In the words of the Court, "Contract A (being the contract arising forthwith upon the submission of the tender) comes into being forthwith and without further formality upon the submission of the tender."[7] The Court described the bidding contract as "Contract A" and the construction contract as "Contract B." Contract A is created upon submission of the bid, and Contract B is created upon *acceptance* of the bid.

There are many Contract A's, but only one Contract B in each contract process. Each bid that complies with the terms and conditions of the call for tenders creates a Contract A between the buyer and the bidder. However, the only Contract B ever formed is between the buyer and the successful bidder. For example, Figure 10-5 on the following page shows that Bidders 1, 2, 3, and 4 each submitted compliant bids, bids that fully complied with the call for tenders; but Bidder 5's bid was non-compliant because it was submitted late. Thus, Contract A's were formed with Bidders 1, 2, 3, and 4, but not with Bidder 5. Ultimately, the buyer chose Bidder 2 as the successful bidder based on the conditions in the call for tenders; and Contract B was formed between Bidder 2 and the buyer.

However, this definition begs the question of what the terms of Contract A might be. The Court in *Ron Engineering* offered some guidance on that point. The following terms for Contract A's in *Ron Engineering* were either express or implied, based on the invitation for bids:

- The contractor has the right to recover the tender deposit 60 days after the opening of tenders if the owner does not accept the tender.
- The contractor cannot revoke the bid.
- Both parties are obliged to enter into a contract, Contract B, upon the acceptance of the tender.
- Subject to the terms and conditions of the call for tenders, the buyer must normally accept the lowest tender.

Thus, the overriding policy consideration behind the Court's decision was that the "integrity of the bidding system must be protected where under the law of contracts it is possible so to do."[8]

---

[6] *Ontario v. Ron Engineering & Construction (Eastern) Ltd.,* [1981] 1 S.C.R. 111.

[7] See previous note at p. 121.

[8] See previous note at p. 121.

FIGURE 10-5 Example of Bids and Contract Formation

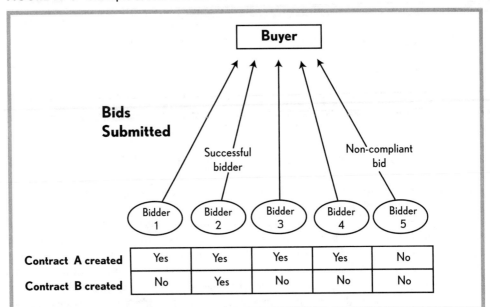

	Bidder 1	Bidder 2	Bidder 3	Bidder 4	Bidder 5
Contract A created	Yes	Yes	Yes	Yes	No
Contract B created	No	Yes	No	No	No

In 1999, the Supreme Court of Canada modified the law in *Ron Engineering* in the case known as *MJB*.[9] In *MJB*, both parties accepted the *Ron Engineering* interpretation described above, since it had been the law in Canada for almost 20 years. However, the Court in *MJB* decided that there may be circumstances where submission of a bid does not create Contract A, and it emphasized some of the language in *Ron Engineering*:

> However, other passages suggest that Estey J. [the judge who wrote the decision in *Ron Engineering*] did not hold that a bid is irrevocable in all tendering contexts and that his analysis was in fact rooted in the terms and conditions of the tender call at issue in that case. . . .
>
> Therefore it is always possible that Contract A does not arise upon the submission of a tender, or that Contract A arises but the irrevocability of the tender is not one of its terms, all of this depending upon the terms and conditions of the tender call. To the extent that *Ron Engineering* suggests otherwise, I decline to follow it. . . .[10]

To summarize, *MJB* held that Contract A does not automatically come into existence when a bidder submits a bid. Rather, the parties must have also intended to create a contractual relationship. *MJB* provided two alternate approaches for owners:

1. They can carefully define the parameters of the relationship between the owner and the bidders; or

2. They can specifically indicate in the tender documents that no relationship exists between the owner and the bidders.

---

[9] *MJB Enterprises Ltd. v. Defence Construction (1951) Ltd.*, [1999] 1 S.C.R. 619.

[10] See previous note at p. 630.

One danger of the second approach is that a court may ignore this wording and determine that Contract A still exists and then imply the terms that it sees fit.

This principle was confirmed in the recent Tercon,[11] in which the Supreme Court held that Contract A came into existence in an RFP because, for example, the bid was irrevocable, there was bid security, proposals had to be received before a set closing time, there were detailed evaluation criteria, a specific form of agreement was attached, pricing was fixed and non-negotiable, and the proponents agreed to enter into a contract if they were chosen.

The other important ruling in *MJB* was that the buyer was not entitled to accept bids that did not comply with the tender call (non-compliant bids). Hence, buyers must now reject bids that do not materially comply with the call for tenders. Buyers and their consultants should therefore minimize the number of mandatory criteria to limit the number of non-compliant bids.

(b) The Bidding Process

Contract formation (offer and acceptance) in the purchase of goods or services is most often accomplished through the bidding process. Bidding is used in many different industries. The process involves a number of stages: preparation of the bid documents; solicitation or invitation of bids; pricing by the potential sellers; submission of bids; evaluation by the buyer; and award of the contract.

The law relating to bidding recognizes rights, obligations, and liabilities at each stage in the process. As a result, bidding sometimes leads to lawsuits—by unsuccessful bidders, by successful bidders who have made mistakes in their bids, by suppliers to the sellers whose pricing information has been used by sellers to "*bid shop*," by sellers against suppliers who withdrew their prices after those prices were incorporated into sellers' bids, and by buyers and sellers against each other seeking to keep, or protect, a bid deposit or bid bond.

A buyer may invite bids in one of three ways: by invitation only to a select group of pre-qualified potential sellers, by open invitation, or through a bid depository. The method used may have an impact on the rights of the parties, particularly if there are allegations of bid shopping.

(i) **PREPARATION OF THE BID DOCUMENTS** The first stage of the bidding process is preparation of the bid documents. These documents typically include the drawings, specifications, general conditions, special or supplementary conditions, instructions to bidders, and the invitation to bidders. On larger and more complex projects, a lawyer may be contracted to prepare the contract documents and review the non-technical aspects of the specifications and general conditions.

In general, the tender documents must clearly describe the process, the required elements, the terms, and most importantly, the process and criteria for awarding the contract. Ideally, from the buyer's perspective, the tender documents should also contain acknowledgements from the bidders that they have no claim against the buyer or the buyer's consultant for damages resulting from the tender process. However, such clauses are not always enforced; for example, when the buyer or the consultant acted in bad faith.

The drafting of the invitation is critical in determining the terms of the agreement between the buyer and the successful bidder. For example, the invitation usually requires the bidders to conduct independent investigations. The invitation also allows the buyer to allocate risks to the seller. Buyers should draft invitations with care and with the benefit of legal advice. The invitation should also define the buyer's right to reject bids and to determine the ground rules for acceptance. The invitation is the buyer's opportunity to set out the factors to be used for acceptance and rejection.

---

[11] *Tercon Contractors Ltd. v. British Columbia*, [2010] S.C.C. 4.

In the private sector, almost all invitations to bidders contain a statement that the lowest or any bid may not be accepted, and that the buyer reserves the right to accept or reject any bid in its sole discretion. These statements, often called *privilege clauses*, are similar to disclaimers in that they are designed to protect the buyer against liability for improper rejection of a bid. In addition, public bodies, such as municipalities, may be required by statute or charter to follow certain rules in the evaluation and acceptance of bids, which usually require that the lowest conforming bid from a qualified seller be accepted. However, the statement that the lowest bid may not necessarily be accepted cannot always be taken at face value, because that statement in and of itself does not permit the buyer to choose any bidder (see also Case Study 8.1, p. 66).

While in some cases the privilege clause itself can provide the buyer with the "privilege" of awarding to any contractor, the better interpretation of the law is to ensure that the terms and conditions of the tender documents clearly spell out all criteria to be used in evaluating and assessing bids. Thus, any buyer policies governing the exercise of the privilege clause should be disclosed. For example, if a buyer has a policy of contracting with local contractors in preference to foreign ones, that preference should be explicitly revealed in the tender documents. If instead the buyer employs undisclosed criteria in the evaluation of bids, an unsuccessful bidder may sue the buyer for damages for a breach of Contract A.

Many procurement documents also contain disclaimer clauses. These will be treated the same way as any other disclaimer clause (see Chapter 15, p. 171). In *Tercon*,[12] the Court in a split decision (meaning that the judges were not unanimous and were, in this case, split 5-4) held that the disclaimer clause was not effective because it did not contemplate the breach of Contract A that was committed by the Province. This reinforces the view that disclaimer clauses must be very specific to be effective.

(ii)   SUBMISSION OF BIDS After the buyer posts or delivers the invitation, potential sellers must prepare their response. The time period between the invitation and the closing of the bid period is often very short, and potential sellers must work quickly to estimate all of the labour, material, and equipment. In many cases, sellers have to rely on other suppliers of goods and services not only to perform the work but also to price it. Similarly, the suppliers often obtain prices from a further level of suppliers before submitting their prices to the ultimate bidder.

These supplier quotes can cause problems. In order to prevent the bidder from using supplier's prices against each other, suppliers tend to submit their prices as late as possible. As a result, many suppliers deliver their prices to the seller on the last day and often in the last hour before bids are due. Thus, the seller often cannot investigate or evaluate supplier prices before deciding whether to use those prices. For example, if a bidder receives a very low price from an unfamiliar electrical supplier, the seller is faced with a dilemma: use the price and face the risks that the supplier is unreliable or that the price contains an error; or don't use the price, in which case a competitor may use it and be awarded the contract. Bidders must make these critical decisions in the hours or minutes before the deadline.

The seller then submits the bid in compliance with timing, form, and content instructions. If the bid is late, the buyer may have to reject it. In addition, form and content requirements usually restrict the bid from containing substantial exclusions or qualifications. For example, the bidder may want to add conditions to its bid that are inconsistent with the invitation, such as an extension to the delivery time. However, the buyer may be prevented from accepting such a nonconforming bid, because statutes, special rules for public bodies, rules incorporated into the call for tenders

[12] See previous note.

(such as the rules of the local construction association's bid depository), and the terms of the call for tenders can all disqualify nonconforming bids. A private owner can retain the discretion to either accept or reject a nonconforming bid by including language in the invitation which informs bidders that technical nonconformance *may* be considered grounds for rejection. However, some courts have held that materially nonconforming bids can never be accepted.

(iii) RECEIPT AND EXAMINATION OF BIDS After receiving the bids, the buyer usually examines them with the assistance of a consultant. The buyer considers whether the most favourable conforming bid is within the project budget and evaluates alternatives proposed by all bidders. Alternatives should only be considered if permitted by the wording of the call for tenders.

Determining the most favourable bid is not always a straightforward process. For example, consider these two bids: Bidder 1 submits a very favourable bid for the supply of computer equipment for a price of $100 000, while Bidder 2's price was $125 000 for effectively the same equipment. In addition, as requested by the buyer, each bidder also submits an alternative price based on reusing the buyer's existing file server. Bidder 1's alternative price is $5000 lower than its base bid, while Bidder 2's alternative price is $40 000 lower than its base bid. If the buyer decides to follow the alternative approach, Bidder 1's price is $95 000, while Bidder 2's price is $90 000. In addition, Bidder 1 submits a further alternative that was neither requested nor permitted by the call for tenders, consisting of a further price reduction of $10 000 if the buyer arranges for disposal of the replaced computer equipment.

The buyer rates both bidders equally on all of the other criteria listed in the call for tenders. The buyer could either choose Bidder 1 at a price of $100 000, if the buyer chose to have its existing file server replaced, or Bidder 2 at a price of $90 000 if the buyer decided that it was better to use its existing file server. But the buyer is not permitted to accept Bidder 1's unsolicited alternative respecting disposal.

Other rules can also affect acceptance of bids. For example, the call for tender also frequently incorporates the reputation of bidders and their ability to perform into the criteria upon which Contract B is awarded. In addition, private buyers often inappropriately attempt to negotiate with bidders during the evaluation phase. If a buyer intends to negotiate with any bidders, that intention must be made very clear in the invitation; yet even if it is clear, a court may decide that negotiation was improper. Otherwise, that buyer's choice of bidder may end up being contested in court. Furthermore, all the requirements of the tender documents must be rigorously observed. For example, late bids must not be accepted, unless the tender documents clearly permit such. Finally, the tender documents should explain whether bids are to be opened publicly or privately and whether reasons for the award will be provided. By law, the buyer has no obligation either to open bids publicly or to provide reasons, even for public buyers. But some public buyers have regulatory requirements or policies that do require such disclosure.

(iv) CONTRACT AWARD The award of the contract constitutes the acceptance of an offer and thus determines contractual rights between the buyer and the successful bidder. It may also determine rights, obligations, and liabilities of other parties, including suppliers and sureties.[13] In order to avoid liability for wrongful award of a contract, buyers and their consultants should adhere to the following general rules:

1. All bidders should be treated fairly, in good faith, and consistent with the terms and conditions of the tender documents.

2. Bids that do not fully comply with all mandatory requirements of the tender documents should be rejected.

---

[13] For a description of the surety's position upon acceptance of a bid, see Chapter 17, p. 192.

3. Except in exceptional circumstances, a bidder should not be asked or permitted to clarify an ambiguity in his or her bid or provide further information after tenders have closed.

4. Negotiations before award should not occur, since the tendering process replaces negotiation, unless the bid process clearly indicates that negotiations will occur. If negotiations are going to be part of the process, the rules regarding such negotiations should be clearly defined.

5. If changes must be made to the contract, they should generally be made and discussed only after the bid has been accepted and written into the contract.

6. The award should be made in accordance with the criteria stated in the tender documents, although price is often considered the most important criterion.

Even if the contract documents seemingly permit the buyer to "breach" any of these rules, legal advice should be sought if the buyer intends to do so.

In addition, where there is a privilege clause in the tender documents (see p. 95), the buyer may be entitled to reject all bids if all are above the budget for the project, or if changed circumstances negate the viability of the project. The buyer can later re-tender the project, provided that there are legitimate and significant changes to the tendered work so that it is clear that the buyer is not simply trying to bid shop.

(v) **THE CONSULTANT'S ROLE** Award of the contract is generally the buyer's decision. However, the buyer's consultant can be sued over tendering issues either by the buyer (for failing to provide appropriate advice) or by an unsuccessful bidder. Some examples of potential claims against a consultant by an unsuccessful bidder include participating in post-tender closing negotiations, disqualifying a bidder, and acting in bad faith.

The consultant may recommend that a particular bidder be rejected, based either on the reputation of the bidder or upon past dealings. In some cases, rejected bidders have sued consultants for giving such advice, basing the action in defamation. However, the consultant may be able to defend such an action if the recommendation is made in good faith, without malice or negligence.

Some specific issues respecting tenders are discussed in Figure 10-6.

(c) Bid Shopping

Bid shopping refers to the practice of using a price submitted by one party to obtain a lower price from another party. Bid shopping may or may not create a legal cause of action, depending on factors discussed below, but it is generally considered unethical. Both buyers and bidders have been known to bid shop, either prior to submission of a bid or afterwards. Courts have struggled with the issue and have had difficulty fitting it into traditional legal frameworks, with the result that in many jurisdictions, particularly outside of Canada,[14] there is no remedy for a party whose price has been shopped. In Canada, a plaintiff must fit the claim into a legal theory, such as breach of contract, promissory estoppel, or unjust enrichment.

As stated, bid shopping may be unethical, but it is not necessarily illegal, meaning in this context that it gives rise to no obvious legal remedy.[15] Bid shopping occurs regularly, despite its unethical nature. In the 1990 *Ron Brown* case,[16] the plaintiff tried to argue that, by virtue of custom and practice in the industry, bid shopping was considered so repulsive that it had become an implied

---

[14] Outside Canada, there is no "Contract A / Contract B" analysis of the bid process.

[15] *Stanco Projects Ltd. v. B.C. (Ministry of Water, Land and Air Protection)*, 2004 BCSC 1038.

[16] *Ron Brown Ltd. v. Johanson and JCL Ventures Ltd.* (unreported), B.C.S.C., Vernon registry No. 337/87, Aug. 10, 1990.

FIGURE 10-6 FAQs about Tendering Law

1Q. **What are the buyer's options if all tenders exceed the budget, or if intervening reasons make the project unfeasible?**

1A. A buyer will generally be able to use the privilege clause to reject all of the bids and cancel the project. If the buyer wants to reinstate the project later on, significant changes will have to be made to the project before it is negotiated or re-tendered in order to demonstrate that the buyer was not bid shopping. The buyer's other alternative is to award the contract in accordance with the terms and conditions of the tender call and then, after the contract is signed, issue a change order deleting a portion of the work. It is generally not open to a buyer to negotiate with any bidders after tenders have closed but before contract signing, since this would be a breach of Contract A. However, buyers may be able to pre-define a hybrid process in the call for tenders.

2Q. **When can a bidder withdraw its tender?**

2A. A bidder can withdraw its bid any time before the tender contract has been formed. A bidder who refuses to enter into a contract after this point will be subject to an action for damages. However, some buyers take the view that they would rather not force a contractor who has clearly indicated that it no longer wants the contract to attempt to do the work as that may lead to the contractor's bankruptcy (which always leads to a spoiled project) or to a contractor who will try to improve its financial position through claims or "cost savings." Some bid depositories specifically permit subcontractors some time after tenders close to withdraw their subcontract bids. A bidder may also be able to withdraw its bid if the bid contains an obvious and material error.

3Q. **Can a buyer accept an unsolicited alternate?**

3A. Often, buyers include a request for alternate pricing for different options. For example, the buyer could ask for a wooden bridge but request pricing as well for a steel bridge. This is a solicited, or requested, alternate. A solicited alternate price can be considered if the tender call permits it. An unsolicited alternate is one that was not requested by the buyer. Using the same example, a buyer requested tenders for a wooden bridge, but the contractor also put in pricing for a steel bridge. The buyer can only consider the steel bridge alternate if the tender call permits it.

4Q. **Can amendments be made to submitted tenders?**

4A. The tender call should answer this question. Some buyers permit amendments to previously submitted tenders as long as the amendment is made before tenders close. Some buyers accept these changes by fax, and some do not. It is important to make sure that the tender call is clear on this point.

term of every bid that the bid would not be shopped. That argument failed, because every expert who gave evidence at trial confirmed that bid shopping was still a regular occurrence. To be characterized as custom and practice, the custom must be almost universal.

In order for a claim based on breach of contract to succeed, a plaintiff must demonstrate that Contract A has been created. Assuming that can be proven, the plaintiff must then show that it was an implied term of Contract A that the recipient of the price would not use the price for bid shopping. A variation of this argument was attempted unsuccessfully in the *Ron Brown* case. In *Ron Brown*, the plaintiff was the only subtrade bidder for the mechanical portion of the work in a construction project. The prime contractor, the bidder, considered the subtrade's price too high and decided to do the mechanical work himself. The Court found that the prime contractor had used Ron Brown's bid price as a component in the calculation of its bid to the buyer, although that bid itself did not indicate that Ron Brown would be used for the mechanical work. Ron Brown argued that the prime contractor had in effect shopped the price to itself. The Court rejected this analysis.

Although many courts have been unable or unwilling to provide a remedy to the victims of bid shopping, the practice is considered sufficiently unethical that bidders who have attempted to bid shop suppliers' bids will jeopardize their right to hold those suppliers to their quoted price.[17]

Where a statute requires that the contract be awarded to the lowest qualified bidder, the public buyer is precluded from bid shopping. The primary purpose of these statutes is to ensure the prudent use of public funds and to prevent public servants from abusing their position; but a secondary effect of the legislation is to prevent unethical activity by public buyers.

(d) Mistakes

A mistake must be both obvious and material in order for it to render a bid incapable of being accepted, even if the bid is irrevocable. Of course, if the matter relates to a risk that has been knowingly accepted by the bidder, and the "error" is a matter of judgment by the bidder, it may not be considered a mistake.

For example, suppose that a bidder is bidding on a large geotechnical investigation. All of its competitors base their bids on the assumption that they will perform the investigation using expensive drill rigs. The successful bidding company bases its bid instead on the assumption that it can use backhoes. If that method proves impossible or ends up being more expensive than anticipated, the assumption on which the bid was based does not qualify as an error for the purposes of escaping the contract.

If a bid is irrevocable, and the bidder catches an error after the time for submitting bids has closed but before the bid has been accepted, a bidder who revokes its bid forfeits any bid bond or bid deposit unless the error is obvious and material.[18] The bidder must prove obvious and material error in order to avoid the formation of a contract with the buyer, or else it must prove that Contract A was never formed based on another theory.[19]

While a buyer may be able to hold a bidder to a bid even where the bidder has made an error, many buyers elect to let the bidder withdraw the bid and forego the damages that could otherwise

---

[17] For an excellent discussion of the case law on bid shopping, bid chopping, and bid chiseling, see Richard Oertli, *Construction Bidding Problem: Is There a Solution Fair to Both the General Contractor and Subcontractor?* (St. Louis: U.L.J., 1975) at 552.

[18] This was the consequence in *Ontario v. Ron Engineering & Construction (Eastern) Ltd.,* [1981] 1 S.C.R. 111.

[19] For example, if the bid is non-compliant, there may be grounds to withdraw.

be covered by the bid deposit or bond if the contractor can prove the error is genuine.[20] The rationale for doing so is that a bidder forced to complete an underbid contract will likely try to recoup the losses through other contractual means or may in fact go bankrupt attempting to perform. In either case, the project will likely be difficult and unsuccessful for all parties.

Some bidders have argued that a calculation error resulting in a large spread in bids entitles them to withdraw their bids. However, because a calculation error on the face of the bid may be obvious but not material, it may or may not qualify a bidder to withdraw the bid. Further, the courts have not established any simple rules that govern all cases. A difference of more than 10 percent in some areas (*e.g.*, concrete work) may be material, whereas a 25 percent spread in other areas (*e.g.*, painting) may not be unusual. A spread that results from an obvious error in one situation may not be obvious in another. It is therefore risky for a bidder to rely on an error to withdraw from a bid.

For example, in *Graham Industrial Services Ltd. v. Greater Vancouver Water District*,[21] the lowest bidding company made a mistake in its bid, but the mistake was not obvious on its face. Therefore, the only tactic available to the low bidder to escape without being forced to perform the contract or pay the buyer damages was to argue that the bid was non-compliant. But the Court held that as a precondition to the formation of Contract A, the terms of the invitation required the bid to be "materially" compliant. In other words, the bid had to meet all of the important mandatory requirements of the invitation. As a result, the Court found for reasons not connected with the mathematical error that there was material non-compliance, allowing the bidder to avoid the contract and obtain return of its bid bond.

(e) Bid Depositories

A bid depository is a system used by buyers, sellers, and suppliers to receive sealed bids from trade contractors in the construction industry. It is designed to facilitate bidding according to specified rules and to enable bidders to obtain firm, written quotations and adequate time to compile their tenders completely and accurately. Bid depositories are run by provincial construction associations. They are the result of industry demands for a less chaotic and more predictable system, as well as inadequacies of the common law in dealing with improper bidding practices.

If a buyer decides to use a bid depository, the buyer may owe all users, including subcontractors, a duty of care to treat everyone fairly in accordance with the rules of the bid depository. A buyer may be liable in damages to a subcontractor if this duty is breached, whether through failure to adjudicate disputes between contractors and proposed subcontractors as contemplated by the rules, failure to make reasonable efforts to require a contractor to enter into a required agreement with a subcontractor, or failure to protect a subcontractor from the improper conduct of a general contractor. The buyer should seek legal advice if any user of the bid depository complains that the rules have been violated.

Normally, only members of a depository can use it. Members agree to abide by its rules or risk losing membership privileges, including the right to bid on projects handled by the depository.

The buyer deposits plans, specifications, and other bid documents in the bid depository for bidders to review, and in turn, bidders must submit their bids through the depository. For contractors

---

[20] The CCDC Document 23, *A Guide to Calling Bids and Awarding Contracts*, recommends that no penalty should be assessed in the case of serious and demonstrable error. The AIA approach is similar. The Public Construction Council of British Columbia, in its September 1989 publication entitled *Procedures and Guidelines Recommended for Use on Publicly Funded Construction Projects*, states that irrevocable tenders "should be used only when there is good reason for doing so. In normal circumstances a bidder should be allowed to withdraw his bid at any time before it is accepted. If a tender is to be irrevocable, the fact must be clearly stated on the Form of Tender. Legally, it must be supported by consideration or the tender must be sealed."

[21] [2004] B.C.C.A. 5.

who do not want to pay for a set of plans and specifications for the purpose of bidding, the depository allows members to use the depository's facilities to perform calculations. The frantic compilation of subcontractor prices by general contractors at the eleventh hour of open bidding processes has led most depositories to include a requirement that subcontractors' bids be deposited at least a minimum period (*e.g.*, one day) prior to the deadline for submission of the general contractor's bid.

Virtually all depositories prevent bid shopping in a number of ways. For example, subcontractors are given the opportunity to give bids only to those general contractors with whom they are willing to work. They must place their bid in sealed envelopes in the contractor's mailbox. The contractor then has the choice of taking the envelopes of those subcontractors with whom the general contractor is willing to work. However, some bid depositories insist that the general contractor use the lowest conforming bid from the envelopes accepted and to contract with that subcontractor if the contractor's bid is accepted. Failure to do so may result in a claim for breach of contract, on the theory that the members have contracted with each other to abide by the rules of the depository.

Some contractors consider using their own staff for a particular subcontract but would use a subcontractor if a low-enough price were available. To deal with that situation, the bid depository may require the contractor to submit a subcontractor price *to itself* in a sealed envelope and to treat it as it would the other bids. Only if its own bid were lowest would the contractor be entitled to use its own staff for that work.

(f) Subcontract Formation in the Bidding Process[22]

In the absence of a bidding process, a contract is not considered formed until acceptance of an offer is communicated to the offeror. But when subcontracts are involved in the bidding process, this rule becomes modified. Other quirks and unusual rules relating to subcontractors change the rules applying to contracts and bids.

(i) **USING A NAMED SUBCONTRACTOR** Because the basic principles of *Ron Engineering* were modified by the Supreme Court of Canada's decision in *MJB*,[23] submission of a bid does not now always create Contract A, and it may be possible for an invitation for bids to exclude the formation of Contract A. In fact, one court has already determined that an owner can "contract out" of Contract A through the express language of the call for tenders.[24]

In deciding when a contractor must use a particular subcontractor, the courts must determine whether a contract with the subcontractor has been formed through the bidding process. Legal analysis of this question does not produce clear-cut answers.

For example, if a prime contractor has named a subcontractor in a bid and used that subcontractor's price, and the buyer has accepted the prime contractor's bid, then does that contractor have a contract with the subcontractor? The *Westgate Mechanical Contractors Ltd. v. PCL Construction Ltd.*[25] case suggests that immediately upon the buyer's unconditional acceptance of a bid, a series of subcontracts crystallize between the prime contractor and each of the subcontractors named. However, the *Westgate* case contains very little analysis of this issue: this point was simply accepted by both plaintiff and defendant as well-established law.

---

[22] Much of this section was presented by Brian Samuels in a paper for the C.L.E. Society of British Columbia, in a seminar on Construction Law held in November 2004. Those portions taken from that seminar are reproduced with the kind permission of the C.L.E. Society.

[23] *MJB Enterprises Ltd. v. Defence Construction (1951) Ltd., [1999] 1 S.C.R. 619.* See also pp. 94–95.

[24] *Maple Ridge Towing (1981) Ltd. v. Maple Ridge (District)*, 2001 BCSC 1328.

[25] *(1989), 33 C.L.R. 265 (B.C.C.A.).*

The same principle may not hold true in other circumstances:

- if the prime contractor has used the subcontractor's price but has not identified the subcontractor in the bid to the buyer
- if the prime contractor has used a price that is an average between two subcontract bidders and has named neither or both subcontractors
- if the buyer's acceptance of the bid occurs after the prime contractor's bid has expired
- if communication of the buyer's acceptance has occurred after the subcontractor's bid has expired.

An example of the first scenario occurred in the *Ron Brown* case.[26] The Court examined the issue on the basis of mutuality, by asking the question: if the prime contractor wanted Ron Brown (the plaintiff subtrade bidder) to act as its subcontractor, and Ron Brown had refused, could the prime contractor have forced Ron Brown to do so? The answer was no. Because Ron Brown was not identified in the bid, Ron Brown was under no obligation to the prime contractor; therefore, there was no mutuality of obligation. As a result, Ron Brown could not force the prime contractor to accept his bid.

Similarly, in the 2001 *Naylor Group v. Ellis-Don Construction* case,[27] the Court affirmed the principle that a subcontractor named by a general contractor has rights under Contract A. However, that finding appears to be limited to a bid depository situation. The Court made no ruling whether the same result would apply outside a bid depository.

The bidding process is somewhat of a cat-and-mouse game between subcontractors and prime contractors. Subcontractors generally submit their prices within an hour of bid closing to avoid being shopped. Prime contractors continually dream up new ways to gain a price advantage by attempting to lower the subcontract prices after bid closing through negotiation. However, based on the reasoning in the *Stanco* case,[28] if a plaintiff can demonstrate that the procedure rendered the bidding process unfair, a claim for breach of Contract A might succeed.

Prime contractors try to shop bids by refusing to name a subcontractor in the bid. Sometimes the prime contractor sets a price in the bid that is an intermediate amount between the bid prices of two subcontractor bids. In other cases, the contractor will simply insert "own forces" instead of the name of a subcontractor. Then, after the bid is accepted, the prime contractor approaches two or more subcontractors for that trade and plays them off against each other.

Sometimes a buyer attempts to accept a bid price after the bid has expired. When such delayed bid acceptances occur, the seller must first ensure that the named subcontractors will hold their prices before agreeing with the buyer to accept the late contract award. The buyer has no right to accept an expired price. In effect, the buyer's purported acceptance after expiry is a new offer. Just as the prime contractor's price has expired, so have all of the subcontract prices. In a heated market where subcontract prices are rising or where material prices are unstable (as with steel supply prices across North America in 2004), a prime contractor would bear the risk of price increases unless that prime contractor has locked in all subcontractor prices before agreeing to a fixed price with a buyer.

However, if the buyer's acceptance occurred within the prescribed time (*e.g.*, 60 days), but that acceptance has not been communicated to the named subcontractors until well after the bid period, should the subcontractor be bound? Not much case law bears directly on this topic; but common sense dictates that if the communication to the subcontractor is made within a reasonable

---

[26] *Ron Brown Ltd. v. Johanson and JCL Ventures Ltd. (unreported), B.C.S.C., Vernon registry No. 337/87, Aug. 10, 1990.*

[27] *2001, SCC 58.*

[28] For an excellent discussion of the law in Canada on bid shopping, see *Stanco Projects Ltd. v. B.C. (Ministry of Water, Land and Air Protection), 2004 BCSC 1038.*

time, the subcontractor should be bound. Certainly if one accepts the proposition in *Westgate* that the subcontracts are crystallized at the moment of the buyer's acceptance, then the subcontractor is bound. Communication of acceptance generally must be received by the offeror before the offer expires in order for there to be a contract. However, *Westgate* holds that buyer acceptance of a bid crystallizes a contract between two parties (the prime contractor and a subcontractor) before any communication of the acceptance has been received by the subcontractor.

The subcontractor should therefore be entitled to know within a reasonable time whether it must order material and mobilize its staff. The prime contractor cannot wait too long before communicating acceptance to the subcontractor and still expect the subcontractor to be ready at a moment's notice (although that is precisely how some contractors treat their subcontractors).

(ii)   HOLDING A SUBCONTRACTOR TO A PRICE While mutuality is a requirement for enforceability of contracts, contract law is not the only legal theory that applies to holding a sub-contractor to a bid price. Obviously, a prime contractor who uses a subcontractor's bid price in the formation of the prime contractor's bid price has in fact relied on that subcontractor's price. Reliance is one of the key elements in a claim of detrimental reliance based on estoppel. These issues of detrimental reliance and estoppel also come into play when a subcontractor seeks to withdraw a price after the prime contractor has submitted a bid to the buyer. The subcontractor may have made a mistake, experienced a supply price increase, or wanted to take advantage of better busi-ness opportunities. The presence of a mistake that is both obvious and material on the face of a subcontractor's bid, as with any bid, entitles the bidder to withdraw the bid without penalty; and a prime contractor should therefore not use a bid that contains an obvious and material error. In theory, this is clear; however, in practice, what constitutes an obvious and material error in a subcontractor's price is not always clear.

In general, except for an obvious and material error or some other material non-compliance, a subcontractor is not allowed to withdraw a price that the prime contractor has relied upon in a bid. Otherwise, the prime contractor could suffer a significant loss.

(iii)   QUANTIFICATION OF DAMAGES FOR WITHDRAWAL OF SUBCONTRACTOR BID When a subcontractor has no justification for withdrawing a bid or otherwise rendering the bid incapable of acceptance, refusal to provide services constitutes grounds for demanding damages. Quantification of damages generally depends on two factors: first, the spread in prices between the two lowest prime contractor bids; and second, the difference in price between the two lowest subcontractor bids.

For example, suppose that the lowest prime contractor bid is $1 million, and the next lowest bid is $50 000 higher. If the prime contractor withdrew the bid because a subcontractor refused to supply services quoted, the prime contractor would be subject to a claim for damages by the buyer of $50 000 for breach of Contract A. The prime contractor would also lose the anticipated profit on the project. Assume that lost profit is $30 000. Therefore, if the subcontractor withdrew a bid, the prime contractor would likely claim damages of $80 000 against the subcontractor.

On the other hand, if the subcontract price being withdrawn were $200 000, and the next sub-contractor bid was $240 000, then in order to mitigate the loss, the prime contractor might hire the second subcontractor rather than refuse to contract with the buyer. The prime contractor would then sue the withdrawing subcontractor for the $40 000 loss.

In looking at these options, a subcontractor considering withdrawing a bid must also consider the losses involved in performing the subcontract. If the reason is a cost that is not large (for example, $25 000), the subcontractor would be better off performing the work and absorbing a $25 000 loss, rather than facing a lawsuit for $40 000 (or $80 000). The existence of bid bonds from the prime contractor or subcontractor may affect the best decision in this scenario. Subcontractors considering withdrawal of a bid must weigh options carefully before deciding on a course of action.

10.5 Choosing the Best Process and Delivery Method for a Project

The contract process most beneficial to a project is intrinsically related to the delivery method that best suits the project. For some projects, negotiating a contract is better than tendering it; but the strength of this decision depends on whether the right delivery system was chosen in the first place. The key is to match the process and the system to the project. A simple four-wall, one-roof warehouse demands much different choices than a 40-storey, mixed commercial and residential high-rise. Other factors can impact the project, such as governmental approvals, criticality of timing, and the labour and material markets. Thus, project administrators should consider all options and variables before deciding on a contract process.

Currently, some Canadian buyers in the construction industry favour negotiated contracts or guaranteed maximum price contracts with shared cost savings, because the labour and material markets are highly variable, both in terms of cost and quality. A market and industry where labour is inexpensive and materials abundant might fuel a preference for tendered fixed-price contracts.

In addition, project administrators must view each project as a new venture, without preconceived notions about how to proceed. The same level of thought and research should go into choosing processes and delivery methods as went into the business decision to proceed with the project in the first place. All too often, the buyer and consultant simply recycle methods from other projects without considering the peculiarities of the new project.

Traditional construction contract models, such as design-build, design-bid-build, and construction management (see pp. 105–109), have benefits and detriments that are well known. One constant is the adversarial relationship between buyers and sellers that appears to be the rule rather than the exception in the construction industry. In a twisted sense, this antagonism forces buyers and their consultants to develop creative methods for relieving or reducing the adversarial atmosphere.

One approach to improving contractual relationships has been to utilize traditional contractual models while limiting the pool of potential participants to just those whom the buyer trusts. Some buyers have in fact retreated to simply contracting with one contractor for all of their projects. In this case, the buyer would not use an open tender process.

Another approach is the partnering model. However, this model often consists of little more than good intentions and handshakes. Alliancing takes the good intentions of the partnering model and adds an enforceable agreement.

Table 10-3 on the next page shows the contractual models compared in terms of the various risks and obligations. Because these models are highly dependent on specific contractual language, generic contract models are presented for comparison purposes. The scale for risks and obligations is broken down for buyers (owners) and sellers (contractors) with risk and obligation increasing from none to minimal, some, most, and all. Although many potential models for the "contractor" exist (*e.g.,* joint venture or single-purpose entity), a generic contractor model was used.

The following section explains the strengths of each contractual model from owner/buyer and contractor perspectives, as they relate to risks and obligations.

(a) Design-Bid-Build

In a design-bid-build process, the buyer and consultant conceptualize and design the project in its entirety; the buyer then hires a general contractor to build it (generally through a tendering process);

TABLE 10-3 Comparison of Project Delivery Methods

Risks	Owner's Perspective						Contractor's Perspective					
	Design Bid Build	Design Build	Design Build Operate	PPP & Design Build Finance Operate	Const. Manage	Alliance	Design Bid Build	Design Build	Design Build Operate	PPP & Design Build Finance Operate	Const. Manage	Alliance
Warranty	None	None	None	None	None	Most	All	All	All	All	All	Minimal
Changes in Law	Some	Some	Some	Some	Most	Most	Some	Some	Some	Some	Minimal	Minimal
Site Conditions	Minimal	Minimal	Minimal	Minimal	Most	All	Most	Most	Most	Most	Minimal	None
Project Revenue	All	All	Minimal	Minimal	All	All	None	None	Most	Most	None	None
Cost	Minimal	Minimal	Minimal	Minimal	Some	Some	Most	Most	Most	Most	Most	Some
Limitations of Liability	Some	Some	Minimal	Minimal	Some	Most	Some	Some	Most	Most	Some	Minimal
Termination Rights	Minimal	Minimal	Minimal	Minimal	Minimal	Some	Some	Some	Most	Most	Some	Some
Environmental Impacts	Most	Most	Some	Some	Most	Most	Minimal	Minimal	Some	Some	Minimal	Some
Design	Most	Some	Minimal	Minimal	Most	All	Minimal	Most	Most	Most	Minimal	None
Warranty	None	None	None	None	Minimal	Some	All	All	All	All	Most	Some
Real Property Value	All	All	All	Most	All	All	None	None	None	Minimal	None	None
Operations	All	All	Minimal	Minimal	All	All	None	None	Most	Most	None	None
"Acts of God"	Some	Some	Minimal	Minimal	Some	Most	Most	Most	Most	Most	Some	Minimal
Time Requirements	Some	Minimal	Minimal	Minimal	Some	Most	Most	Most	Most	Most	Some	Minimal
Obligations												
Performance	None	None	None	None	Minimal	Some	All	All	All	All	Most	Some
Authorities and Permits	Some	Some	Minimal	Minimal	Most	Most	Most	Most	Most	Most	Minimal	Minimal
Insurance	Some	Some	Minimal	Minimal	Most	Most	Most	Most	Most	Most	Minimal	Minimal
Design Control	All	Minimal	Minimal	Minimal	All	Most	None	Most	Most	Most	None	Minimal

FIGURE 10-7 Design-Bid-Build Process

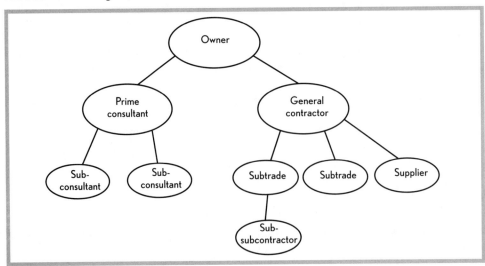

the consultant inspects the construction; and the buyer pays for the project and puts it into service. A diagram of the contractual framework is shown in Figure 10-7 on page 106. Table 10-4 lists the advantages and disadvantages of this method from the buyer's perspective. Some of these disadvantages can be reduced or eliminated by negotiating rather than tendering. Yet this in turn eliminates some of the advantages.

TABLE 10-4 Advantages and Disadvantages of Design-Bid-Build

Advantages	Disadvantages
• Well known	• Generally requires choice of lowest bidder
• Allows the buyer to shift risks and define obligations	• Generally forces bidder to accept risks and obligations as dictated by the buyer and the costs associated with these risks and obligations
• Separates the designer/inspector from the builder	• Reduces innovation
	• Can be time consuming

(b) Design-Build

In a design-build model, the buyer conceptualizes the project, and then engages one party to design and build it (including inspection); then the buyer pays for the project and puts it into service. However, the buyer's need for control over the design process conflicts with concerns over the integrity of the inspections. In many cases, the buyer wants far more control over design than is considered in the pure design-build concept. In addition, the buyer often must engage a separate consultant to ensure quality, thereby distancing the buyer from the decision-making process. A diagram of the contractual framework is shown in Figure 10-8 on page 108, and the advantages and disadvantages to the buyer/owner are listed in Table 10-5.

(c) Design-Build-Operate

The design-build-operate model is effectively the same as design-build, except that it reduces some of the buyer's concerns over intervention and integrity of inspections by making the contractor responsible for operations for a length of time after the design-build phase is complete. The rationale is that a contractor will not cut corners where ability to profit from operations would be impaired by poor design or construction. The operations component of the contract sometimes includes a portion of the contractor's profit and overhead from the design-build phase, such that the contractor's profits are at risk throughout the operation period.

TABLE 10-5 Advantages and Disadvantages of Design-Build

Advantages	Disadvantages
• Increasingly well known	• Concern over level of permissible buyer intervention in design
• Allows the buyer to have "one-stop shopping"	• Concern over integrity of inspection process
• Integrating designer and builder can foster cooperation	• Possible conflict of interest on the part of the consultant, since the consultant is paid by the contractor whose work the consultant is reviewing

FIGURE 10-8 Design-Build Process

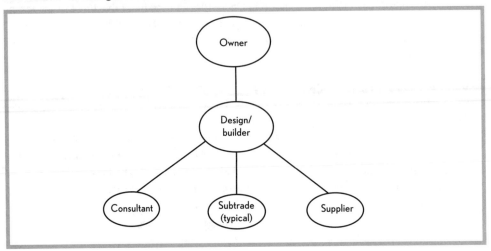

(d) Design-Build-Operate-Finance

The design-build-operate-finance model resembles the design-build-operate except that it includes the additional element of finance. The contractor invests in the project as part of the contract. This investment has minimal impact on the model. However, the contractor's risk increases because of the capital at risk. Public-private partnerships generally follow the design-build-operate-finance model. A diagram of the contractual framework is shown in Figure 10-9.

(e) Construction Management

Construction management is a process through which an owner contracts directly with the trades instead of hiring a general contractor. The owner also hires a construction manager to manage the trades. The construction manager does little or no physical work on the site and is supposed to simply manage the process. Construction managers can be employed in many of the traditional contract scenarios; but in Canada, construction managers are generally hired either as the buyer's agent (wherein the contracts are entered into between the buyer and the trades) or as an "at risk" construction manager (wherein the construction manager is contracted directly to the trades). A diagram of the contractual framework is shown in Figure 10-10.

The most significant problem with the construction management model is that the buyer cannot simply hold one or two parties responsible for a problem on the project. Problems usually involve several trades. In addition, construction managers are generally not prepared to accept risks associated with cost increases and the time for completion of the project, which means these risks generally remain with the buyer, the consultant, or the tradespeople. Table 10-6 lists the advantages and disadvantages of this method.

10.6 Effects of Interprovincial and International Trade Agreements

International and interprovincial trade agreements can alter the normal processes for procurement and related disputes. An **international trade agreement** is an agreement between two countries

FIGURE 10-9 Public-Private Partnerships (P3s)

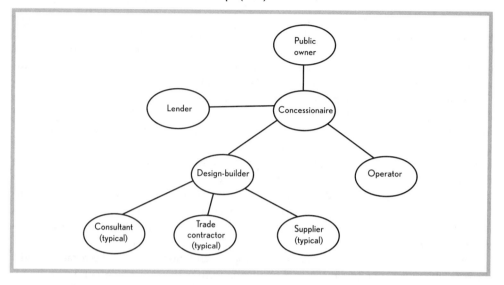

FIGURE 10-10 Construction Management

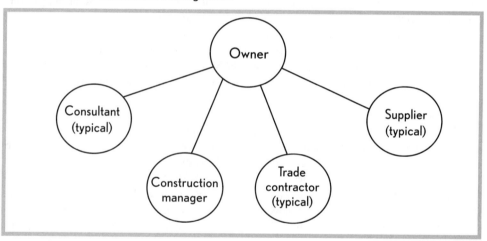

TABLE 10-6 Advantages and Disadvantages of Construction Management

Advantages	Disadvantages
• Cuts out the "middle man" profit normally made by the general contractor • Allows the buyer to have a closer relationship with the trades	• Requires relatively sophisticated parties • Separates liability unless done "at risk" and with schedule and cost risk on the construction manager

that sets rules about the transfer of goods and services between two countries. An **interprovincial trade agreement** is an agreement between the federal government and the provinces that sets rules about the transfer of goods and services between the provinces. The most important international trade agreements include the North American Free Trade Agreement (NAFTA), the World Trade Organization (WTO) Agreement on Government Procurement, and the Canada-Korea Agreement on the Procurement of Telecommunications Agreement. Currently, the only important interprovincial agreement is the Agreement on Internal Trade (AIT).

All international and national treaties are only applicable to countries that agree to adhere to the treaty. For example, if a country does not adhere to the Kyoto Accord, then the provisions of that treaty are not applicable in that country. Where Canada adheres to a treaty, as with NAFTA, all provincial or municipal governments within Canada are bound to follow that treaty, subject to potential arguments about whether the federal government can, without agreement of the provinces, make treaties that impact areas where the provinces have exclusive constitutional jurisdiction.

Generally, international and interprovincial treaties require procurement processes to be transparent, accessible, and fair. Each agreement details the rights and remedies of government and bidders in a given procurement. They also detail exclusions from the general rules, such as projects related to national security.

Each agreement also includes some form of dispute resolution process and generally excludes the rights of parties to litigate in court. The common form of dispute resolution is series of processes that ends in a trade tribunal. For Canadian government procurements, a bidder must take complaints to the Canadian International Trade Tribunal (CITT). The time-lines for filing a complaint with the CITT are very short. In fact, the supplier (bidder) must file the complaint within 10 days of the moment when the supplier knew or ought to have known about the breach.

In addition, each treaty has monetary limits for each type of procurement. For example, the 2005 threshold for services contracts under NAFTA was $84 400. This means that contracts for services less than $84 400 do not fall within the NAFTA procurement rules. The AIT threshold for services is $100 000.

Under these treaties, except in a very narrow set of circumstances, government departments and agencies must put their projects out for tender. If a government uses technical specifications that are too restrictive, includes unclear evaluation factors, or provides unequal access or any form of local preference, it will be in breach of the applicable treaties.

The AIT was supposed to do within Canada what NAFTA has done between Canada, the United States, and Mexico. To date, only the federal government has adhered to the AIT. Therefore, the AIT is only binding on the federal government. However, many provincial governments have requirements and dispute processes that follow the AIT for provincial government procurements. As such, professionals need to understand the applicable rules when dealing with a government procurement, both when creating or responding to a tender and when considering, making, or responding to a complaint.

CHAPTER QUESTIONS

1. Which of the following actions can a buyer in a tendering situation do without risking legal liability?

   a) It can accept bids that do not comply with a material requirement of the call for tenders

   b) It can negotiate with the bidders after tenders close

   c) It can use criteria other than price in deciding on the successful bidder, as long as the criteria are disclosed

   d) It can use undisclosed criteria in deciding on the successful bidder

2. Describe the buyer's advantages and disadvantages of using a familiar procurement process.

3. When preparing contract documents for tender purposes, what should an engineer include?

   a) Only that information which the engineer personally believes should be necessary for a competent contractor

   b) All relevant information which may assist a contractor in preparing a bid

   c) Only that information which the buyer, his client, agrees that the engineer should include and disclose

4. Discuss the statement: "It is always best to hire consultants who have the lowest price."

5. If tender documents state that submitted bids are irrevocable until a specified date, what does this mean?

   a) A bidder can never withdraw his bid.

   b) A bidder can withdraw his bid before tenders close, but not after.

   c) A bidder can withdraw his bid any time before the specified date.

6. Describe the nature and effect of the privilege clause.

7. If an engineer who is supervising a contract tender receives inquiries from prospective bidders before the close of bids, which of the following actions should he take?

   a) Refuse to answer any questions, because the bidders must base their bid only on the written tender documents

   b) Answer questions as fully and fairly as possible, keeping a written record of his advice in case a future controversy develops

   c) Answer the question fully and fairly and pass on to all other bidders the same information by issuing an addendum

8. Describe the impact of international treaties on government tendering practices.

9. List and discuss the advantages and disadvantages of the design-build contract model.

10. During the tender period, if the engineer discovers deficiencies or errors in the tender documents, which of the following actions should he take?

    a) Immediately notify the buyer and all tenderers of the potential difficulty, providing amending information by way of addenda, where possible

    b) Say nothing and deal with any problems which may arise by way of change orders after the contract is awarded

    c) Cancel the entire bid and re-tender the project after correcting the errors and deficiencies

ANSWERS

1. C

2. The buyer's familiarity enables him or her to be comfortable with the chosen delivery system and, hopefully, to understand it. On the negative side, using the same system without evaluating its appropriateness may lead to use of an inappropriate or less appropriate system.

3. B

4. Price is only one factor in making any choice to purchase goods or services. The quality of the service provider, especially for a professional service, is crucial to the success of a project.

5. B

6. The privilege clause permits an owner to accept or reject all of the bids, but it does not permit an owner to utilize undisclosed criteria in awarding a bid.

7. C

8. International treaties can impose additional requirements on government to follow certain rules or not include criteria, such as local preferences, in a tender call.

9. One advantage is that design-build is becoming increasingly well known, so that contractors and professionals are used to working with it. In addition, this method allows the buyer to have one-stop shopping. As well, because the designer and builder are integrated, this method fosters cooperation. However, there are concerns about the level of permissible buyer intervention in the design. In addition, there are concerns about the integrity of the inspection process. Finally, the design-build method creates a potential conflict of interest on the part of the consultant, since the consultant is paid by the contractor whose work is being reviewed.

10. A

CHAPTER 11

SPECIFIC CONTRACTS AND CLAUSES

Overview

Contracts come in all shapes and sizes. From a complex agreement respecting a significant transaction to a simple contract for the supply of goods or services, every contract contains important provisions. Many industry groups have created standard form contracts for broad use in their industries. These contracts usually represent good starting points, but they can be risky if the contract is not modified to fit the needs of the situation.

In the construction industry, the method of payment defines most of the different contract forms. A **fixed-price** or **stipulated sum contract** is a contract in which the goods are provided and services are performed for a lump sum amount, including overhead and profit. A **cost-plus** or **cost reimbursable contract** provides the contractor with payment for the labour, equipment, and material expenses incurred on the project, plus a percentage for profit and overhead, sometimes to a guaranteed maximum. A **unit price contract** provides for payment for each unit that is provided, regardless of the cost of actually providing the unit (*e.g.*, $X per tonne of gravel placed on the road).

Other construction contracts are not based on method of payment. A *design-build contract* is based on the concept that the contractor will perform all the design and construction work with little or no input from the owner. An **alliance agreement** creates a close relationship between all parties on a construction project. In this kind of agreement, the level of compensation for each party depends on the success of the entire project, and there is a specific provision prohibiting litigation. In addition, **public-private partnerships** are used for public projects, such as highways: the public agency contracts private companies to design, construct, finance, and operate these projects on a long-term basis. Finally, a **licensing agreement** provides a grant of rights to a licensee to use a right held or owned by a licensor. Licences can be exclusive or non-exclusive. Licensing agreements are particularly important in the software industry.

However, the field of geoscience sometimes involves other geoscience-specific forms of contract, including syndication, grubstake, royalty, and option agreements. In most cases, the objective of these contracts is to raise sufficient funds to conduct exploration and provide for a return on the investment of such funds.

In addition, *professional service agreements* are used to engage the service of a professional or firm of professionals in a project. Such agreements can be made on a cost-plus, fixed-price, or percentage basis. As in all agreements, professionals negotiating a professional service contract must pay particular attention to the clauses that define the relationship between the parties, especially those relating to the scope of services provided, the limitations of liability, the level of field services, and the pricing. Other important clauses are those defining the time for performance, dispute resolution options, termination, provision of insurance, and ownership of the work product. Professionals entering a contract must carefully examine each of these clauses.

Many projects also require the involvement of lenders, such as banks and investors. In order for a project to gain the support of a lender, those organizations must be able to see that the contract outcome is achievable. Moreover, they must also see the risk/benefit balance for the borrower. The lower the risk of the party seeking financing, the more likely that financing will be available.

11.1 Standard Form Contracts

The complexity of the business world has led many industry groups to create standard form construction contracts that can be used in their specific field. In a perfect world, these contracts would fit all projects without modification. However, standard form contracts usually require some changes to reflect the needs of each project.

Fitting the contract to the project is no easy task. In some ways, choosing a contract for a construction project resembles a clothing purchase: the budget determines in large part the level of tailoring—from an off-the-rack suit with no tailoring (which rarely fits well) to an off-the-rack suit with some tailoring (which fits reasonably well) to a made-to-measure suit (which fits very well). Likewise, budgets determine the starting point for contract planning in a construction project and drive most of the decisions.

Standard form contracts are generally created by industry groups. Each of these documents represents a trade-off between industry players, with risk and obligation shuffled to satisfy the politics of each industry group. That is not to say that standard form documents are without benefit. They represent an amalgam of good and bad, depending on one's perspective. However, at the very least, they also provide a common starting point for creation of a contract.

The single largest drawback of standard form contracts is that they are used too often without attention to the specific needs of the project in question. Sometimes budgets force a project to use the standard form as is, since spending more time and money on a contract than a project is worth makes little fiscal sense. However, weakly drafted contracts are the most likely to end in costly litigation.

11.2 Construction Contracts

Industry groups such as the Canadian Construction Document Committee (**CCDC**), the American Institute of Architects (AIA), RAIC, ACEC, the Design-Build Institute of America (DBIA), and the International Federation of Consulting Engineers (FIDIC) all have their own standard construction contracts for all the major contract types: fixed price, construction management, design-build, alliance agreement, and public-private partnership.

(a) Fixed-Price Contracts

A fixed-price contract is a contract in which the goods and services are provided for a lump sum amount, including overhead and profit. Such contracts are also referred to as *stipulated price contracts* or *lump sum contracts*. Under these contracts, the owner usually has no right to direct the contractor on the means and methods of construction and no right to inquire about the actual cost of the work. The contractor is entitled to keep any additional profit earned as a result of cost-saving measures but is also responsible for overruns.

The myth that fixed-price contracting is always the best choice for an owner still pervades much of the North American construction industry. Many owners still believe that transferring all risk and obligation to a contractor not only results in a low price that does not change throughout the project but also results in an on-time, on-quality project.

Certainly, a fixed price rarely stays fixed throughout the project. The reality is that change orders resulting from changes in scope, improperly detailed design drawings and specifications, changes in conditions, and other contractual entitlements almost always result in a price that looks very different from the original contract price. The biggest risk is that some owners believe that the price is fixed and budget accordingly, often facing financial crises by the end of the project.

However, owners that understand the concept that a stipulated or fixed price is just a starting point can use this pricing technique effectively. These owners understand that making the contractor include in the price contingencies for unknowable risks (*e.g.*, subsurface conditions) is not always wise, because in the long run, the contract price for those unknowns is likely to exceed the actual cost of the risk.

The most common form of stipulated price contract in Canada is the Canadian Construction Document Committee 2 Contract (CCDC 2). It has been revised many times, most recently in 2008. Because CCDC 2 is the best-known construction contract in Canada, it provides a stable platform for a stipulated price contract. The composition of the CCDC documents committee is found on the CCA website. Users of the document should be aware of the composition of the committee.

Other forms of stipulated price contract are available. These include the Canadian Construction Association (CCA) L-1 and S-1, the Institute of Civil Engineers (ICE) A, the American Institute of Architects (AIA) 101, the American Consulting Engineers Council 24, the International Federation of Consulting Engineers (FIDIC) Conditions of Contract for Works of Civil Engineering Construction, and the Associated Owners & Developers (AOD) contracts. Each of these standard contracts has strengths and weaknesses that must be evaluated before the form is used. However, in Canada, industry-wide familiarity with CCDC 2 leaves it as the *de facto* leader in stipulated price contract forms.

(b) Cost-Plus Contracts

A cost-plus contract is a contract in which the contractor is paid the actual cost of the goods provided and services performed, plus a fee for overhead and profit that may be fixed or may vary with the costs. In such agreements, the costs to be reimbursed must be accurately defined.[1] Definitions of cost have been the source of many disputes, particularly regarding overhead items.

Owners need to be aware that cost-plus agreements may provide the contractor an opportunity to use *transfer pricing* as a way to increase profits by transferring costs to the owner. For example, a contractor may need a trailer as an office for a 12-month project. However, rather than renting from an outside source, the contractor could purchase the trailer, then rent it to the project as part of the cost-plus agreement. After 12 months, the owner has effectively paid for most of the purchase price of the trailer, but the contractor owns it. Such results occur when the owner fails to exercise control over costs or has not provided a contractual means of control.

---

[1] The CCDC 3 cost-plus agreement, Article A-9, contains such a definition.

One drawback of the cost-plus contract is the lack of built-in price control. A contractor operating under a cost-plus contract with a fixed fee has no incentive to save money for the owner, other than the prospect of repeat business and a sense of ethics. The contractor's profit is fixed percentage regardless of the overall cost of the project. Moreover, a fee based on a percentage of the costs of construction may provide the contractor with an incentive to increase the cost in order to increase the fee.

To deal with these problems, owners often use incentive formulas. One such formula is known as a **target price contract**, which ensures that the cost is fixed if the cost of construction exceeds a certain amount (the target); but the owner and contractor share any savings below the target, according to a predetermined formula, such as 50/50 or 60/40. Some target price contracts also require the contractor to absorb all costs of construction above the target. These target price contracts are hybrids of the fixed-price and cost-plus agreements and are commonly known as *guaranteed maximum price* (GMP) contracts. They give the owner the primary advantage of a fixed-price contract, which is a guarantee that the total price will not exceed the agreed amount. GMP contracts also have the attributes of cost-plus contracts in that they allow the owner to audit the contractor's costs if the contractor appears to be claiming any portion of the cost savings.

One key issue with cost-plus contracts is responsibility for rework (*i.e.,* the correction of deficiencies or the replacement of defective equipment). A cost-plus agreement should clearly define whether the contractor is entitled to be reimbursed the cost of correcting work and whether this includes work that was negligently performed. Thus, all parties should give careful consideration to the interaction between the requirement to perform the work competently and the cost-plus nature of the contract.

In addition, cost-plus contracts place many of the costing risks back on the owner. A cost-plus contract is generally used when the scope of the work is uncertain yet the owner must move forward with a project. For this reason, the risks that a fixed-price contract ascribes to the general contractor are retained by the owner in a cost-plus contract. These risks include the cost and time of performance of the work. The most significant issue is the cost of rework on deficient work. Some contracts, like the DBIA design-build contract, divide the risk by requiring the contractor to pay for negligently performed rework, and the owner to pay for all other rework. All parties can offset or negate their risks associated with rework and inefficiency by defining appropriate sharing of risk through guaranteed maximum prices and cost-saving bonuses.

The CCDC 3 is the most popular cost-plus contract form in Canada and resembles the CCDC 2 fixed-price contract form. However, the CCDC 3 has significant weaknesses as a cost-plus contract because it does not adequately allocate responsibility and risk. For example, clause GC 2.4 of the CCDC 3 requires the contractor to pay for all rework howsoever caused. If a project were priced according to this clause, a contractor would likely include very large mark-ups on rates and equipment to offset the risks of undefined costs related to future rework orders. Thus, this rework clause needs careful definition. Another weakness is that the CCDC 3 requires a construction schedule, even though the rationale for using a cost-plus contract in the first place is usually the undefined scope of work. For these reasons, the CCDC 3 needs revision before it would be useful for most projects.

In contrast, American standard forms of cost-plus contracts are more practical. The AIA forms balance the roles and responsibilities of the parties. Even so, these contracts must still be modified to suit particular circumstances, and the issue of nonconforming work should still be reviewed carefully.

Figure 11-1 suggest ways to use cost-plus contracts effectively.

FIGURE 11-1 Tips on Using Cost-Plus Contracts

> • Using a combination of payment plans may be of benefit. Using a general contractor as a construction manager for a fixed fee to obtain a budget followed by negotiation for a lump sum with the same general contractor may be beneficial where there is an established relationship of trust with that general contractor.

(c) Unit Price Contracts

A **unit price contract** is a contract that requires the owner to pay a stipulated amount for each unit or quantity of work performed, such as $100 per cubic metre of concrete in place. These contracts are common for road-building, earthmoving, and pipeline projects. In theory, such contracts could be used for almost any component of construction, such as square footage of drywall or lineal footage of conduit installed; however, they rarely are.

While unit price contracts do not guarantee the final cost, they may be advantageous to the contractor and owner for different reasons. In projects where the quantity of work varies but where the contractor has to bid on a lump sum basis, the price must include a contingency to protect the contractor against the risk of rising requirements and changing amounts. Under a fixed-price contract, the owner would end up paying a premium; but with a unit price contract, the owner can minimize this risk by requiring unit cost pricing for unanticipated or extra work. However, the contractor must be careful not to inflate the unit prices in these situations; otherwise, the owner might delete rather than add work, and the contractor would be bound to provide an inflated credit. As an example, if a contractor is building a 10 000-seat arena and has been asked to provide unit pricing on a per-seat basis in case the owner decides to add or deduct seats, the contractor should provide an accurate number. If the contractor provides an exorbitant unit price per seat, the owner could simply delete all of the seats and have a third party install them later at a reduced cost.

In addition, although unit price contracts require the owner to pay the agreed price per unit, this requirement is not absolute. Many unit price contracts contain provisions that adjust the compensation per unit if the total quantity varies from the estimate by more than a given percentage. The logic behind this adjustment is that when quantities increase or decrease substantially from the original contract, the job becomes materially different. For example, a contractor's mobilization costs remain unchanged throughout most projects. Thus, if a project ends up requiring a significantly smaller quantity of units than originally estimated, the contractor ends up having fewer units through which to recover the cost of mobilization. On the other hand, a large increase in quantity can reduce the contractor's per-unit overhead; but it can also require the contractor to perform that part of the job in inclement weather or cause other increased costs. If no formula is included in the contract to deal with such changes in quantity, a significant change could result in a claim based upon changed conditions.

The CCDC 4 is the most notable standard unit price contract in Canada. This form is a reasonable starting point for a unit price contract; however, because such contracts are not used much in the industry, the CCDC has not recently updated it.

(d) Construction Management Contracts

In fixed-price, cost-plus, and unit price contracts, the owner generally contracts with one party, a general contractor, to provide all of the goods and perform all of the services. However, a general contractor sometimes only provides 15 percent to 20 percent of the actual goods and services and contracts the balance to subcontractors.

An alternative to contracting with a general contractor is to hire a construction manager to deal with all the trades. The construction manager does little or no physical work on the site and simply manages the process. The construction management model can be employed in many of the traditional contract scenarios; but in Canada, construction managers are considered the owner's agent, such that the owner contracts directly with both the trades and the construction manager.

Construction manager contracts must take into consideration whether the construction manager takes the risk of increased cost or time for performance. At-risk construction managers are construction managers that are at risk for increased cost and time.

The CCA 5 construction manager contract is an acceptable standard contract for small projects, where the construction manager is acting solely as a manager. However, general contractors sometimes provide construction management services as well as a relatively large portion of the construction work itself under a construction management contract. This dual role can lead to a complex scenario in which many trades are contracted to the owner with lump sum pricing, and the construction manager provides construction services on a cost-plus basis. Moreover, the CCA 5 form is not suited to situations where the construction manager is at risk for time or cost or for performance of some of the construction work. In general, when using the CCA 5 form, parties should ensure that the role of the construction manager is appropriately described. They should also include clauses that determine the impact of a time extension on cost and carefully review the impact of a termination.

In contrast, the AIA has forms specifically designed for placing a construction manager at risk for time and cost, as well as a separate form for doing so where the construction manager will be performing part of the work as a contractor. This form references the AIA general conditions of contract; but wording from the CCDC 2 can be used in its place with some modification.

Generally, in a pure construction management scenario (where the construction manager does not perform construction or take on cost or time risk), the owner enters contracts directly with the individual trades, with the construction manager signing these contracts on behalf of the owner. A CCA 17 trade contract is appropriate for contracting with individual trades through a construction manager.

(e) Design-Build Contracts

In a design-build contract, the owner generally contracts out the design, construction, and inspection, often to a single party; but often the owner wants to retain some control over the design process and wants to monitor the integrity of the inspections. In these cases, the owner contracts a separate consultant to help ensure quality.

The most popular Canadian form of design-build contract is a CCDC 2–based contract known as CCA 14. However, one drawback of the CCA 14 is that it fails to recognize the fundamental differences between a design-build contract and a fixed-price contract. In addition, the CCA 14 does not deal directly with the issues of owner control in any detail.

American forms can provide tools for dealing with these shortcomings. The Design-Build Institute of America (DBIA) has many forms of design-build contracts, provided the parties modify them to suit the project. Designed from the bottom up for the design-build application, the DBIA contract reflects the roles and responsibilities of project participants. However, even though it is a good starting point, this form does not deal adequately with builders' liens, safety, and other Canadian issues. Other useful non-Canadian design-build contracts include forms from the American Institute of Architects (AIA), the American Consulting Engineers Council, the International Federation of Consulting Engineers (FIDIC), and the Associated Owners & Developers (AOD).

(f) Alliance Agreements

By its nature, an alliance agreement is intended to diffuse the historically adversarial relationship of parties in a construction project. It typically involves all the major parties, including the owner, significant contractors, and design professionals. The key terms of such an agreement are that all parties agree that there will be no liability for poor performance or defective work, and that all parties will share in the success or failure of the project from an economic perspective.

From the contractor's perspective, an alliance agreement theoretically removes most of the traditional risk/reward (often boom/bust) from the relationship by limiting both the contractor's potential liabilities and the potential for large windfalls that can result from good bidding and efficient work. Instead of large risk/reward, the contractor's exposure arising from poor performance or defective work is typically limited to a combination of overhead and profit. The owner pays the contractor on a cost-plus basis with a loss/profit or painshare/gainshare available at the end based on performance.

Meanwhile, the owner offsets increased risks through cost savings on an individual project basis and, more importantly, through reduced legal claims. The owner also reduces risks by selecting only contractors that are trustworthy and competent.

One legal commentator has defined alliancing as follows:

Essentially alliancing is a collaborative, incentive driven method of contracting where all participants work co-operatively to the same end, sharing the risk and rewards of bringing the project within time and under cost, whilst respecting principles of good faith and trust.[2]

Thus, the crucial elements of an alliance agreement are the following:

1. a "no blame" clause (perhaps with some limited exceptions), which operates between all of the parties, including the owner;
2. a clearly defined gainshare/painshare; and
3. a clear understanding of the roles of the participants.

In addition, most alliance agreements deal with the following important issues:

- A clear definition of the scope of the arrangement
- An exclusivity clause providing that all work within the scope of the arrangement (with some exceptions) will be done under the alliance agreement
- Other commercial aspects including insurance, third party liability, indemnities, and dispute resolution
- A time frame for the agreement
- The composition, voting rights and requirements, and procedures for an alliance board which acts in a similar fashion to a corporate board of directors.

---

[2] Presentation by Tony Abrahams, Director, Construction and Infrastructure, KPMG Legal, as reported by Juliet Pratley, "Project Alliancing: Does it work?" (1999) 15(2) *Building Australia* 33.

(g) Public-Private Partnerships

Strain on governmental resources has led to innovation in the contractual framework for public projects. Projects involving hospitals, transportation systems, water and wastewater treatment facilities, and public buildings are often public-private partnerships.

The Government of British Columbia, Ministry of Competition, Science and Enterprise, produced a Public Private Partnerships (PPP) Best Practices Guide which included this definition of a PPP (or P3):

> Simply put, a P3 is a partnership between the public and private sectors where there is a sharing of risk, responsibility and reward, and where there is a net benefit to the public. Specifically, a P3 is a partnership for some combination of design, construction, financing, operation and/or maintenance of public infrastructure which may rely on user fees or alternative sources of revenue to cover all or part of the related costs of capital (debt servicing and principle payment and return on equity if applicable), operations and capital maintenance.[3]

This definition shows that public-private partnerships share some but not all of the features of alliancing, while also dealing with other issues, such as financing and maintenance of the public infrastructure.

In one popular form of public-private partnership, the private sector partner provides the capital for the project. In exchange, the private partner receives a lengthy operating contract through which it earns profit over the long term. For example, in a new highway project, the government and a consortium consisting of design professionals and contractors work together to conceptualize and design the highway. The consortium then constructs the highway with little or no government funding. While the government retains ownership of the new highway, the consortium obtains a lengthy lease and permission to charge a toll. This toll pays for the capital cost of the highway and ongoing maintenance and also provides the consortium's return on investment. The government gets a highway at no cost, and the users pay for the benefit of using the new highway.

11.3 Professional Service Agreements

All professional service agreements should clearly define the scope of work and expected outcomes. As with other contracts, the price for performing the services, whether expressed as a fixed price, cost plus, or some percentage of construction cost, should be clearly defined.

One important clause in a professional service agreement is the *limitation of liability clause*, which defines the limits of available claims against the professional. These clauses often limit claims by the party hiring the professional to the limits of the professional's insurance for any claim. However, there is no guarantee that the professional will have insurance available at the time of a claim or that an exclusion clause in the insurance policy will not exclude the type of claim being brought.[4] Hence, some professional service agreements include an alternative remedy should there be no available insurance, such as the re-performance of the services or a payment (either a specifically predetermined amount or the amount of fees, or a portion thereof, paid to the professional under the agreement). Another critical element of a limitation clause is a limitation on the time in which parties can commence legal action.

---

[3] Government of British Columbia, Canada, Ministry of Competition, Science and Enterprise, *PPP (Public Private Partnerships) Best Practices Guide* (January 1998).

[4] See Chapter 16 on insurance.

For services provided in professional service agreements, clauses dealing with field reviews are often problematic. The client expects the professional to be the guarantor of the work being reviewed by the professional, while the professional believes that he or she is responsible only for carrying out reviews to determine if the work is in general conformance with the contract. Even the language used in these clauses is problematic. The client wants inspection to determine that the work performed is in accordance with the contract; in contrast, the professional wants to conduct periodic field reviews just to determine general conformance to the contract terms. Because these opposing objectives often lead to disputes, all parties must deal with issues regarding field reviews and inspections before signing the contract. Realistically, neither the client nor the professional is being reasonable in maintaining such positions, since the client cannot expect the professional to guarantee performance, for that would require an enormous level of review and cost, even if it were possible; while the professional cannot expect that he or she can choose the level of review needed and then only be responsible for what he or she happens to notice at the time. The balance is somewhere in the middle.

Other clauses of importance include those defining time for performance, dispute resolution, termination, provision of insurance, and ownership of the work product of the contract. Many of these topics are discussed in greater detail in section 11.6 on page 123.

(a) Engineering Service Agreements

The most common standard form of engineering service agreement in Canada is the Association of Consulting Engineering Companies (ACEC) 31. The principal clauses needing attention in this form are those regarding the level of inspection service, defined in the agreement as being in the sole discretion of the engineer; those assigning copyright to the engineer; and those limiting liability of the engineer in both time and dollar amount.

All parties must understand the contract terms before signing: the services, and level thereof, that are included and excluded; the limitations of liability; the impact of termination; and the role of the engineer in provision of contract administration services.

(b) Architectural Service Agreements

The principal agreement used in Canada for architectural services is the Royal Architectural Institute of Canada (RAIC) Document 6. The 1997 version of this document was modified in 2002. While the 2002 version creates more flexibility, it does so at the expense of simplicity. Document 7 is a shorter form of this contract, updated in 2005 to more closely follow Document 6. In addition, the AIA offers a form of contract for architectural services, but it is rarely used in Canada.

All of the issues discussed for the ACEC 31 are also relevant to the RAIC document forms. One additional drawback is that the RAIC agreement provides no way for the owner to terminate the project (unless the project is cancelled); and the termination "expenses" to pay to the architect include significant percentages above the actual cost. These costs apply even if the termination is the result of a fundamental breach by the architect. Thus, owners must examine the RAIC contract carefully.

(c) Geoscience Service Agreements

The American Association of Petroleum Geologists (AAPG) has a form of contract entitled "Contract for Geoscience Services." Two drawbacks of this form are the clauses defining the scope of services, which include the potentially ambiguous phrase "and other duties that customarily are a part of the geoscience function," and the clause that assigns responsibility for workers' compensation insurance to the geoscientist. In addition, the form fails to address the outcome of a termination and the liability of the geoscientist.

Both parties must ensure that they define and understand the scope of services, the limitations of liability, the impact of termination, and the non-disclosure and noncompetition obligations of the geoscientist.

11.4 Licensing Agreements

A licence is a grant of rights. An **exclusive licence** grants the licensee (the party gaining the rights) the exclusive right to use the rights held by the licensor (the party granting the rights).

Licences can be granted in many areas of the law. These include intellectual property[5] and real property.[6] For example, a licence can be granted to use a copyrighted song or to travel over private land.

The most important clause in any licence is the grant provision that defines the scope of the rights given to the licensee. The grant provision should define the exclusivity of the licence, the geographical extent and time of the grant, the purposes for which the rights can be used, and the conditions under which the grant can be terminated. A licensing agreement should also make clear whether the licensee can sublicence or whether the licence is only for the benefit of the licencee.

Particularly in fields related to technology, the licensor wants the licensee to use the licence, because compensation for many licences is based on a percentage of sales (*e.g.*, royalties). Hence, many licences attempt to enforce a "use it or lose it" type of obligation. However, forcing someone to use a licence is difficult to define in a contract and generally results in a "best efforts" type of obligation, which often becomes the subject of dispute. The parties should carefully define the use obligations.

11.5 Geoscience Agreements

Geoscience agreements can take many forms. Often, joint ventures[7] and limited partnerships[8] are used for oil and gas and mining explorations. Geoscience projects provide also some unique alternatives.

One such alternative is a grubstake agreement. Historically, a **grubstake agreement** was an agreement between a store owner (who provided grub) and the prospector so that the prospector did not starve while searching for a big strike. Later, prospecting syndicates evolved into grubstake agreements with units sold to participants, which raises prospecting money for the prospector in exchange for a share of the outcome. Many grubstake agreements are now subject to securities legislation.[9]

In addition, geoscientists also use **option agreements**, which provide the optionee (the party purchasing the option) with the right to buy or not buy land, goods, or rights from the optionor (the party granting the option). Mining law options are especially intricate because they involve the transfer of money and/or shares together with a promise to do work on the property, all in exchange for the right to earn an interest in the property. Option agreements are similar to licences; thus, the considerations that apply to licences also apply to option agreements.

---

[5] See Chapter 4, pp. 28–31.

[6] See Chapter 4, pp. 24–28.

[7] See Chapter 5.

[8] See Chapter 5.

[9] See Chapter 25.

11.6 Standard Clauses

All contracts attempt to define and control the same aspects of work on a project: the scope of the work, the contract time, the procedures for changes, and terms for damages and bonuses, warranties, termination, indemnification, exclusion, and dispute resolution. By knowing the requirements of these standard clauses and reviewing them in the contract before signing, a professional best protects his or her interests.

(a) Scope of the Work

A contract must clearly define the scope of the services and materials to be provided. This is often done through detailed specifications and drawings. Many difficulties arise on projects because of poorly defined scope of work. The more accurate and complete the definition of the scope of work, the lower the probability of subsequent claims.

Depending on the contract form, the scope of work definition may be very detailed (a stipulated sum contract) or functional (defined by a general description of the scope such as a 20 000 m^2 warehouse) or may simply specify the performance criteria (for example, a design-build water treatment facility contract stating the required quantity and quality of the water to be treated). In each instance, the key is to adequately define the scope of work to reduce ambiguity.

(b) Contract Time

Almost every contract contains a "time is of the essence" clause. The first thing to understand about this language is that it applies to both parties. Hence, each party must perform obligations in a timely manner.

Secondly, contracts often contain obligations to meet a particular schedule. Scheduling[10] is an art, not a science. Thus, a poorly planned schedule may end up being more of a hindrance than a help to the project. Professionals should pay close attention to schedule requirements and ensure that adequate contingency time is included. In addition, the definition of what components are to form part of the schedule is very important. Major tasks generally get most of the attention in a schedule; yet smaller tasks often determine the timing and feasibility of major tasks and as such should be defined and included as well. In general, all parties must pay particular attention to the feasibility of a project schedule.

(c) Changes[11]

Construction contracts need to define a method for modifying general terms and scope of work as circumstances change. Many contracts contain only a simple clause that permits changes to be agreed upon by the parties in writing. Other contracts contain extensive provisions dealing with ways changes must be priced and incorporated into the revised contract.

Changes to a contract can cause problems from a number of perspectives. First, parties often ignore the change procedures in the contract by issuing oral instructions or by making oral agreements for changes. This in effect creates a new contract through estoppel, thereby undermining the applicability of the original contract. Second, introducing changes during the project often requires modifying the work program; meanwhile, each party has a very different perspective on the impact

---

[10] See Chapter 19, p. 217.

[11] See also Chapter 9, pp. 71–73.

of such changes on the cost of the work and the time to complete. Thus, change can create conflict. Where changes are necessary, all parties should ensure that they are carried out in accordance with the contract, including providing appropriate documentation and notice.

(d) Damages and Bonuses

In Canada, damages clauses are sorted into different categories. *Penalty clauses* for late performance are unenforceable. Penalty clauses are distinguished from *liquidated damages clauses*, which *are* enforceable, by the way in which the amount is determined. Liquidated damages are an accurate pre-estimate of the damages suffered as a result of the late performance; whereas a penalty is simply an arbitrary punishment cost chosen to give the party a significant incentive to perform on time.

If a liquidated damages clause is included in a contract, both parties should perform and keep a calculation showing the rationale for the calculated amount. Whether the damages actually incurred are greater or less than the estimated liquidated damages, the general rule is that the liquidated amount will be awarded.

Some professionals believe that every clause that gives damages to one party for late performance must be accompanied by a clause that provides a bonus for early performance. However, no such principle exists in Canadian law. Contracts may provide bonuses for performance that exceeds the expected level, usually early performance; but there is no requirement to do so even where there is a liquidated damages clause.

(e) Warranty

A **warranty** is a contractual promise to repair defects in the goods and services provided for a specific period of time after the goods and services were provided. For example, a new home may come with a one-year warranty stating that the builder will repair any defect in the home for one year after it is substantially complete. In addition to being limited in time, warranties often contain other important limitations, such as the location of service, the obligation of the purchaser to provide notice, and certain costs that must be borne by the purchaser.

However, Canadian law implies statutory warranties in addition to those provided in contracts. Most jurisdictions have *sale of goods statutes* that provide implied warranties that the goods sold are fit for their intended purpose, are of merchantable quality, and are free of charges or encumbrances. However, terms and conditions written into the contract can contradict and negate these implied warranties.

In fact, in some cases, contracted warranties are actually just limitations of liability and provide little or no real benefit to the purchaser. For example, if a new home has a defect, the purchaser has a set period (either two or six years) during which he or she can commence action against a builder for defects or deficiencies. However, the builder may argue that any warranties normally implied by law are negated by the wording of the warranties in the contract, and that the limitation period during which the injured party could commence action is the stated warranty period. In other words, some warranty providers try to make the warranty in the contract the only remedy for the other party. The likelihood of success of such an argument depends on the clarity of the wording of the contract. Many contracting parties add language to warranties indicating that the warranty is in addition to, rather than in place of, their rights at law.

Buyers should also know who is providing the warranty. Often, the builder's contract warranty is only one piece of the warranty, since manufacturers' warranties on particular elements of the construction project may in fact be longer than the contract warranty. Buyers need to ensure that such warranties are assignable and then must obtain their assignment. In either case, the warranty is only as good as the financial strength of the party providing it, because obtaining an extended warranty from a company that will not be in existence to back it up is of no value.

(f) Termination

Most contracts are terminated when the parties have each performed their obligations under the contract. However, there are situations where a contract is terminated prior to completion. The contract must clearly set out the circumstances under which the parties can terminate the contract. In most cases, the contract will only permit early termination for a fundamental breach of contract.[12] Even then, in most cases, termination can occur only after a period of time after notice of the breach is given and only if the breaching party receives a reasonable opportunity to remedy the breach.

Some contracts state that a party will be in breach of the contract for particular actions. However, these clauses should be read with caution. While failure to provide a progress report on time may technically be a breach of the contract, a court is unlikely to regard that act as grounds for terminating the contract early. Generally, only a fundamental breach can become grounds for terminating the contract. However, contracts can provide appropriate wording to ensure that multiple unremedied minor breaches also provide grounds for termination.

Contracts can also provide for termination for convenience, which means that a party can terminate simply because they want the contract to end. However, most such clauses require reasonable compensation for the other party because that party was not at fault.

In general, early termination of a contract is a very serious action. If the terminating party is found to have acted prematurely, then that party will be found in breach of contract. Professionals should always seek legal advice before terminating a contract and before giving notice of a breach.

(g) Indemnification

Many contracts contain indemnification clauses (see Chapter 9, pp. 70–71). In most cases, each party indemnifies the other against their negligence or breach of contract. For example, a software developer generally indemnifies the client for breaches of copyright of the software program created by the software developer, such that if a third party brings claims against the client for breaches of copyright regarding that software program, the software developer protects the client from financial harm, both in terms of damages and legal costs.

In some cases, especially in the United States, indemnities are used to shift risk. For instance, the software developer in the previous example may try to obtain an indemnity from the client, so that the client ends up protecting the software developer from actions for breach of copyright committed by the software developer.

As with warranties, an indemnity is only as good as the financial strength of the party providing the indemnity. Professionals must read and understand the scope and consequences of an indemnity, while recognizing that it may provide very little protection if the party that provides it is not financially sound.

(h) Exclusion, Limitation, or Waiver Clauses

An **exclusion clause** is a clause that purports to completely exclude the damages or remedies available to the innocent party upon the occurrence of specified events. Courts have referred to exclusion clauses variously as "exemption," "exculpatory," "exception," "escape," and "protective"

---

[12] See Chapter 7.

clauses. They are sometimes referred to as "weasel" clauses (although not by the courts). To be enforceable, an exclusion clause must clearly define the exact contingency or event to be exempted or limited. Generally, the courts construe these clauses strictly against those who drafted them and resolve any ambiguities in favour of the other party.

Limitation or limiting clauses are a subcategory of exclusion clauses. Instead of completely excluding liability for specified events, **limitation clauses** purport to contain or limit the damages or remedies available to the innocent party upon the occurrence of specified events. Some courts have taken the position that limitation clauses are different in quality from exclusion clauses and should be given a more liberal interpretation. In all other respects, the principles governing court interpretation of exclusion and limitation clauses are the same.

In *B.G. Linton Construction v. C.N.R.*,[13] the majority of the Supreme Court of Canada held that if an exclusion clause is drafted clearly enough to leave no doubt about its meaning, then full force and effect will be given to it. An exclusion clause will only be interpreted against its creator where there is ambiguity about the meaning of the clause in the contract.[14]

Whether a clause is effective to exclude or limit liability depends upon its wording and the particular circumstances of the case. The Supreme Court in *Tercon*[15] recently set out a three-stage test in analyzing an exclusion clause:

1. Does the exclusion clause apply to the circumstances of the case?

2. If the exclusion clause does apply, was the exclusion clause unconscionable at the time the contract was made?

3. If the exclusion clause is valid and applicable, should the Court override the clause for public policy reasons?

If you are seeking to draft an exclusion clause, it is therefore critical that the clause deals with the circumstances that you are trying to exclude.

Limitation clauses can be used to re-allocate any type of risk. These risks can include restricting one party's ability to make certain types of claims, limiting a party to claims against insurance, limiting the time within which a claim may be brought, or limiting the right to claim certain types of damages. One very common example of the last category is a limitation against claims for consequential damages. *Consequential damages* refer to lost profit and other damages that are unrelated or only indirectly related to the claim being made. The reason to exclude consequential damages is to control prices from bearing absurd costs. Consider the purchase of seals for the bottom of a ship. A $5.00 seal may cause the loss of the $50 million vessel. From an economic point of view, for the seal manufacturer to be able to bear the risk of failure of its $5.00 seal, it would have to charge considerably more than $5.00 (such as $10 000 per seal). However, this price change would greatly increase the cost of building and maintaining ships. Thus, ship owners would likely be better off buying $5.00 seals and monitoring them carefully, rather than having the seal manufacturer bear that risk. For this reason, contracts often define the limits to consequential damages.

Care should be taken in understanding how risk is being re-allocated whenever an exclusion clause is used. For instance, where damages are limited to those recoverable from insurance, the party must understand what type of insurance is being carried and the conditions on that insurance.

---

[13] [1975] 2 S.C.R. 678.

[14] See also *McClelland and Stewart Ltd. v. Mutual Life Assurance Co. of Canada*, [1981] 2 S.C.R. 6, where the Court held that the doctrine of *contra proferentem* (see Glossary) did not come into play at all as there was no ambiguity in the contract.

[15] *Tercon Contractors Ltd. v. British Columbia*, [2010] S.C.C. 4.

Case Study 11.1

Coles v. Clarenville Drydock Ltd. (1998), 170 Nfld. & P.E.I.R. 17 (N.S.C.T.D.)

Facts

The defendant Clarenville was a marine service centre operator. The plaintiff Coles had stored his boat at the defendant's marina. While Clarenville's employee was in the process of lifting Coles's vessel, a strap placed around the hull malfunctioned, causing the boat to incur damage. However, the Statement of Acceptance of Responsibility signed by Coles provided that Coles was to assume all liability for any damage which occurred during the lift, launch, and storage of the vessel.

Question

1. Does the exclusion clause bar Coles from pursuing Clarenville in negligence?

Analysis

1. The Court stated the following basic principle: courts regard such exclusion clauses "with a critical or jaundiced eye, approaching the interpretation of such a clause strictly, applying the ordinary rules of construction, including the *contra proferentum* rule, which says the clause, particularly in a standard form contract, is to be strictly construed, against the party who drafted it."[16]

    The Court also stated that the negligent dropping of the crane, if it occurred, would be a fundamental breach of the contract, and "This would have the effect of depriving the Plaintiffs of substantially the whole benefit which it was the intention of the parties that they should obtain from the contract, namely, the safe lifting, launching and storage."[17]

    Thus, the Court concluded that in these circumstances, it would not be fair and reasonable to enforce the clause.

(i) Dispute Resolution

Nearly every contract has some form of dispute resolution provision. A **dispute resolution provision** is a clause that defines the process to be followed in resolving disputes between the parties to a contract. The principal methods of dispute resolution are discussed in Chapter 14. From a contractual perspective, it is important to understand the nature of the dispute resolution process (DRP) incorporated into the contract.

Almost all contracts contain some form of notice of claim provision. A **notice of claim provision** is a clause that defines the process to be followed for giving notice of and information in respect of a claim under a contract. Notice must generally be given within a reasonable time once the claimant has determined that a potential claim exists. The reason for providing prompt notice is that many contracts contain strict time limits for bringing claims, because the breaching party wants the opportunity to remedy or mitigate the breach at an early stage, rather than being told

---

[16] *Coles v. Clarenville Drydock Ltd.* (1998), 170 Nfld. & P.E.I.R. 17 (N.S.C.T.D.), at para. 15.

[17] See previous note at para. 28.

after the contract work is complete that a claim is being brought. Professionals must understand the time constraints for providing notice, ensure that the notice given contains the proper information, and deliver the notice in accordance with the requirements of the contract.

After notice of a claim is given, many contracts then require the parties to negotiate, generally with some time constraints. The contract may or may not require the parties to negotiate in good faith. If the parties agree to settle the dispute, the settlement will be binding. However, if they fail to agree, the dispute moves on to the next stage.

The next stage may involve a referee process, which may or may not be binding. If a claim has not been resolved, the DRP clause may require binding arbitration. Note that the parties can always agree to enter arbitration even if it is not contractually required; but if it is contractually required, in most Canadian provinces, the matter cannot skip arbitration and proceed to litigation.

Before including a mandatory arbitration provision in a contract, all parties must give the idea serious consideration. Many parties and lawyers dislike arbitration because most disputes involve more than the two contracting parties. A mandatory arbitration provision in a contract may force the dispute to be resolved in multiple forums. This approach rarely functions well. For this reason, parties should consider whether mandatory arbitration is right for their project and their situation. They must carefully scrutinize these provisions and determine whether such a clause is in their best interest, especially since parties can mutually agree to arbitrate a dispute at any time.

11.7 Project Finance

In a traditional project, the owner finances the project from cash reserves, a lender, or, most commonly, a combination of the two. In order to obtain lender financing, the owner has to satisfy the lender that the increased value of the asset, the cash flow generated by the project, or both will assure the lender of repayment. For this reason, the lender wants to minimize risk for both the lender and the owner so that the project will be successful.

The project consultant facilitates the financing process by doing the following:

- assisting the owner in development of a feasibility study
- providing a design that meets the project's needs
- working with the owner and lawyers to create contract documents that are "bankable"
- providing appropriate contract administration services, which may well include reporting to the lender.

A **bankable contract** is one with terms acceptable to the lender financing one of the project participants. *Bankability* is an issue when the borrower does not have sufficient assets or assured cash flow to secure a loan from the lender. If the borrower had sufficient assets or cash flow, the lender would simply take security on that property or cash flow. Bankability applies to not only the contract between the lender and borrower but also all other contracts on the project.

Bankability is achieved when the lender is satisfied that the project will be successful such that the borrower will profit from the project and be able to repay the loan plus interest. The lender must be satisfied that the contractual allocation of risk between the project parties is such that, even if difficulties are encountered, the debt will be protected so far as reasonably possible. Where possible, lenders like to see risks allocated to insurers[18] and bonding companies.[19]

---

[18] See Chapter 16.
[19] See Chapter 17.

For public-private partnership (P3) projects,[20] the principal asset of a P3 project is the public authority's (generally the government's) promise to make payments over time. This type of long-term payment obligation is called a *concession*. Because a concession is over a long term, any risk to the flow of funds to the private party that built and maintains the project places both the private party and their lender in jeopardy. A reasonable risk allocation by the public authority in the contract between the private company and the public authority makes the project more bankable.

CHAPTER QUESTIONS

1. Which statement best describes a standard form contract?
    a) It is perfect for every situation.
    b) It represents the best practices for an industry.
    c) It is a useful starting point.
    d) It is completely balanced.

2. A contractor enters into a fixed-price contract with an owner based on a well-defined scope of work. Midway into the contract, the contractor determines that its estimating department missed portions of the work. Is the contractor entitled to seek extra compensation? Why or why not?

3. Is a cost-plus contractor at risk for the cost of the work? Why or why not?

4. Describe the circumstances under which a contractor is able to seek increased unit prices.

5. Why is it not advisable to have construction managers perform construction work?

6. Define a P3 contract.

7. Describe why professionals should not guarantee the work of other parties.

8. Can liquidated damages be assessed against a party if the contract does not contain a bonus clause?

ANSWERS

1. C

2. The contractor is not entitled to additional compensation because the contract is a fixed-price contract.

3. A cost-plus contractor is not at risk for the cost of the work because the inherent risk allocation of such a contract is for the owner to pay all of the costs (as defined in the contract).

4. The contractor can seek increased unit prices where there is a significant change in quantities from the estimate provided by the owner.

---

[20] See p. 120.

5. When construction managers perform some of the construction work, they are in a conflict of interest, in that the construction manager is supposed to oversee the work of contractors on the project. This means that the construction manager is left to manage his or her own work on the owner's behalf. However, the owner may agree to this arrangement, assuming that the owner understands the implications.

6. A P3 (public-private partnership) is a partnership between the public and private sectors where the parties share risk, responsibility, and reward, and where there is a net benefit to the public.

7. Professionals should not guarantee the work of other parties because it is physically impossible to provide an absolute assurance that every piece of that other party's work is perfect.

8. Yes

CHAPTER 12

TORTS

Overview

A **tort** is a breach of a duty to care for another party where the breach causes injury or loss to that party, independent of whether the two parties involved have a contract for which the law provides a remedy.

The modern law of negligence developed out of a famous English case called *Donoghue v. Stevenson* in 1932. In that case, Donoghue was injured by the contents of a bottle of ginger beer, which included a decayed snail. The bottle was purchased by a friend of Donoghue's from a shop that had in turn purchased it from the manufacturer, Stevenson. Hence, there was no contractual relationship between Donoghue and Stevenson and therefore no legal claim. However, the English House of Lords (the equivalent of the Supreme Court of Canada) chose to create the law of negligence in large part by the following statements from the Court in the *Donoghue* case:

> You must take reasonable care to avoid acts or omissions which you can reasonably foresee would be likely to injure your neighbour. Who, then, in law, is my neighbour? The answer seems to be—persons who are so closely and directly affected by my act that I ought reasonably to have them in contemplation as being so affected when I am directing my mind to the acts or omissions which are called into question.[1]

In effect, the Court in *Donoghue* held that even though there was no contract between the ultimate consumer of the contents of the bottle and its manufacturer, the manufacturer ought to have considered that consumer when it filled the bottle and ought to have taken reasonable care to avoid causing that consumer injury.

Torts are divided into two categories: unintentional and intentional (see Figure 12-1 on page 132). The largest category of unintentional torts and the primary focus of this chapter is negligence. Negligence claims are of particular interest to architects, engineers, and geoscientists, because a large percentage of claims against such professionals are framed in negligence. Intentional torts also include fraud and trespass.

Since *Donoghue v. Stevenson*, the law of negligence has continued to evolve. However, the basic elements of negligence are consistent from one jurisdiction to another. In order to succeed in a negligence claim, the plaintiff must prove the following:

---

[1]*Backman v. Canada*, [2001] 1 S.C.R. 367.

FIGURE 12-1 Types of Tort

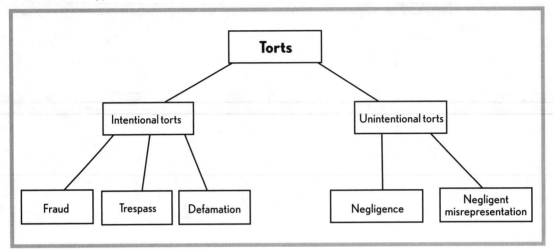

1. The defendant owed the plaintiff a duty of care.

2. The defendant breached that duty.

3. The plaintiff suffered loss or damage.

4. The breach was the proximate cause of the plaintiff's loss.

These elements also form the basis for important defences against negligence claims. A defendant need only prove that any one of the elements does not exist to defeat the claim. If there was no duty of care owed, no breach, no damage suffered, or no causation, there can be no recovery. For example, even if an architect is careless and leaves an important element out of the design, it is entirely possible that no loss will be suffered by the client (*e.g.*, if the error is caught before pricing is complete), in which case a claim against the architect in negligence would fail.

Of these elements, duty of care is the least settled area of the law. In general, professionals have a duty to care for everyone who is so closely and directly affected by their acts that they ought reasonably to have them in contemplation. In general, this duty includes those members of the public whom such professionals might foresee would be affected by their acts or statements, as well as the clients and employers of such professionals. Breach of duty is a legal concept that involves the establishment of a standard of care.

12.1 Duty of Care

(a) The General Rule

Early cases established that a duty of care must be owed before a negligence claim can succeed. **Duty of care** is an element of negligence based upon reasonable foreseeability. If at the time he or she committed the negligent act, the defendant could have reasonably foreseen that the plaintiff might suffer loss or damage, then a duty of care is owed to the plaintiff. For example, an engineer designing a building should reasonably foresee that if the design is performed negligently, a tenant might be injured in the event of a collapse. Therefore, a tenant injured in such a collapse would be owed a duty of care by the engineer. On the other hand, if a geoscientist prepares a confidential report for a client, and that report is subsequently relied upon by a third party, such as an investor, that investor will have difficulty proving that the geoscientist owed him or her a duty of care, even if the report was prepared negligently. That is because foreseeability would be difficult to prove in light of the confidential nature of the report.

Because the courts have feared that too much litigation might occur as a result of the tort concept, they have tried to limit the categories where a duty of care is owed. Courts have limited the scope of duty of care in this way with respect to the following issues:

- Does an auditor preparing financial statements owe a duty to public investors?
- Does an architect owe a duty to a contractor to point out construction errors?
- Does a design professional owe a duty to a contractor for errors in the plans and specifications?
- Does a designer owe a duty of care to construction workers with respect to safety issues?
- Does a contractor owe a duty to subsequent purchasers for construction defects?
- Does a non-contracting party owe a duty for pure economic loss, where no physical damage or injury has occurred?

The duty of care in the situations described above has not been decided consistently from one jurisdiction to another. Duty of care is more easily established if there is a pre-existing relationship between the plaintiff and defendant. For example, if the parties have a contract, duty of care can be more easily established, absent any contractual language to the contrary. The following sections provide a brief summary of duty of care issues that have been addressed by the courts.

(b) Duties of Care for Architects and Engineers

Both architects and consulting engineers owe a duty of care to protect owners, even if there is no contract between them. In the case of *Surrey (District) v. Carroll Hatch and Associates*,[2] the structural engineer had no contract with the owner; rather, the architect had hired him. The engineer became concerned about the soil conditions and warned the architect. However, he failed to tell the owner and was found liable for failure in his duty of care. American courts have reached the same conclusion.[3] Thus, liability may exist with or without the existence of a contractual relationship.

Another duty of care question is whether a designer who prepares plans and specifications for a bid owes a duty of care to the successful bidder for losses caused by errors in those plans and specifications. The Supreme Court of Canada considered this issue in the case of *Edgeworth Construction Ltd. v. N.D. Lea & Associates Ltd.*[4] and concluded that a duty of care was owed by the design firm to the bidder. This conclusion was based in part on the fact that the bidding period is too short to allow bidders to conduct a thorough review of the accuracy of the engineering work; furthermore, duplication of the work would be costly.[5] Thus, bidders must be able to rely on those who supply information to them.

Note that a negligent contractor cannot generally blame a designer for failure to notice and point out the contractor's mistakes. However, the owner can fault the designer for the same failure. The difference is that the designer almost always owes a contractual duty as well as a duty in tort to

---

[2] (1979), 101 D.L.R. (3d) 218 (B.C.C.A.).

[3] In the United States, intended third party beneficiaries of a contract can also sue to enforce that contract. The owner is an intended third party beneficiary of the contract between the prime consultant and the subconsultant. Therefore the owner could sue the subconsultant in both contract and negligence. This intended third party beneficiary rule is not the law in Canada.

[4] [1993] 3 S.C.R. 206.

[5] This result is also consistent with the approach taken in the United States. For example, see *C.H. Leavell & Co. v. Glantz Contracting Corp. of Louisiana, Inc.*, 322 F. Supp. 779 (E.D. La. 1971).

the owner; whereas the obligation of the designer to the contractor is not as clear.[6] Those rare cases where the courts have recognized such a duty have been criticized, since they allow the contractor to escape responsibility for negligence and breach of contract and instead place that responsibility on the architect.[7]

Another important duty of care issue is whether a design professional can be liable to a worker for unsafe work practices. Generally, design professionals do not control the construction work itself. Industry custom and practice dictates that contractors are responsible for means, methods, and techniques of construction, absent contractual language to the contrary. The CCDC 2 agreement between owner and general contractor reinforces these responsibilities.[8] However, if the designer accepts responsibility for supervising the construction, he or she may owe a duty of care to workers for safety hazards.

However, even if a design professional is found to owe a duty of care to workers, a claim in negligence by the worker against the design professional may be statute-barred (prohibited by law). Typically, lawsuits by injured workers against their employer are barred by virtue of the applicable workers' compensation legislation. Some workers' compensation legislation bars lawsuits against anyone registered as an employer under the statute, and design professionals may be registered as employers in some circumstances. For example, in Ontario, the *Workers Compensation Act* states:

> The provisions of this Part are in lieu of all rights and rights of action, statutory or otherwise, to which a worker or the members of his or her family are or may be entitled against the employer of such worker, or any executive officer thereof, for or by reason of any accident happening to the worker or any occupational disease contracted by the worker on or after the 1st day of January, 1915, while in the employment of such employer, and no action lies in respect thereof.[9]

Thus, third parties who are not part of the construction process (*i.e.*, excluding contractors, subcontractors, suppliers, workers, owners, and consultants) and who are not barred from doing so by statute may make claims for physical loss or injury against owners, occupiers, designers, and builders, as long as reasonable foreseeability can be established.

However, courts do not always determine whether a duty of care exists strictly on the basis of legal analysis. Policy considerations may also be a factor. Consider the problem of a contractor who is aware of latent defects[10] in a recently completed project. The contractor discloses all of these defects to the purchaser, who demands and receives a substantial reduction in the purchase price on account of the defects. The purchaser then resells the property to a subsequent purchaser without disclosing the defects, despite a legal obligation to provide full disclosure. The subsequent purchaser eventually discovers the defects, cannot locate the developer who sold the property to him, and sues the contractor in negligence. Strict legal analysis might dictate that a duty of care is owed, because at the time the negligent construction was performed, the contractor could reasonably foresee that a subsequent purchaser might suffer loss. However, a court might decide in the interest of fairness not to accept this analysis, thereby exonerating the contractor. This example is given to illustrate the point that some duty of care issues are difficult to determine.

---

[6] Even in the United States, it is difficult to find cases where architects and engineers have been found to owe such a duty to the contractor. One notable exception is the California case known as *U.S. v. Rogers and Rogers*, 161 F. Supp. 132 (S.D. Cal. 1958).

[7] See Lee, Byrum C., Jr. *Architect-Engineer Liability Under Colorado Law* (Bienna CA: The Cambridge Institute, 1987) at pp. 39–40.

[8] Clause GC 2.2.5.

[9] R.S.O. 1990, c. W.11, s. 16.

[10] A latent defect is defined as one that is not discoverable through reasonable inspection.

(c) Claims for Purely Economic Loss

Historically, courts in Canada have barred negligence claims for pure economic loss. A **pure economic loss** is one in which the only damages are lost money: there is neither personal injury nor property damage. For example, an engineer issuing instructions to stop work on a worksite might cause the contractor to suffer delay, with its accompanying financial implications. However, this is a pure economic loss and is therefore not recoverable. Pure economic losses are recoverable for claims of breach of contract; however, some courts still exclude tort claims for pure economic loss.

But there are exceptions to this general rule. For example, claims for negligent misrepresentation are in effect claims for pure economic loss but are not barred. The rule has also been modified as a result of the decision in *Winnipeg Condominium*:[11] the Supreme Court of Canada established a basic rule that allowed recovery for pure economic loss but only in circumstances where there is a substantial danger to the health and safety of the occupants. However, the courts still exclude negligence claims for purely aesthetic defects.

Although there are a few well-recognized exceptions to the rule barring claims for pure economic loss, that rule is still very much alive in Canada. For example, because claims by a subcontractor against an owner cannot be asserted directly as a breach of contract claim, because there is no privity of contract, such a claim might be attempted as a negligence claim, as in the *Design Services* case.[12]

In that case, the Supreme Court of Canada was mindful of the fact that, in the construction industry, there are multiple layers of potential claimants, including sub-subcontractors and employees of sub-subcontractors. It was the indeterminacy of the class of potential claimants that caused the court to apply the rule excluding claims for pure economic loss, fearing that allowing such claims might open the floodgates to a very large number of claims:

> … a recognition of an owner's duty of care towards subcontractors could lead to a multiplicity of lawsuits in tort, an undesirable result.[13]

(d) Reducing the Risk of Negligence Claims

Professionals cannot eliminate every risk of a negligence claim. However, they can help prevent a breach from occurring. In addition, they can try to prevent a duty from existing, although this is considerably more difficult.

In general, contractual limitation clauses and exclusionary language can restrict duties owed to the party with whom they have contracted. For example, many agreements between architect and client state that the architect has no duty to the client to supervise construction. Similarly, a contractor can negotiate a provision into its contract with the owner excluding responsibility for the adequacy of the design.

Note that exclusion or limitation clauses are only effective between the parties to the contract and do not impact liability to third parties. However, a well-defined scope of obligations can influence whether certain duties are owed to third parties not privy to the contract. For example, where a method of construction has caused injury to a third party, contract clauses that limit the scope of an architect's duties regarding contractors' means and methods of construction can help a court conclude that a duty is not owed by the architect to third parties. The reverse is equally true: if the contract includes supervisory duties in clauses defining the scope of work, a court may hold that a duty is owed to third parties.

---

[11] *Winnipeg Condominium Corp. No. 36 v. Bird Construction Co.*, [1995] 1 S.C.R. 85.

[12] *Design Services Ltd. v. R*, 2008 SCC 22.

[13] See previous footnote, at para. 64.

Therefore, professionals must take care must during contract negotiation to reduce or exclude risks and responsibilities that properly belong to others. A well-drafted contract may assist in establishing a duty of care defence.

12.2 Breach of Duty

Breach of duty is another way of saying that the standard of care was not met. **Standard of care** means the level of skill and care required of a person. This standard is generally based on the level of skill and care expected of a reasonable and competent member of that profession or occupation. The same rules regarding standard of care apply to all occupations, even if they are not considered "professions."

The rule about breach of duty quoted below applies to all common law jurisdictions, including Canada and the United States, and to engineers and contractors (even though only architects are mentioned):

> An architect undertaking any work in the way of his profession accepts the ordinary liabilities of any man who follows a skilled calling. He is bound to exercise due care, skill and diligence. He is not required to have an extraordinary degree of skill or the highest professional attainments. But he must bring to the task he undertakes the competence and skill that is usual among architects practising their profession. And he must use due care. If he fails in these matters and the person who employed him thereby suffers damage, he is liable to that person. This liability can be said to arise either from a breach of his contract or in tort.[14]

Thus, the mere fact that an error has occurred is not sufficient to establish negligence. However, in most cases, if that error is of such a kind that reasonable skill, care, or diligence would have prevented it, liability in negligence can be established.

Lawyers typically establish this applicable standard by the use of expert evidence. Each party to the lawsuit customarily calls an expert witness, who is a practitioner of the same profession as the defendant. The expert witness provides evidence that the conduct of the defendant was or was not consistent with what is expected in the profession and therefore met or did not meet the standard of care and skill of a reasonable practitioner.

However, standards for duty of care can change or may be different in different jurisdictions. The relevant standard used in each case has to be the one that was in place at the time the alleged negligence occurred, not at the time of trial. Changes and improvements in the state of the art may have occurred during the intervening period; but requiring the defendant to have predicted these future improvements would be unfair. It may be impossible to say what will be generally accepted in the profession in the future and what was the general practice of years past. In addition, an entire profession may not adopt a new method on the same date. At times, two or more schools of practice may be in use simultaneously. For example, in a 1957 English medical malpractice case, the Court expressed this principle, which applies equally to all professionals:

> [A doctor] is not guilty of negligence if he has acted in accordance with a practice accepted as proper by a responsible body of medical men . . . Putting it the other way round, a doctor is not negligent, if he is acting in accordance with such a practice, merely because there is a body of opinion who takes a contrary view. At the same time, that does not mean a medical man can obstinately and pig-headedly carry on with some old technique if it has been proved to be contrary to what is really substantially the whole of informed medical opinion.[15]

---

[14] *Voli v. Inglewood*, [1963] A.L.R. 657 (Aust. H.C.).

[15] *Bolam v. Friern Hospital Mgmt. Committee*, [1957] 2 All E.R. 118 at p. 122 (Q.B.).

In rare cases, a court may reject entirely some expert evidence regarding standard of care. If the entire profession seems to adhere to a standard of practice that is careless or unsafe, the court may exercise its own judgment. In the *T.J. Hooper* case, the Court stated:

> Indeed in most cases reasonable prudence is in fact common prudence; but strictly it is never its measure; a whole calling may have unduly lagged in the adoption of new and available devices. It may never set its own tests, however persuasive be its usages.[16]

Moreover, if professionals hold themselves as experts or specialists in a field, then the courts apply a higher standard of care. Specialists are measured against other specialists, not against average professionals. Therefore, specialist engineers cannot defend themselves with claims that their design was as good as what one should expect from an average engineer. What constitutes a specialist? In law, anyone who represents himself or herself to the public as a specialist or expert, even if he or she does not possess special expertise, is held to the specialist standard.

The vast majority of breaches of standard of care fall into two categories: incompetence (lack of skill) and lack of care. Duty of care cases involving lack of skill are relatively rare compared with those involving lack of care. That is because most professionals have the integrity to avoid doing work which they are unqualified to do. Yet architects, engineers, and geoscientists do occasionally get into duty of care trouble when they take on work for the first time in a new field. Any venture into a new field should be preceded by additional training; and ideally, the first project in the new area should be done in partnership with or under the guidance of an experienced practitioner in that field.

Claims caused by lack of care happen more frequently than those caused by lack of skill. While lack of skill is preventable through additional training and association with others, lack of care is equally preventable. One proven method of reducing claims is to establish a list of procedures. Professionals can create checklists that set out procedures and then follow them. Professional associations and liability insurers often have standard checklists or practice guidelines for their members. In addition, professionals can ensure that calculations are correct by hiring someone to check them. Finally, peer review is useful in many situations, and the cost of such review is more than offset by the savings in claim reductions. Professional associations and liability insurers are good sources of advice on peer review and other claims reduction programs. In fact, some associations require members to join claims reduction programs after they have been subject to a claim.

Professionals should also make a point of taking part in continuing education. Many professional associations run courses for their members. Some professions have mandatory continuing education. Even if continuing education is not mandatory in order to maintain registration as a professional, a professional has an obligation to remain current as standards evolve. Failure to keep up with changing standards is likely to result in negligence.

A final method of reducing the likelihood of such claims is to ensure that the contract governing performance of the work does not set an unreasonable standard of care or include a guarantee of results.

12.3 Proximate Cause

In order for an event or act to be considered legal cause of a loss, it must satisfy the "but for" test: *but for* the act, the result would not have occurred. The "but for" test simply asks the question: "Would the loss or damage or injury have occurred without the negligent action of the defendant?"

---

[16] *T.J. Hooper v. Northern Barge Corp.*, 60 F 2d 737 (1932).

If the answer is *yes*, then there is no causation. Once the claim meets this test, the court analyzes it to determine whether it is proximate.

A **proximate cause** is a cause that plays a significant role in producing the result, compared to one that is too remote. Consider the example of a contractor who leaves a trench excavated and installs barricades around it. A municipal crew removes the barricades and fails to replace them. A motorist then drives into the trench, resulting in injury and loss. It is true that "but for" the contractor's act of performing the excavation, the injury and loss would not have occurred. However, it is clear that act was not a proximate cause of the loss.

In the majority of cases, courts seem to have little difficulty determining causation. In a negligence case, causation need only be determined on a balance of probabilities, not beyond a reasonable doubt.

Case Study 12.1

University of Regina v. Pettick (1986), 51 Sask. R. 270 (Q.B.)

Facts

University of Saskatchewan ("the U of S"), which later became the University of Regina, retained Joseph Pettick as architect to design a physical education building at its Regina campus. Pettick hired Haddin Davis & Brown Co. Limited as its structural engineers on the project. After the project was complete, failure of the roof system occurred due to the inability of the hub assemblies to resist the forces applied to them. The architect and engineers had incorrectly assumed that the fabricator had designed a compensating camber into the roof frame. The U of S sued the contractor, the architect, and the engineers. Each denied liability.

Question

1. For the sake of discussion and analysis, assume that the architect was negligent in failing to ensure that the camber had been built in to the trusses. Does it automatically follow that the architects are liable?

Analysis

1. The Court found that it did not:

   Failure to ensure camber and thereby minimize roof deflection ... was cited by the University as indicators of the negligence of several of the defendants. I need not consider these suggestions further because even if these omissions constituted negligence, that negligence did not contribute to the deterioration of the joint assemblies which was the root cause of the University's damage.[17]

   In other words, even if the architect was negligent, that negligence was not a proximate cause of the loss.

---

[17] *University of Regina v. Pettick* (1986), 51 Sask. R. 270 (Q.B.) at p. 278.

12.4 Loss Caused by the Breach

When a negligent act or omission takes place, sometimes no loss or damage occurs as a result. If there is no loss resulting from the breach, the negligence claim fails.

For damages for breach of contract,[18] the proper measure of damages for breach of contract is the amount of money needed to put the innocent party in a position as if no breach had occurred. The same principle applies in negligence cases. The difference is that in a breach of contract, the contract has been breached; whereas in a negligence case, the common law duty of care has been breached. In breaches of both contract and duty of care in tort, the courts must decide whether to award diminution (a reduction in value) or the full cost of repair. For instance, if a car is damaged and it will cost more to repair than the car is worth, the proper measure of damages is the reduction in value.

12.5 Tortious Misrepresentation

A **misrepresentation** is a false statement. Misrepresentations fall into three categories: innocent, negligent, and fraudulent. Depending on the category of misrepresentation, different remedies may be available to the victim of the misrepresentation.

The most serious type of misrepresentation is fraudulent misrepresentation. Fraud is discussed in detail on pp. 140-141.

Negligent misrepresentation is a misrepresentation of facts that breaches a duty of care, and as such, it can be considered to be a form of negligence. However, negligent misrepresentations differ slightly from pure negligence cases. First, a negligent misrepresentation involves a statement, rather than an action. For example, a car driver who causes an accident by going through a red light commits an act of negligence but not a negligent misrepresentation. Second, the injured party must have reasonably relied on the negligent statement in order for the claim to succeed. Simply, a person who provides erroneous information to another person in the performance of his or her profession may be liable if the recipient of the information suffers loss as a result of reasonably relying upon that information. These cases typically involve pure economic loss.[19] Negligent misrepresentation is one of the best-recognized exceptions to the rule that pure economic losses are not actionable in tort.

In the seminal 1964 case on negligent misrepresentation by professionals, *Hedley Byrne & Co. v. Heller & Partners Ltd.*, the Court described the law of negligent misrepresentation by the Court as follows:

> if someone possessed of a special skill undertakes, quite irrespective of contract, to apply that skill for the assistance of another person who relies upon such skill, a duty of care will arise.[20]

Lack of causation can be an effective defence in negligent misrepresentation claims, because the reliance requirement of a misrepresentation claim is related to causation. If there is no reliance, then the claimant would have followed the same course of action with or without the erroneous information; therefore, the "but for" test fails.

An innocent misrepresentation involves no intention to deceive. Yet it may still give rise to a remedy. Where there has been an innocent rather than a fraudulent or negligent misrepresentation, the plaintiff may still be entitled to the remedy of **rescission**. Rescission means the cancellation of the

---

[16] See Chapter 7, pp. 59–61.

[19] *Hedley Byrne & Co. v. Heller & Partners Ltd.*, [1964] A.C. 465 (H.L.).

[20] See previous note, p. 694.

contract. However, in the absence of negligence or fraud, the plaintiff is not entitled to additional damages. As well, the mistake must be fundamental or substantial in nature to warrant rescission.[21]

Misrepresentations can take the form of a statement that is overtly false or one that is false because of what it does not say. The ancient Roman statesman Cicero used the following example to illustrate the second type of misrepresentation. A merchant from Alexandria had transported a large stock of corn to Rhodes, where a famine was in progress. The merchant knew that more merchants with an even larger stock of corn were just behind him over the horizon. During bargaining over price, the merchant was asked whether more corn was on its way. Rather than answering that question, the merchant simply replied by asking whether or not the citizens saw more merchants and imploring them to trust their eyes. The merchant's statement is a misrepresentation: it does not conceal facts, but rather avoids disclosure.

The field of geoscience has had many cases dealing with disclosure of material facts, including the notorious BRE-X scandal in the 1990s. In another well-known 1990s case, *Ontex Resources Ltd. v. Metalore Resources Ltd.*,[22] the Ontario Court of Appeal held that failure to disclose a favourable report in the course of negotiations over a contract should result in the rescission of the contract. In *Ontex*, the parties had entered into a "lease option" agreement whereby the defendant was to carry out exploratory work in return for an interest in the plaintiff's mining claims. One of the terms of that agreement was that the defendant had an obligation to disclose information to the plaintiff. The parties entered into negotiations approximately two years after the initial agreement, and the plaintiff agreed to reduce its interest, unaware that the defendant had obtained a favourable report. The Court held that the defendant's failure to disclose the report meant that the second agreement should be rescinded. Hence, geoscientists must clearly understand the obligations associated with information and its disclosure (see also section 12.7 on fiduciary duty, p. 141).

12.6 Fraud

Fraud, also known as deceit, is an intentional tort. It is also known as fraudulent misrepresentation, because in almost all cases fraud involves statements (representation) by one party to another. Most Canadian courts adopt the following definition of fraud:

> A representation in order to be fraudulent must be one (1) which is untrue in fact; (2) which defendant knows to be untrue or is indifferent as to its truth; (3) which was intended or calculated to induce the plaintiff to act upon it; and (4) which the plaintiff acts upon and suffers damage.
>
> It is not, however, necessary that a misrepresentation to sustain an action of deceit should be made in express terms; it is sufficient if the words used are intended to convey a false inference.[23]

Some also use the following definition:

> A fraudulent misrepresentation is a statement known to be false or made not caring whether it is true or false. A person induced to enter into a contract by such a statement is entitled, *prima facie*, to damages for fraud . . . and to rescission.[24]

Fraudulent misrepresentation can also occur by omission, rather than by an express statement. Where the courts find that there is a positive obligation to disclose information, failure to disclose

---

[21] *Ennis v. Klassen* (1990), 70 D.L.R. (4th) 321 (Man. C.A.).

[22] (1993), 103 D.L.R. (4th) 158 (Ont. C.A.).

[23] *Greenglass v. Rusonik*, [1983] O.J. No. 40 (Ont. C.A.), quoting Kerr, W.W. *Law of Fraud and Mistake*, 7th ed. (London: Sweet & Maxwell, 1952) at p. 25.

[24] *1018429 Ontario Inc. v. FEA Investments Ltd.* (1999), 179 D.L.R. (4th) 268 (Ont. C.A.). Quoted from Waddams, Stephen. *The Law of Contract*, 4th ed. (Aurora, Ontario: Canada Law Book, 1999) at p. 299.

that information may result in a finding of fraud. For example, in some cases, courts have held that an owner soliciting bids who is in the possession of a soils report must disclose that report to all bidders.

12.7 Fiduciary Duty

A **fiduciary duty** is a heightened duty to care for the interest of another party in priority to one's own interest and a duty not to act against the interest of the other party. Courts have found a fiduciary obligation in circumstances involving a trust relationship. While there is no firm rule describing when such a duty exists, the following guidelines generally apply:

- One party (the "fiduciary") has the ability to exercise some discretion or power.
- The fiduciary can unilaterally exercise that power or discretion so as to affect the beneficiary's legal or practical interests.
- The beneficiary is peculiarly vulnerable to or at the mercy of the fiduciary.
- A fiduciary relationship does not normally arise between arm's length commercial parties.

One case with extensive analysis and explanation of fiduciary relationships in Canada is *Lac Minerals Ltd. v. International Corona Resources Ltd.*[25] In that case, a junior mining company was negotiating with a larger mining company to develop a mine in Ontario. The junior company provided confidential information to the larger company during the negotiation process. The larger company secretly purchased adjacent properties. The Court found that conduct not to be a breach of fiduciary duty, because it involved two commercial arm's length parties in negotiation of a contract. However, it held the action to be a breach of an implied obligation of confidentiality. Thus, the Court awarded damages to the junior mining company. The basic principles of fiduciary relationships in Canada were clarified as a result of the Court's reasons in that case. The discussion in *Lac Minerals* described breach of fiduciary duty as "carrying a stench of dishonesty." It is similar to a claim in fraud.

The relationship between an architect, engineer, or geoscientist and client is not inherently a fiduciary relationship. To determine whether a fiduciary relationship exists in any particular case, the court look for the hallmarks of such a relationship, such as vulnerability and power, as well as factors described above.

Case Study 12.2

Canadian Transit Co. v. Girdhar (2002), 17 C.L.R. (3d) 1 (Ont. S.C.)

Facts

Canadian Transit operated a bridge crossing between Windsor and Detroit.

The defendant Girdhar was a consultant hired by Canadian transit to work on a project related to this crossing. However, the defendant was also working as a consultant for the Peace Bridge Authority, who operated a crossing several hundred miles away. Canadian Transit asserted a claim that because Peace Bridge Authority was a direct competitor of Canadian Transit, the defendant's work with both companies on similar projects was a breach of fiduciary duty.

[25] [1989] 2 S.C.R. 574.

Question

1. Did a fiduciary relationship in fact exist in this case?

Analysis

1. The Court ruled that a fiduciary relationship did not exist in these circumstances. The Court specifically stated that there is no broad principle that professional advisors are fiduciaries because such a relationship is inherent in the professional relationship.

12.8 Trespass

Trespass is the unauthorized entry onto the land of another person. It is a **strict liability** tort, meaning that in order to prove liability, the plaintiff need only prove that the defendant entered or encroached onto the plaintiff's land without authorization. The plaintiff does not need to prove negligence or moral transgression.

Property refers to more than the land's surface. Property lines are usually marked on the surface of the land, if they are marked at all. However, in theory, each property line extends from the centre of the earth up through the atmosphere. Subsurface and airspace encroachment rights are of practical importance for architects, engineers, and geoscientists. In construction, encroaching onto neighbouring property is sometimes necessary for carrying out the construction in an efficient manner. This means that rock anchors or soil anchors might need to be installed into neighbouring properties to stabilize the embankments until backfilling is completed; and tower cranes might swing over small portions of adjacent properties. The rock anchors are a permanent encroachment, even though they are harmless, whereas the over-swing of the crane is a temporary encroachment.

If a company cannot obtain consent of neighbours, then it must find alternative means of construction. Instead of rock anchors, it could use sheet piling, but at far greater expense. Instead of a tower crane, it could use mobile cranes, again at greater expense. However, extra cost is no excuse for trespassing. In one case, a developer unable to obtain consent of the neighbour chose to install rock anchors anyway. The court awarded punitive damages to the neighbour equal to the amount of money the developer saved by not using the more expensive method.[26]

To facilitate these construction situations, neighbours often negotiate reciprocal encroachment agreements, such that the neighbour granting permission to the developer is guaranteed the same courtesy in the future. The party carrying out construction also normally provides assurances, indemnities, and agreed levels of liability insurance.

In addition, geoscientific activities create situations that create a risk of trespass. Collecting samples, doing airborne surveys, and drilling wandering boreholes can all result in actions for trespass. Geophysical trespass has been defined as a "wrongful entry on land for the purpose of making a geophysical survey on the land."[27] Geoscientists must ensure that they have appropriate permission to enter onto land (above or below grade) to conduct any form of investigation prior to doing so.

[26] *Austin v. Rescon* (1987), 18 B.C.L.R. (2d) 328 (S.C.).

[27] Williams, H. and C. Meyers. *Manual of Oil and Gas Terms*, 9th ed. (New York: Matthew Bender, 1994).

Trespass is not the only land-related example of a strict liability tort. Another related example is the tort of *nuisance*, which originated in an old English case known as *Rylands v. Fletcher*.[28] In that case, the defendant's underground water reservoir caused an old mine shaft owned by the plaintiff to collapse. Although the Court found that the defendant was not negligent, he was still liable for damages. Canadian courts find liability in nuisance where the plaintiff has proven that the following:

1. The defendant made a non-natural use of his land.

2. The defendant brought onto his land something which was likely to do mischief if it escaped.

3. The substance in question escaped.

4. Damage was caused to the plaintiff's property (or person) as a result.[29]

12.9 Duty to Warn

A special concept in tort is the duty to warn. In addition to obligations in contract and more generally in tort, professionals owe a duty to warn of impending damage to persons or property. Especially for risk of personal injury, the law holds professionals to a very high standard and requires that they take reasonable—and in some cases more than reasonable—actions to prevent such injury.

For example, if an electrical engineer on a worksite has become aware through experience that high, vertically unsupported trenches could collapse, then he or she has a duty to warn if he or she notices any such thing on a worksite. Even if the unsafe trench is not within the electrical engineer's scope of work, he or she is not relieved of the responsibility to take steps to prevent that trench from collapsing. Defining precisely what steps are necessary is difficult, since these steps depend on the circumstances. It is possible that the engineer could discharge this duty by alerting the contractor of the danger, followed by the owner and regulatory authorities, or perhaps taking the immediate step of suspending work on the site.

For example, in *Surrey v. Carroll-Hatch*,[30] the prime consultant was an architect who in turn had hired a structural engineer. The structural engineer was concerned about the lack of sufficient geotechnical information and notified the architect, with whom he had a contract. However, he did not notify the owner. The building incurred damage as a result of uneven settlement. The Court found that he owed a duty to warn the owner.

12.10 Product Liability

Product liability cases in general have resulted from the decision in *Donoghue v. Stevenson* (see pp. 131–132). **Product liability** is generally understood to mean the liability of the manufacturer to a consumer for a defective product. Where the consumer has purchased the product directly from the manufacturer, the consumer can claim for breach of contract if the product fails and causes injury. However, in practice, the vast majority of products purchased by consumers are not purchased directly from the manufacturers of those products. Instead, the products pass through wholesalers, distributors, and retailers before they reach the consumer. For this reason, the consumer must assert the claims using tort law, rather than contract law.

---

[28] (1868), L.R. 3 H.L. 330.

[29] See, for example, *John Campbell Law Corp. v. The Owners, Strata Plan 1350*, 2001 BCSC 1342.

[30] (1979), 101 D.L.R. (3d) 218 (B.C.C.A.)

Defects in products occur for reasons that can be divided into two categories. First, defects may occur because the product has not been manufactured in accordance with the specifications. Second, the product was manufactured according to specifications, but the specifications were inadequate. In either case, the ultimate user of the product may have a claim directly against the manufacturer.

Moreover, a claim against the manufacturer may be based as well upon a duty to warn. If the manufacturer is or should have been aware that the product poses a danger to the public, warnings should have been prominently displayed on the product packaging. For pharmaceuticals, the warning should have been conveyed to key intermediaries, such as a college of physicians, who would typically provide access to the product through prescriptions.

Note that Canadian laws governing claims about defective products differ from those in the United States, and this difference causes some confusion in the general public. In the United States, a consumer can prove a claim against the manufacturer without proving negligence. This is known as the doctrine of *strict liability*. However, in Canada, where there is no direct contract between the consumer and the manufacturer, the consumer must prove negligence. In other words, all the general requirements that apply to claims in negligence also apply to product liability cases. These requirements include establishing a duty of care, a breach of that duty, causation, and damages resulting from the breach.

There are, however, some differences between an ordinary claim in negligence and a negligence claim based on a defective product. For example, in a simple negligence claim, such as a claim against an architect for defective design, the plaintiff has the burden of proof for each element of the claim. In a product liability claim, once the plaintiff proves that the product is defective, the burden of proof falls on the product's manufacturer to disprove that its conduct fell below the applicable standard of care.

In product liability cases, there is also a defence known as the "learned intermediary" defence. In many situations, architects, engineers, and geoscientists may be characterized as learned intermediaries. A **learned intermediary** is a person positioned between the manufacturer of a product and the consumer, a person who has specialized technical knowledge and upon whom the law imposes a duty to warn the consumer. Professionals must be vigilant to warn clients of risks and dangers regarding the systems and products that they design and specify. They must communicate risks and dangers that have been communicated to them by manufacturers, as well as those they recognize based on their training and experience.

CHAPTER QUESTIONS

1. An architect negotiates a term into her contract with the owner that states: "The architect shall not be liable to third parties for damage caused by error in design." Later, an error in the architect's design causes a parapet to collapse, injuring a pedestrian. The pedestrian sues the architect in negligence. Is the architect liable? Explain.

2. While conducting a periodic inspection required by the contract, the architect advises the contractor to change the design of the falsework and shoring below the formwork in preparation for an impending concrete pour. The contractor complies, and an accident occurs. A piece of formwork falls to the street and injures a pedestrian, who sues the architect and contractor. The architect argues that the contractor is responsible for means and methods of construction. Who should prevail? Why?

3. A professional engineer contractually responsible for structural design of a building hires an engineer-in-training (EIT) to do the calculations for a column. The engineer checks the calculations but fails to catch an error. A collapse results. Ignore any issues of contractual liability. Is the engineer liable in negligence? Is the EIT liable in negligence? Draft an argument presenting the basic principles to be used in each of their defences.

4. An architect has prepared plans for a new office tower. The municipal building authority approves the plans, subject to changes to the emergency egress and fire suppression systems. These changes will cost $30 000. The owner is annoyed and wants to know why the architect did not design these systems according to municipal requirements. The architect responds that the municipal authorities have very recently changed their requirements. What defences would be available to the architect if sued in negligence?

5. An engineer designing the boiler for an industrial installation determines that a particular design complies with all applicable codes and standards, but just barely. Based on experience and knowledge, the contractor suggests minor enhancements that would increase the safety factor by a significant margin at modest cost. The engineer declines to act on the suggestion. The boiler fails, and the owner sues the engineer. What defences, if any, are available to the engineer? Describe the strengths or weaknesses of those defences.

6. What is the difference between the remedies available for innocent misrepresentation and fraudulent misrepresentation?

7. What is a key difference between the remedies available for a claim in negligence and a claim in negligent misrepresentation?

8. An owner carrying out construction activity on his own property mistakenly believes that his property line is one foot west of where is actually lies. Because of that mistaken assumption, he puts up a fence that sits on his neighbour's property. Does that constitute trespass, or is mistake a defence?

9. A geoscientist obtains conflicting results in a geophysical mineral survey. The geoscientist knows that her report will be used by prospective purchasers of the company that she is working for; so she chooses only to disclose the positive aspects of her results. Is she potentially liable to the prospective purchasers? Why or why not? What if she chose just to disclose the negative results?

10. To what level of standard of care do courts hold professionals?
    a) Perfection
    b) A reasonably competent professional practising in the same field
    c) The professional with the lowest competence practising in the same field

ANSWERS

1. The pedestrian is not a party to the contract in which the limiting term is found. Therefore, the limitation does not protect the architect from claims by third parties.

2. The architect would be unlikely to succeed in that argument. By instructing the contractor as to means and methods, the architect has assumed some of the liability. The problem

should be broken down into two parts: the claim against the contractor, and the claim against the architect. In the claim against the contractor, the plaintiff would have to establish three elements referred to above: duty of care, breach, and damage. Duty of care and damage would be easy to establish. The contractor's defence, if any, would likely be on the basis of breach of duty. The contractor could show that he met the standard that would normally be expected of a reasonably competent and prudent contractor in those circumstances. If the evidence at trial established that there was a breach of that standard, the contractor would be liable.

The architect could assert the same defence as the contractor—that there was no breach of the duty of care. In addition, she could argue that a duty of care was not owed. Means and methods and techniques of construction are traditionally outside the scope of the architect's duties, and the contract documents reflect that fact. However, the facts in this case indicate that the architect voluntarily assumed responsibility by recommending methods of construction. Even though the architect was not obliged to offer this advice, she chose to do so. It is therefore likely that the architect would be found to owe a duty of care, and that the architect would be held liable in negligence.

3. The defence of the EIT in this case would be based on whether there was a breach of duty; that is, did the engineer-in-training meet the standard of care that would be provided by an average engineer-in-training exercising reasonable prudence and competence? Even in the case of a fully qualified and experienced professional engineer, a standard of perfection is not expected. As long as the engineer exercises an average degree of competence and care, the claim in negligence fails. In the case of an engineer-in-training, the amount of care expected might be the same as that of a fully qualified professional engineer; but the level of skill expected would be lower. Furthermore, the engineer-in-training expects a supervisor to check his work, which was in fact done in this case. In summary, the engineer-in-training performed to the level of skill and care that was required, and the mere existence of an error is not sufficient grounds for establishing negligence. If the plaintiff is unable to prove the EIT's lack of skill or care, then the claim in negligence is defeated.

The argument for the professional engineer's defence is based on standard of care. Did the engineer, by supervising the EIT and checking the calculations, meet the standard of care that was expected and required of him? No law prevents an engineer from delegating work to junior employees. Delegation of duty, however, does not relieve the engineer of his obligation to use proper skill and care. The fact that the engineer in this case approved the drawings means that he has taken responsibility for the work performed by his junior employee. The fact that some of the work was delegated cannot create a higher standard (*i.e.*, perfection) than would be imposed upon the engineer-in-training if no such delegation had taken place. Thus, by checking the calculations, the engineer exercised a level of care that a reasonably prudent and competent engineer would normally be expected to exercise. Unless the error was such that it should not have been missed by a reasonably competent and prudent practitioner, no liability should be found.

4. The architect could argue that the design met the standard of care in existence at the time that the plans were first created.

5. Compliance with applicable codes and standards is always a strong defence. The weakness occurs if the defendant is, or should be, aware that the standard is inadequate. A professional who is aware of problems or possible problems with the codes and standards has a duty of care to go beyond the codes and standards. In the Ford Pinto cases, which involved several fatal accidents due to explosions from minor collisions, the designers involved needed to move the gas tank forward to improve the vehicle's ability to withstand rear impact. The fact that this design had met existing codes and standards did not exonerate Ford from negligence in causing injury and death. Ford designers knew or should have known that the placement of the gas tank was risky.

6. A claim in innocent misrepresentation can only result in rescission of the contract, whereas a claim for fraudulent misrepresentation can result in an award of damages.

7. In a claim for negligent misrepresentation, claims for pure economic loss are allowed. In a claim for negligence, claims for pure economic loss are barred unless they fall within a recognized exception.

8. Mistake is not a defence.

9. The geoscientist's failure to fully disclose the results could well result in claims both by her own company, if she chooses to only disclose the negative results, and the purchaser, if she chooses to only disclose the positive results.

10. B

CHAPTER 13

COMMON ISSUES IN CONTRACT AND TORT

13.1 Concurrent Liability in Contract and Tort

Historically, contractual remedies were available for economic losses, and tortious remedies were used for personal injury and property loss.[1] For many years, the courts held that where a contract existed between two parties, they could not owe each other a duty of care in tort; their only obligations were contained in the contract.

In the landmark decision in *Central Trust v. Rafuse*,[2] the Supreme Court of Canada held that parties to a contract could also owe a duty of care in tort as long as the contract did not define the nature and scope of the duty of care. The use of tort actions could not be used to circumvent the terms of a contract.

The importance of this concept is that parties should define their legal obligations to each other as much as possible. Wherever the contract is silent on a particular matter, the court will either imply a contractual term or impose a duty of care in tort. For instance, in some poorly constructed contracts, the contract fails to specify the terms of payment. In such a case, the court either imposes a reasonable set of payment terms or holds that the paying party breached its duty to care by not paying on time. Thus, specifying the payment terms would be better for the paying party than letting a court determine them later.

13.2 Limitation Periods

Each province has legislation limiting the time in which an aggrieved party can commence legal action. For example, in Ontario, Alberta, and Saskatchewan, the time limit for commencing a legal action is two years from the breach; while in British Columbia and other provinces, the time varies with the type of action: injury to persons or property is two years, while breaches of contract and tort causing economic loss is six years

---

[1] Barnett, Ian Hunter, "Concurrent Liability in Construction Law: The Australian Experience" (1988), 27 *C.L.R.* 63.

[2] (1986), 31 D.L.R. (4th) 481 (S.C.C.).

(see also Chapter 14, p. 159). When a breach of contract or tort occurs, professionals should check the applicable statute in the province where the action will be commenced, because new legislation occasionally amends these limitation periods.

But determining when the applicable limitation period starts can be difficult. In cases where a plaintiff could not reasonably have known that he or she had a cause of action, the legislatures and the courts have found that it is unjust to an aggrieved party to deprive them of the ability to sue before they are aware that they can do so. For example, in *Kamloops v. Nielsen*,[3] the Supreme Court of Canada held that the time on a limitation period did not begin to run until the plaintiff knew of the existence of the damage or ought to have known through reasonable diligence. Ontario and Saskatchewan negligence statutes provide that the time period commences when the cause of action arises; but a plaintiff may be able to show that it should run from the time of *actual discovery* (as opposed to the *Kamloops reasonable discoverability* standard). In the remaining provinces, the rule in *Kamloops* continues to be the law.

This distinction may be important depending on whether the case involves a latent or a patent defect. A *patent defect* is an obvious defect that can be identified easily upon reasonable inspection. A *latent defect* is often described as a hidden defect. Examples of latent defects are software "bugs" that remains undiscovered until the plane crashes or building foundation defects that remain unknown until the building collapses. For latent defects, the law maintains that the limitation period does not commence until the plaintiff knew or ought reasonably to have known that it had a cause of action. For many architectural and engineering works, this time period could stretch for the life of the architect or engineer.

However, professionals have access to three potential protections from liability that would otherwise extend indefinitely into the future:

1. The parties to a contract can generally agree to a shorter (or longer) time period than is contained in the statute by agreeing to limit liability for claims to the following:
   - a defined time period (*e.g.*, six years from substantial completion of the construction project or delivery of the software);
   - a defined financial limit (*e.g.*, $250 000);
   - insurance available at the time;
   - the re-performance of the services; or
   - a combination of the above.

2. Each limitation statute contains an "ultimate" limitation period, beyond which claims cannot be brought even for latent defects (as short as 10 years or as long as 30 years).

3. The defendant can purchase insurance, recognizing that this must generally be maintained throughout his or her lifetime, because professional liability insurance is issued on a claims-made basis (see Chapter 16, pp. 180–181).

The standard form architect's agreement (RAIC Document 6) and engineer's agreement (ACEC 31) both contain clauses defining the above contractual limitations. Contractual limitations are usually the most significant and difficult point of negotiations in engineering and architectural contracts. Professionals generally want some assurance that their potential liability will come to an end in the foreseeable future and that they will not be at risk for their work forever. But owners and clients want the engineer or architect to be responsible for their work and do not want to become at risk for the engineer's or architect's errors. Of course, this issue is equally applicable

---

[3] [1982], 1 W.W.R. 461 (B.C.C.A.).

to third party contracts, such as the contract between an owner and a contractor for building construction[4] or between an engineer and client for software "licences."[5] However, this issue should be clearly addressed so that each party understands how the limitation period is to work.

Note that the contractual protections described here only apply to contracting parties. Professionals can do little to protect themselves from liability to third parties.[6]

13.3 Joint and Several Liability

Each province has legislation permitting the courts to apportion liability among the parties that caused the loss. But what if one of the parties found negligent and liable for a portion of the loss is unable to pay its share because of bankruptcy or lack of assets? At common law, the historical rule was that each party was only responsible for its portion of liability unless a common action caused the loss. But legislation in each province[7] has changed that rule such that all parties found partially responsible for a part of the loss can under some circumstances be held responsible for the whole loss.

Some jurisdictions distinguish between cases where the plaintiff is partially at fault and those where the plaintiff is not at fault at all. Where the plaintiff is partially at fault, the parties are responsible only for their own portion of the loss. This is called *several liability*. But where the plaintiff is not partially at fault, the plaintiff can recover the entire amount from any of the at-fault parties; and then the at-fault party must seek contribution from the other at-fault parties.

For example, assume a building collapses, and the court apportions fault with 70 percent to the contractor and 30 percent to the engineer. The owner is entitled to recover 100 percent of the damages from the engineer, and then the engineer is forced to try and collect the 70 percent balance from the contractor. If the contractor is defunct, the engineer is left paying 100 percent of the damages even though it was only 30 percent responsible.

Alternatively, the courts may decide that the contractor is 69 percent, the engineer 30 percent, and the owner 1 percent at fault. This 1 percent constitutes *contributory negligence*, or negligence by the plaintiff. In provinces with statutes that eliminate joint and several liability in cases with *contributory negligence*, the engineer would now be responsible only for its 30 percent portion of the damages; and the owner must bear the defunct contractor's 69 percent. This example highlights the dramatic effect of contributory negligence.

Many states in the United States used to have joint and several liability legislation, but most have repealed it. The original rationale for the joint and several liability rule was that in claims between a blameless plaintiff and another at-fault party, the at-fault party should bear the risk of impecunious parties. But some state legislatures have decided that this places too much blame for the whole situation on the at-fault party. The new rationale is that since the plaintiff usually chooses the parties, the other parties should not bear the risk of the plaintiff having chosen to enter into a relationship with an impecunious or high-risk party.

---

[4] CCDC 2 contains a limitation clause in section 11.1.1.

[5] See Chapter 11, p. 122.

[6] See Chapter 12, p. 136.

[7] For example, in Manitoba, see the *Tortfeasors and Contributory Negligence Act*, C.C.S.M. c. T90.

13.4 Vicarious Liability

An employer is vicariously liable for the actions or inactions of employees as long as the employees are acting within the course and scope of their employment. *Vicarious liability* is the liability of one party for the fault of another. This liability is based on the premise that those who profit from an activity should also be liable for losses that result from that activity.

Employers want to ensure that limits are placed on the employment relationship so that they are only liable for their employees' actions while the employees are doing the tasks that they were assigned. For example, they do not want to be responsible for their employees' after-hours bar-room activity, their drunk driving, or their consultation on the side. Therefore, employers want clearly defined limits on employee's duties.

Employees believe that employers have deeper pockets, such that any action against them by third parties for their breaches should be taken against their employer. But for vicarious liability to apply, the employee had to have been acting within the course and scope of his employment when the event occurred. Events occurring outside the course and scope of employment fall within the employee's personal liability; moreover, any such action taken against the employer will likely fail. In addition, the employer's insurance (professional liability insurance or certified general liability insurance) would also likely be exempt from liability for non-work-related actions. Thus, employees who engage in consulting activity separate and apart from the employer's business place themselves and their assets at risk, without the employer being vicariously liable or the employer's insurance providing coverage.

For vicarious liability to exist, there must be an employer/employee relationship. Courts recognize a difference between employees and independent contractors, as discussed in Chapter 20 (p. 228). Generally, the law does not hold someone vicariously liable for the actions of independent contractors.

Note that vicarious liability applies only to damages sustained by third parties. It does not address the question of whether the employer who has been found vicariously liable has the right to recover those damages from the negligent employee. In the employment relationship, the employee owes the employer a duty of competence; but in practice, employers very rarely sue their employees for negligence.

13.5 Codes and Standards

Many industries have written and unwritten codes and standards. These cover many aspects of architecture, engineering, and geoscience, including safety, construction requirements, software development, geophysical reporting, and consumer product testing. While governments write and enforce many of these codes and standards, many codes and standards have also been developed independently by technical societies or professional organizations.

Architects, engineers, and geoscientists should keep current with the codes and standards applicable to their practice. While knowing the 1960 electrical code is interesting, it is not particularly relevant to today's practice. Codes and standards evolve over time, and the newest codes and standards must be followed.

Legislated codes, such as building codes, are mandatory for everyone in the relevant industry, whereas codes required by contract are mandatory between the contracting parties. All other codes are recommendations only, and compliance with them is not mandatory. However, for practical purposes, in the event of a failure or accident, any existing, relevant code can become a minimum mandatory standard in court. In other words, a professional has the right to exercise independent judgment and may need to determine a special way to obtain a specific result. But if problems occur and the professional is required to justify a design that does not meet code, the resulting presumption against the professional would be very difficult to overcome. Even though violation of the standard is not conclusive evidence of negligence, it can be persuasive evidence. However, violation of a code mandated by legislation or by contract is almost always a breach of contract. A statement of this principle is found in the New Zealand case of *Bevan Investments Ltd. v. Blackhall and Struthers (No. 2)*:

> bearing in mind the function of codes, a design which departs substantially from them is *prima facie* a faulty design, unless it can be demonstrated that it conforms to accepted engineering practice by rational analysis.[8]

The more difficult question is whether a professional in full compliance with the code can be held liable. The simple answer is that such a situation is unlikely in most cases.

There are four sets of circumstances where such liability can be found. The first is where the professional has actual knowledge that the code requirements are inadequate. Second, a professional can be found liable if he or she reasonably should have been aware that the code was inadequate. For example, it may have become common knowledge as a result of an accident, a study, or new research that the code needed to be changed. The revisions to the code may not have kept up with knowledge in the industry. Moreover, code revisions could still be in the draft process following a major study. In the meantime, the professional has an obligation to keep up with current and future developments in the relevant field, in particular if that professional is a specialist.

Third, in rare cases, courts have held that an industry cannot be permitted to set its own standard; and professionals should not follow such standards where those standards are careless. For example, courts have found that adherence to an industry standard is insufficient defence to a claim in negligence in automobile design and manufacture cases.

Finally, where the design is leading edge or unique, professionals need to be aware that a different standard may apply. Traditional codes may not be relevant in such circumstances.

If a professional uses a new approach which deviates from common practice, what is the effect on potential liability? This question is relevant only at two stages in the process: at the time of contracting, and in the event of a failure or problem. If a professional is contemplating using a new approach because of anticipated cost savings, that professional must fully inform the client of any potential risks associated with the method, such as the effect it might have on use or maintenance costs or on safety factors; then, the professional lets the client make an informed decision. If the decision to use the new method is passed on to the client in this way, then the client takes the risk. This agreement may afford the professional a defence against the client in the event of a poor outcome, although it will be no defence against a third party who did not accept that risk. Whatever the result, a court will impose a very high standard on the professional, particularly if a tried-and-true method is known and available, or if the risks were not fully revealed to the client.

In addition to liability, safety is another important consideration in codes and standards. Many architectural, engineering, and geoscience failures occur because of a combination of factors. Design error, construction error, poor inspection, poor investigation, and unforeseen use, or some combination of these factors can all contribute to eroding the intended safety factor until failure occurs.

---

[8] [1973] 2 N.Z.L.R. 45 at p. 66.

Normally, the design professional chooses the safety factors for a project. While the law or accepted codes may mandate a minimum standard, the professional is the one who determines whether the project meets, exceeds, or fails to meet the safety factor. For example, in a disciplinary hearing following a roof collapse, the judgment panel made the following comments:

> Codes of practice set forth minimum standards to which there should be adherence. The engineer cannot deviate from the requirements of the Code without adequate grounds to do so. . . . If the Code is in any way ambiguous and requires interpretation, the engineer must ensure that the interpretation is based upon sound engineering principles and is consistent with the intent of the Code.[9]

In general, codes of ethics for professional in both the United States and Canada dictate that public safety is paramount.[10] Courts use this obligation to public safety to establish the duty of care owed by the engineer to members of the public in negligence cases. Although contractors may not be bound by a formal code of ethics, the same duty is likely owed by contractors.

CHAPTER QUESTIONS

1. Assume that it is common practice for an architect to certify progress payments on a construction project. Also assume that legislation in the province requires that the project owner hold back 10 percent from each progress draw until a specified period after substantial completion of the project. Finally, assume that the architect's contract with the owner requires the architect to follow all applicable legislation; but that neither that contract nor the owner's contract with the builder specifically indicates that there must be a 10 percent holdback. Does the architect have an obligation either to advise the owner of the 10 percent holdback requirement or to ensure that the holdback is maintained? Why or why not? If so, can the architect be sued for failing to do so? Why or why not?

2. A geoscientist relies negligently on information from his or her client in preparing a prospectus, and both are found to be liable to a purchaser of shares who relied on the prospectus. But the client is bankrupt. In most provinces, who bears the client's portion of the liability—the geoscientist or the share purchaser? Why?

3. Which statement best describes an employer regarding vicarious liability?
   a) The employer is never liable for the actions of its employees.
   b) The employer is liable to others for the actions of its employees while the employee is acting in the course and scope of his or her employment.
   c) The employer is not bound by labour legislation.
   d) The employer is unable to sue its employees or former employees.

[9] Association of Professional Engineers of British Columbia Inquiry Re: Harrison, Tacy, London and Man (Aug. 1989).

[10] The ASCE Code of Ethics states that "Engineers shall hold paramount the safety, health and welfare of the public in the performance of their professional duties." Similar wording has been used by Canadian engineering associations.

4. In most provinces, when does the limitation period in a negligence case against an engineer about an engineer's negligence begin to run?

   a) When the engineer's work is substantially complete

   b) When the contractor's work begins to run

   c) When the aggrieved party discovers or ought to have discovered the negligence

   d) At the end of the month following the engineer's birthday

5. An environmental engineer becomes aware that the environmental code is being developed by an organization that he believes is insufficiently experienced to be competent. Is it sufficient defence in court if the engineer follows that code? What steps, if any, should the engineer take when he uses the environmental code?

ANSWERS

1. This issue can be complex. In part, the answer depends on whether it is standard practice for architects to provide advice on holdbacks. As well, if the architect is aware that the client is unsophisticated and unaware of the holdback requirements, the likelihood that such a duty is owed increases. Hence, the answer to the question depends on the specific circumstances in which the architect is placed.

2. The geoscientist would bear the bankrupt client's portion of responsibility because of joint and several liability.

3. B

4. C

5. Codes are minimum standards, rather than maximum standards. If the code is insufficient, the engineer must design to the standard that he believes is appropriate. He must also make the client aware of the problem with the code and the rationale for designing to the higher standard.

CHAPTER 14

DISPUTE RESOLUTION AND EXPERT EVIDENCE

Overview

There are four well-recognized methods of dispute resolution: litigation, arbitration, mediation, and negotiation. The appropriate method in each case depends on the nature of the dispute, the amount of money in dispute, the remedies sought, the willingness of the parties to resolve the dispute, and the nature of the relationship between the parties.

Negotiation is a purely voluntary method of dispute resolution. If a party does not wish to negotiate, forcing negotiation is futile. Litigation is at the other end of the spectrum. It is voluntary for the plaintiff but involuntary for the defendant. In other words a defendant's failure to participate results in judgment being awarded against the non-participant; whereas failure to pursue a claim by the plaintiff simply results in no recovery.

Arbitration is voluntary at the contract negotiation phase, in that any party can refuse to accept the inclusion of an arbitration clause in a contract. However, once an arbitration clause—assuming it is a mandatory arbitration clause—is in the contract, then one party can force the other parties to arbitrate. However, arbitration is always an option even if the contract contains no arbitration clause or the parties have no contract.

Like arbitration, mediation may be agreed to before or after a dispute has arisen; but unlike arbitration, mediation is nonbinding. In other words, even if a party takes part in the mediation process, it can still refuse to settle the dispute.

14.1 Litigation

Litigation is the use of the court system to resolve disputes. The litigation process allows an independent party, meaning a judge or jury, to resolve issues of law and fact that are in dispute.[1]

---

[1] In the United States, commercial cases are more commonly decided by a jury than in Canada.

There are inherent advantages to litigation. For example, a court can enforce its own orders and processes. This is particularly important if one of the parties is a reluctant participant. Moreover, not all disputes involve good faith disagreements. Some lawsuits are motivated by a desire by one party to take advantage of the other or to delay the performance of recognized obligations, such as the payment of money. Thus, sometimes litigation is the only way that a party can enforce its rights.

But the litigation process has disadvantages as well. The first disadvantage is the cost. Each party has to hire a lawyer and, in some cases, expert witnesses and incur other disbursements for case preparation and trial. Moreover, the case can take up work time of the party's employees for long periods of time. A second major disadvantage is delay. Years can pass before some matters proceed to trial; and if one party files an appeal, the final result is delayed further. In the end, the personal cost as an investment of both emotion and time is usually much greater than parties had anticipated at the outset.

(a) Pleadings

Pleadings are documents filed in court in a lawsuit or included in a trial record. Pleadings include the statement of claim, statement of defence, reply, demand for particulars, interrogatories, and motions. The pleadings are the only documents that the court sees prior to the commencement of trial. The pleadings give the judge a basic understanding of the nature of the dispute.

To commence the litigation process, the plaintiff drafts a **statement of claim** or, in some provinces, a **writ** and a statement of claim. These documents identify the parties, the allegations that form the basis of the claim, and the nature of the relief sought.[2] The **plaintiff** is the party filing the statement of claim. He or she must file the statement of claim in the appropriate court registry and serve it upon the **defendants**, who are the parties against whom a remedy is sought. There is no limit as to the number of potential plaintiffs or defendants in a case. Upon receipt of the statement of claim, a defendant has a short period of time[3] in which to file a **statement of defence**, which is a document setting out the legal basis for denying the claim.

The body of the statement of claim contains a series of numbered paragraphs, each one containing an allegation. A statement of claim that does not set out a cause of action, such as breach of contract or negligence, will be dismissed. Each element of each cause of action must be proved. For example, the elements one has to prove in a case of negligence include existence of a duty, breach of that duty, and damage or loss flowing from that breach (see Chapter 12). A statement of claim that alleges a duty but not a breach will be struck for failure to allege a cause of action. However, statements of claim are rarely struck. In addition, the statement of claim must also state what relief the plaintiff is seeking. That relief may include general damages, punitive damages, specific performance of a contract, an injunction, or other forms of relief.

The defendant must draft and file a statement of defence, denying all the allegations that he or she plans to contest at trial. Failure to deny an allegation in the statement of defence is considered admission of that allegation, and that admission may be impossible to retract at a later date.

In addition, in many jurisdictions, the plaintiff is entitled to file a reply upon receipt of the statement of defence; but this rarely occurs. Pleadings may be amended after they have been delivered or filed; however, permission of the court may be required.

---

[2] In the United States, the statement of claim is called a "complaint." The statement of defence is called an "answer."

[3] The time limit varies from one jurisdiction to another, but is typically 14 to 21 days. The period may be extended by permission of the court, or, in some jurisdictions, by permission of the plaintiff.

Because of the technical requirements of drafting pleadings, the writing of pleadings is best left to lawyers.[4] As well, there are time limitations for filing and serving documents and other procedural requirements, which make self-representation risky. Self-representation for a corporation is only permitted under certain conditions, and, where permitted, only by an officer or director of the corporation. Courts try to accommodate litigants who are self-represented, but they may be unwilling or unable to assist if failure to comply with the rules has caused prejudice to the other party.

(b) Counterclaims and Third Party Claims

A *counterclaim* is a claim that a defendant asserts against the plaintiff. Counterclaims are frequently made by project owners who have been sued by their design consultants or contractors for payment of fees. There is no rule limiting the counterclaim to the amount of the original claim; and in practice, counterclaims are sometimes much larger than the original claim.

Many lawsuits involve more than two parties. Often a number of project participants are brought into the litigation by the defendant through a **third party claim**, which is a pleading setting out the claim of a defendant against a third party. A **third party** is a party to a lawsuit that is added to the case by the defendant rather than the plaintiff.

Defendants have strong reasons to want their third party claim heard together with the main action. When third party claims are heard separately, the third party is not bound by the findings of fact in the main action. For example, assume that the owner of a mine has sued an engineer for alleged design errors relating to the design of a part of the equipment used in the extraction process. Assume further that the engineer is contractually responsible to the mine owner but has subcontracted the equipment supply and installation to an equipment supplier (the third party). The engineer may be found liable because of errors in the design. However, the engineer has a potential third party claim against the equipment manufacturer. If the lawsuit between the owner and the engineer were to proceed first, without the third party claim, and the engineer found liable, he or she would then have to sue the supplier. But that supplier may have good defences that the engineer did not consider. Moreover, since the supplier was not a party to the first lawsuit, he or she is therefore not bound by the result. Finally, the second trial may have different witnesses and evidence available, which could lead to a result inconsistent with the findings in the main action. To avoid such problems, the engineer should name the supplier as a third party in the original suit, binding the supplier to findings of that lawsuit.

A defendant may also have strategic reasons for keeping a potential third party out of a lawsuit. For example, if the third party becomes hostile as a result of being included in the litigation, then the defendant comes under fire from two sides instead of one.

(c) Costs

The party who loses a lawsuit is usually required to pay a percentage of the other party's legal costs.[5] This rule is a strong disincentive against making frivolous claims. A defendant would not issue a third party notice unless the third party claim had a good chance of succeeding, or unless the potential cost sanctions of failing were dwarfed by the size of the claim. However, costs recovered are generally only a portion of the actual legal costs. For example, in British Columbia, costs recovered by the successful party tend to only be about 33 percent of the actual costs.

---

[4] Small claims courts typically omit these technical requirements or simplify them so that plaintiffs do not need to hire lawyers.

[5] In the United States, subject to some exceptions, the general rule is that each party bears its own legal costs.

(d) Discovery

Discovery refers to procedures available to all parties prior to trial, including but not limited to examination for discovery, interrogatories, inspection of property, and disclosure of documents. All discovery procedures are designed to allow the parties to learn as much about the case as possible. The rationale behind the rules for discovery is that if all parties are aware of the relative strengths and weaknesses of each party's case before trial, there will be no surprises at trial, thereby increasing the likelihood of settlement.

An **examination for discovery** is an oral cross-examination of a representative of the opposing party, conducted prior to trial. Depending upon the complexity of the case, it may take less than one day, or it may take several weeks.

Interrogatories are a series of written questions sent by one party to another, which the other party must answer under oath. Interrogatories have form requirements, but otherwise they are straightforward. The parties send interrogatories shortly after the pleadings phase is complete, in order to obtain preliminary answers and to find out which witnesses are knowledgeable about various aspects of the case.

Inspection of property is another form of discovery, and involves an on-site inspection, usually by the litigant, the lawyer, and often accompanied by an expert witness.

Discovery of documents is the ongoing process of identifying documents that are relevant to the litigation. This process often continues until shortly before trial. Upon being served with a demand for discovery of documents, a party must produce all relevant documents, except for privileged documents. The test for relevance is whether the document tends to prove or disprove a fact in dispute, or whether it may lead to a line of inquiry relating to the dispute.

Construction disputes are notorious for generating massive quantities of documents. They usually include the contract documents (including drawings and specifications), correspondence between the parties and between witnesses, minutes of meetings, diaries, schedules, transmittals, photographs, telephone records, invoices, and internal memoranda. Documents can be in paper or electronic form, including business records and formal and informal documentation, including emails. Emails are considered the new gold mines of litigation because parties tend to write them with an informal attitude (see Chapter 27). Professionals should always treat emails as they would a formal report: if they are not prepared to sign and seal what is in an email, whether internal or external, they should not send it.

Large cases can produce tens of thousands of documents; very large cases can generate hundreds of thousands or even millions. Each party must number its documents and create a list of the documents in a logical sequence. To keep track of the documents, most lawyers enter them into a database, with fields for date, author, recipient, and subject matter, to make them retrievable. Thus, the cost of document control can be substantial for a large case. But failure to spend the time and money on this task early in the process can result in greater expense and tactical disadvantages as the trial approaches.

In an examination for discovery, as the lawyer asks questions and the witness answers them, a court reporter makes a transcript of those questions and answers for use at trial as evidence. If evidence given by a witness at trial is contradictory or inconsistent with the answers given in the examination for discovery, the transcript can be used to impeach (or contradict) that witness. The primary purpose of an examination for discovery is to allow both parties to assess all the evidence before the trial in order to reduce the likelihood of surprises. Although the element of surprise can never be completely eliminated, this procedure helps reduce the number. Moreover, witnesses who change their testimony from the time of discovery to the time of trial almost always damage their credibility, unless there is a very compelling reason for the change.

In addition, in some provinces, as part of discovery, each party must disclose to the other the names of all witnesses they intend to call during the trial. Again, the purpose is to remove the element of surprise and to promote settlement.

Although the examination for discovery process is necessary, it can also be very expensive. For example, for a lawyer with a rate of $350 per hour, a day of examination costs a client approximately $3000 plus expenses, including the cost of the transcript prepared by a court reporter. This amount does not include the time required for preparation, which may exceed the time for examination itself. Thus, professionals involved in a lawsuit must be prepared for the costs of the procedure.

(e) Technical Evidence

Technical evidence often requires special treatment in court. For example, matters of a technical and scientific nature must be explained to the court. The parties can simplify some of the evidence, so that a judge without a scientific background can understand it. But often they hire technical experts to explain the technical evidence to the court in simple terms.

(f) Drawbacks of Litigation

Because litigation is often a winner-take-all proposition, it is always risky. Even if a case looks like a sure winner, there is always the risk that the court may be sympathetic to the other party, or that a key witness may change his or her evidence. Moreover, litigation is very expensive and time-consuming and does not always produce a fair result. Despite the fact that evidence is given under oath, witnesses often contradict each other and in some cases deliberately falsify their testimony. Thus, although litigation is still frequently used to resolve disputes, it is generally considered a method of last resort.

Due to the cost and delay associated with litigation, there has been much interest generated in recent years in **alternate dispute resolution (ADR)**. ADR is any process of resolving a dispute other than litigation. While arbitration and mediation are the most popular methods, other methods are possible. For example, two software company presidents in Australia were reported to have settled their companies' $100 000 dispute through arm-wrestling. One executive was quoted as saying that it was preferable to the court process.

14.2 Arbitration

Arbitration is a private litigation or trial process in which the parties set the rules and choose the judge, called the arbitrator. An **arbitrator** is essentially a private judge and jury. Proponents of ADR claim that arbitration is less expensive and quicker than litigation. It often is, but not always. Arbitration has been used extensively for many years in labour disputes in Canada and in commercial disputes in other countries. It is best suited for disputes involving parties who have an ongoing business relationship, who would like to maintain that relationship, who want the dispute and the outcome to be confidential, and who have a genuine desire to resolve the dispute amicably and quickly. However, if one of the parties prolongs the dispute, arbitration can prove more expensive than litigation.

Parties can pre-select arbitration as their method of resolving future disputes by agreeing to include an arbitration clause in the contract. If a mandatory arbitration clause is in the contract, the process is referred to as *mandatory binding arbitration*. If instead the parties agree to arbitration only after the dispute has arisen, it is called *voluntary binding arbitration*. There is also an option known as *nonbinding arbitration*, which requires the arbitrator to give an advisory opinion.

In addition, a contract clause can specify different types of arbitration: it can call for mandatory arbitration in one party's discretion or for non-mandatory arbitration requiring the consent of both parties at the time of the dispute. However, binding arbitration is by far the most popular method. Thus, all references to arbitration in this chapter refer to binding arbitration unless noted otherwise.

Regardless of the form of arbitration, the parties must answer the following questions before starting the process:

- Who will act as arbitrator?
- What will be the terms of reference of the arbitrator, and what specific questions will the arbitrator have to answer?
- What rules of procedure will the arbitrator have to follow?
- What law will apply?

Of these four questions, the most important is the choice of arbitrator. Arbitrators are chosen for their reputation and expertise. Normally, the parties select a respected professional who has the skills to conduct arbitration, has technical expertise or experience in the field in dispute, and can be counted upon to act impartially. But if the parties cannot agree on an arbitrator, the court can appoint an arbitrator, or the parties can request a list of arbitrators from an arbitration organization.

Several organizations register arbitrators. In the United States, the American Arbitration Association (AAA) is by far the most prominent of these.[6] Arbitration organizations typically provide standard arbitration clauses that can be inserted into contracts or agreed to after a dispute has arisen.

Each association has its own standard rules of procedure. However, if the parties specify the rules in their arbitration clause, then the arbitrator will follow them instead. Each province has its own legislation that governs the conduct of arbitration.[7]

In general, the rules of evidence and procedure in an arbitration are less formal than those in a court of law. Almost all arbitration associations specify in their rules that formal rules of evidence need not be followed. In general, this means that the rules are used but are relaxed to some degree. For example, hearsay or opinion evidence, which might be inadmissible in court, can be heard by the arbitrator, although the arbitrator should be careful to weigh such evidence appropriately. But the arbitrator must follow the fundamental principles of fairness, regardless of whether the arbitrator chooses to follow the formal court rules. This means that each party must receive adequate notice of the hearing, an opportunity to be heard, an opportunity to test the evidence of the other parties, and fair judgment by an impartial and unbiased decision-maker.

The arbitration process is popular because it has considerable advantages. One advantage is privacy. Unlike court proceedings, which are open to the public, arbitration is a private hearing. Thus, if the evidence includes trade secrets that, if made available to competitors, could affect future profitability of a party, or other potentially damaging information, one party may prefer arbitration.

Another advantage is the arbitrator's expertise. The parties can choose an arbitrator with training or experience in the subject matter of the case. As a result, technical evidence can be canvassed more quickly and in greater depth, because there is no need to explain the basics. However, parties must ensure that the arbitrator does not make assumptions based on personal experience rather than on the evidence.

But arbitration has a few disadvantages. For example, while the costs are lower than those of litigation, they are still considerable. Arbitration organizations have set fee schedules for their services. In addition to these fees, which may be a percentage of the amount in dispute, the

---

[6] The AAA has offices throughout the United States and has its headquarters in New York at 140 W. 51st St., New York, NY 10020.

[7] Parties considering the use of an arbitration clause should be aware of the legislation that will govern the arbitration.

arbitrator him- or herself must be paid. Most arbitration agreements provide that the parties bear the costs of the arbitrators equally, unless the arbitrators make an award of costs against one party. Also, the parties may have to pay the cost of rented facilities for the hearing.

In addition, the arbitrator's schedule can cause difficulties, because most arbitrators are practising professionals (often lawyers, engineers, and architects); thus, long hearings may require considerable delays or multiple sessions to accommodate these schedules. Moreover, if the hearing takes longer than scheduled, the parties may have to come back at a time to accommodate everyone else's schedule.

The most serious disadvantage pertains to the third party problem described in the litigation section of this chapter. Because arbitration is a voluntary process, the parties cannot force a third party to participate (except in some instances as a witness), unless that party is signatory to or has agreed to participate in and be bound by the arbitration clause. Consequently, third party claims may result in a second hearing in another forum, with a different judge or arbitrator and a significant risk of inconsistent findings. This problem is so important that many leading lawyers recommend against mandatory arbitration clauses in contracts.

Overall, speed and lower cost are generally touted as benefits of arbitration. Yet in recent years, arbitration hearings are starting to resemble court proceedings, in terms of production of large numbers of documents and examination of witnesses. At the same time, the arbitration process lacks the control that a judge brings to the litigation process.

14.3 Negotiation

Negotiation is discussion aimed at resolving a dispute, usually through compromise. As such, it does not have a formal set of rules. Many books teach professionals how to become effective negotiators; but these techniques and strategies vary from one author to another and from case to case. Thus, negotiation is more of an art than a science. Negotiating skill may be improved through study; but in the final analysis, it is best learned through experience.

Hiring a skilled and experienced negotiator, such as a lawyer, often assists in the negotiation process. Skilled negotiators can often identify important issues and thereby assist a party to a better settlement. In some cases, a skilled negotiator can help parties see that the issues they believed important are really secondary to the main issue.

Negotiated settlements have three clear advantages over litigation and arbitration. First, the parties maintain control over the procedures; second, the process is usually much less costly than litigation and arbitration; and third, it eliminates uncertainty once it reaches a settlement. For example, negotiating to a quick $50 000 settlement is generally preferable to litigating a $100 000 claim that has a 50 percent probability of success, in addition to legal costs and the investment of time.

Finally, the privileged nature of negotiation assures parties that if negotiations do not result in a settlement, those negotiations will not damage their claims in subsequent court proceedings. If negotiation fails, all evidence discussed in the negotiation becomes **privileged**, which means that it cannot be used as evidence in subsequent proceedings.

14.4 Mediation

Mediation is an assisted negotiation process in which a neutral third party facilitates settlement discussions. Mediation is effective in resolving both complex multi-party disputes and small, straightforward ones. In order to succeed, the parties must have a genuine desire to reach a settlement, the mediation must be attended by a representative of each party with authority to conclude a settlement, and the mediator must have the trust of all parties.

The first phase of mediation is usually an exchange of opening statements by each of the parties. The mediator may then meet separately with each party, discussing the strengths and weaknesses of each position, in an attempt to bring the parties closer together. Unlike a judge or arbitrator, the mediator is allowed to meet with one party in the absence of the others to discuss the case. During these separate meetings, the mediator may hear confidential information from each party, including in many cases a bottom line settlement amount. Thus, each party can request that certain information be kept confidential. In turn, the mediator must ensure that such information is not disclosed without consent.

One of the functions of the mediator is to provide a reality check for the parties. A party, and even a lawyer involved in a case for a long period of time, may lose objectivity. The mediator can help that party regain objectivity. In addition, the mediator may be made aware of facts of which that party is unaware and can use that information to move the parties past roadblocks, as long as the information is not confidential.

Like negotiation, mediation provides that if no settlement is reached, all information provided by all of the parties and by the mediator during the proceeding becomes privileged. This allows the parties to be more open with each other and improves the likelihood of settlement.

If the parties reach a mediated settlement, the mediator or one of the parties accepts the task of recording the terms of settlement in the form of a settlement agreement, which the parties then sign.

14.5 Other Dispute Resolution Methods

Other methods of dispute resolution, such as mini-trials and settlement conferences, are similar to mediation. A *mini-trial* is a condensed version of a trial in front of a judge alone, at which parties may present evidence. In some mini-trials, rather than calling witnesses, parties summarize anticipated evidence in written form. The judge makes a nonbinding ruling based on this evidence, which gives the parties a realistic idea of the probable outcome of a complete trial. The parties can then choose whether they want to proceed further with the litigation.

In contrast, a *settlement conference* is an informal meeting or, in some jurisdictions, a formal mediation process with a judge. The judge's role is to assist the parties to settle the case before litigation. Some jurisdictions mandate that parties attempt to settle the case through mediation or a settlement conference before proceeding with litigation.

There are several variations to these methods, including *binding mediation (mediation/ arbitration)*, *court-appointed masters and referees*, *summary jury trials*, and *partnering*. Parties considering these processes should evaluate the following factors before agreeing to proceed:

- the cost
- the time and delay
- whether an independent party will be involved, and if so, the qualifications of the party and the way he or she will be chosen
- whether the process is binding
- the right or lack of right to appeal an unfavourable decision to a higher authority or court.

14.6 Expert Witnesses

An **expert witness** is a person with skill, expertise, training, and experience, who is hired to provide an opinion to the court or to an arbitration tribunal. Expert opinion is one of the exceptions to a general rule of evidence, which stipulates that opinion evidence is not admissible in court. Courts allow opinion evidence in circumstances where subject matter is of a technical nature beyond the knowledge of a nonprofessional.

Architects, engineers, and geoscientists frequently act as expert witnesses in litigation. They are retained by one of the parties to assist in the preparation of the case or to give an expert opinion on some technical matter in issue. In general, the expert witness's obligation is to remain neutral. Expert witnesses must provide their honest and professional opinion, whether or not that opinion advances the client's interest. Moreover, expert witnesses must deal with the grey areas of some technical problems, because not all technical issues have only one correct answer. More than one theory based on scientific principles may fit the facts. In those circumstances, expert witnesses should present the theory they believe to be the most credible; but they must be prepared to agree that it is only one of several plausible explanations. Putting forward an unlikely theory as though it were an equally acceptable alternative is unprofessional conduct and can result in professional sanctions.

Expert witnesses also often testify about standard of care. In all civil cases, including those in which professionals are sued in negligence, the plaintiff must prove on a balance of probabilities[8] that the defendant failed to meet the standard of care required by a contract or expected of a competent professional in his or her field. But in these cases, an expert witness cannot act as a hired gun, despite being hired by one of the parties; rather, the expert's role is to assist the court to understand technical and professional matters.

For this reason, an expert witness cannot charge a fee that is contingent on the outcome of the case. Contingent fees would provide a strong incentive to put the client's interest and the expert's personal interest ahead of the public interest. Moreover, disclosure of such a fee arrangement would likely render the expert's opinion inadmissible. In one well-known case, an engineer entered into a contingent fee agreement to give expert evidence in a lawsuit. He was subsequently disciplined by the APEGBC and suspended for three months. The comments of the Discipline Committee in this case are instructive:

> The function of an expert is often to help the understanding and demystifying of complicated technical matters beyond the immediate comprehension of lay people, including members of a judicial court and whoever hired the expert. For this reason it is incumbent upon the engineer, before accepting such an assignment, to carefully evaluate his/her ability, through training and experience, to provide the necessary expertise in the matter. Expert testimony must be factual, unbiased and non-adversarial. These fundamental requirements can be undermined when fee payment for the expert is contingent upon a successful outcome of the trial.[9]

In short, providing an expert opinion that advances the client's case does not further the client's interest if that opinion will not stand up to scrutiny. The expert will be cross-examined, and any problems with the opinion are likely to be exposed. The expert witness would better serve the client by providing an honestly held opinion in the first instance. If the opinion is unfavourable, the client might be better off settling the case in its early stages, rather than incurring substantial legal and expert fees pursuing a losing cause.

---

[8] The *balance of probabilities test* means that the claim is more likely true than not. In a criminal case, the burden of proof is beyond a reasonable doubt, which is a much more difficult burden to meet.

[9] For further details, see the report "In the Matter of Roger Hill, P.Eng.," October 1991, *The BC Professional Engineer.*

CHAPTER QUESTIONS

1. Which of the following is not at all a voluntary process?
   a) Litigation
   b) Arbitration
   c) Mediation
   d) Negotiation

2. Are photographs and videos considered to be documents for the purpose of document production in a lawsuit?

3. Which of the following is not a pleading?
   a) A reply
   b) A statement of defence
   c) Discovery of documents
   d) Interrogatories

4. Why should a third party claim be decided at the same time as the main lawsuit under which it was filed?

5. Lawyers can be paid a fee on a contingent basis. Why is an expert witness not entitled to be paid this way?

6. Which of the following are not allowed to speak privately with one of the parties in a dispute?
   a) The judge
   b) The arbitrator
   c) The mediator
   d) a) and b)

7. Does the successful party in a lawsuit recover all its legal fees?

8. What documents do judges examine prior to the commencement of a lawsuit?
   a) The statement of claim
   b) Correspondence between the parties
   c) Photographs
   d) Settlement offers

9. If disputing parties attempt mediation and exchange settlement offers but are unable to reach a final settlement, are the amounts of the offers admissible in subsequent litigation?

10. What is usually the most expensive dispute resolution method?
    a) Settlement conferences
    b) Mediation
    c) Litigation
    d) Arbitration

ANSWERS

1. A

2. Yes

3. C

4. By definition, a third party claim is a claim by one of the defendants in a litigation that another party is at fault. If the third party is not part of the litigation, a second lawsuit will have to result, based on the results of the first one. But in the second lawsuit, a different judge may find a different set of facts and may decide the case differently. For this reason, third parties named in third party claims should be incorporated into the original litigation.

5. A lawyer is an advocate for one party in the dispute, whereas an expert witness is supposed to be impartial and unbiased.

6. D

7. No

8. A

9. No

10. C

CHAPTER 15

RISK, RESPONSIBILITY, AND DISPUTE AVOIDANCE

Overview

The ultimate purpose of a contract is to allocate risk and responsibility between the parties: risk of loss in the event of failure, risk associated with unforeseen events, responsibility for making decisions, and responsibility for ensuring certain results. Parties to a contract can reduce or eliminate certain types of risk both at the contract negotiation phase and during performance of the contract by recognizing their risks and responsibilities and taking steps either to accept them or shift them to other parties. The principles of contract law guide parties and courts in interpreting contract risks and responsibilities only up to a point: therefore, the parties must take on the rest of this job themselves.

Responsibilities and risks not adequately addressed in a contract fall where the courts determine they fall. By default, the law makes presumptions and implies terms into the contract.[1] Similarly, a party may inadvertently accept risks because of certain phrasing in a contract. But parties can easily avoid such inadvertent risks if they recognize them and knowingly assign them.

Risk of conflict is greatly increased when a client is dissatisfied. Client dissatisfaction is often unrelated to the quality of service provided or the skill of the professional or contractor providing the service, but may be a function of poor communication.

15.1 Risk Assessment

When parties are about to enter agreements, they often do so in an atmosphere of enthusiasm and hope and with an expectation of profit. However, in order to properly identify risks, both parties must temporarily become pessimists during the contract negotiation stage. Hoping for the best is acceptable as long as one also plans for

---

[1] See Chapter 6, p. 44.

the worst. Contract lawyers are trained to ask themselves what could possibly go wrong and then to find a way to deal with those possible events. Architects, engineers, and geoscientists must learn to adopt this approach during the planning stage of a project, before the contracts have been signed.

When identifying risks, professionals must ask very specific questions, such as the following:

- What losses would be suffered by each of the parties for each day that the project is delayed?
- Who should bear the risk of theft or fire, during construction?
- What if the subsurface conditions are more difficult than expected?
- Who will complete the work if the contractor defaults?

Simply by asking such questions, parties can develop logical solutions. For example, the contract should specify whether insurance should be purchased to cover the risk of fire or theft; whether performance bonds should be put in place to cover the risk of contractor default; and whether bonus or liquidated damages provisions should be added to deal with delays. Once the parties have discussed the risks and solutions, they can decide whether to pass those risks onto third parties, such as subcontractors, suppliers, bonding companies, and insurers; or they may allocate the risks to the contracting parties by mutual agreement.

15.2 Common Law Presumptions

In contract disputes, courts generally view professionals as relatively sophisticated parties. In other words, the law usually presumes that the terms of a contract signed by a geoscientist were negotiated fairly; and if the geoscientist agreed to the inclusion of a specific term, then he or she should be bound by that term. Rarely will a court be persuaded that the professional acted out of economic duress or hardship.[2] Moreover, professionals who argue that had they better understood the terms, they would have refused the contract, which would then have been awarded to someone else, are not likely to be successful. Courts presume in most cases that the parties have freedom of contract and the ability to walk away during negotiations if the terms are unacceptable: they either choose to live with the onerous terms or to let someone else live with them.

The law also presumes that if a risk has been identified in the contract, then the parties had addressed their minds to that issue. Thus, courts are reluctant to substitute their own allocation of risk for the parties' choice. If the risk has been addressed in the contract, courts generally let the contract determine the outcome.

While clients often believe that professionals guarantee the result of their work, the law makes no such presumption. In the absence of contractual language to the contrary, the law provides that a professional must act with the level of competence and care expected of the average practitioner in that field.[3] For this reason, professionals should be extremely careful to avoid any contractual language that places them in the position of guarantor or surety. Perfection in the provision of any professional services is rare, which is one reason why clients have contingencies in their budgets. But if a professional represents to a client in the contract or in a letter, email, or conversation that results are guaranteed, then a court may hold the professional to that standard.

Furthermore, while the law does not necessarily presume that the contract will provide the results desired by the client, it may imply certain warranties. For example, in most jurisdictions, the

---

[2] See Chapter 6, p. 51.

[3] Of course, different standards apply to those who hold themselves out as experts or to those who guarantee results. See Chapter 12, pp. 136–137

law implies that in a contract for design or construction of a residence, the final product will be fit for human habitation. The law will also imply certain minimum standards of quality, workmanship, skill, and care.

Yet these presumptions may be modified by terms of the contract. For example, the CCDC 2 construction contract states:

> The Contractor shall review the Contract documents and shall promptly report to the Consultant any error, inconsistency or omission the Contractor may discover. Such review by the Contractor shall be to the best of his knowledge, information and belief and in making such review the Contractor does not assume any responsibility to the Owner or the Consultant for the accuracy of the review. The Contractor shall not be liable for damage or costs resulting from such errors, inconsistencies or omissions in the Contract Documents which the Contractor did not discover. If the Contractor does discover any error, inconsistency or omission in the Contract Documents, the Contractor shall not proceed with the work affected until the Contractor has received corrected or missing information from the Consultant.[4]

This clause modifies the common law presumption about errors that a reasonably competent contractor would normally be expected to discover. With such a clause in the contract, the contractor would not be liable for failure to discover an obvious problem; whereas at common law, the contractor would probably be liable.

15.3 Shifting Risk

Looking after the client's interest does not mean that the professional must ignore his or her own interest. Certainly, a client's interest is paramount to the personal interest of a hired consultant. Yet the two are not always in conflict; and in many cases, the two coincide. But many professionals end up accepting risk that properly belongs to their client in the mistaken belief that it is somehow improper to leave that risk with the client.

Case Study 15.1
Risk-Shifting

Facts

Assume that a mechanical engineer is hired to design the heating, ventilation, and air condition system for an office building. In assessing what kind of chiller to install, the engineer determines that one make and model would be marginally adequate but might fail in the event of high demand; whereas a larger make and model would be adequate but would cost substantially more. The engineer is aware that the client is on a tight budget and has rejected all requests for upgrade.

---

[4] Clause GC 3.4.1.

Questions

1. What might happen if the engineer selects and installs the less expensive unit?

2. What might happen if the engineer selects and installs the more expensive unit?

3. What might the engineer do to shift the risk?

Analysis

1. If the engineer simply selects and installs the less expensive unit, and it subsequently proves inadequate, the client could sue the engineer. The engineer might be found liable.

2. Similarly, if the engineer selects and installs the more expensive unit, the client might complain once it became apparent that the increased cost was unnecessary.

3. The engineer could write to the client explaining the problem, setting out the possible consequences and likely costs, thereby allowing the client to make an informed decision. The client would then be unable to argue that he or she would have been more than willing to accept the increase in cost in order to avoid the risk.

15.4 Disputes Caused by Client Dissatisfaction

Client dissatisfaction is one of the prime causes of litigation. This dissatisfaction may be justified, or it may be simply a matter of perception. Justified dissatisfaction may result in the client winning the case, whereas perceived dissatisfaction may have the opposite result. Regardless, all parties incur the cost and aggravation of a legal battle.

In some cases, professionals can avoid disputes simply by advising the client of the choices available, outlining the advantages and disadvantages of each choice, and allowing the client to make an informed decision, as in the case above. Even where there is no practical choice, keeping the client apprised of the likely results can avoid disputes as well.

Case Study 15.2
Client Dissatisfaction Dispute

Facts

Assume that the client, a developer of a shopping centre, is constructing a parking garage. The client advises the structural engineer that the cast-in-place concrete slabs must be flat, so that water will not pond on the surface. The engineer knows that over time slabs develop creep deflection; and unless camber, or curvature, is designed and built into the structure, a slab that is flat right after it is poured will result in ponding over time. Furthermore, removal of the formwork after pouring causes short-term deflection, such that measurements taken before stripping the forms differ from those taken afterwards. The amount of deflection is even greater if live load is imposed. For this reason, the timing of measurement largely determines whether or not it appears flat.

Question

1. How can client dissatisfaction be reduced?

Analysis

1. In these circumstances, if the client measures the flatness of the slab after pouring, the client will likely be dissatisfied. The engineer can avoid this problem by informing the client beforehand about the changes in measurement. If the client is informed, any ponding of water before the forms have been stripped will not become cause for complaint. But without having been informed of the explanation of camber and deflection, the owner is extremely unlikely to be satisfied upon finding water puddles after the first rain. He or she may consider suing the engineer for providing a design that did not accomplish the requested result.

Thus, offering explanations before rather than after a result is always advantageous. Explanations after the fact are often perceived as self-serving and are therefore less reliable in a court of law. Moreover, statements made after a dispute has developed and especially after litigation has commenced do not receive the same weight as statements made before litigation.

In some cases, a client tries to pressure a professional into making a design choice that should properly be made by the client. The owner might simply say, "Do whatever you consider appropriate." If possible, the professional should avoid being drawn into making the choice. If that is not possible, the professional should inform the client in writing what choice is being made, for what reason, with what possible consequences, and at what cost. The letter should also advise the client that if he or she is not happy with that choice, the professional must receive immediate notification so that the decision can be reconsidered.

A significant portion of claims against professionals are precipitated by claims for unpaid fees. The typical scenario is as follows: At the end of the project, the professional requests the last instalment of the fee. The client then tells the professional to waive or reduce the fee because the project exceeded the budget due to alleged errors and omissions. The professional files a claim for the fee, and the client counterclaims for losses caused by alleged negligence. It is not unusual for the counterclaim to be many times greater than the claim for fees.

This scenario raises the question of how to avoid such counterclaims. Oftentimes, refusals by clients to pay the final fee instalment stem from a poor understanding of the requirements and costs of the project. For example, a client may not understand the extent of an architect's involvement in changes in a construction project. An unsophisticated client may expect that once the detailed design phase is complete, the architect's involvement will be minimal. But in the construction process, architectural changes are often required for a variety of reasons, including contractor error, owner and tenant requests, unforeseen conditions, and designer error. Thus, a client may not have budgeted adequately for architectural costs. In addition, a client may not understand the need for inspection of work. For example, an engineer may have included in the contract only one or two visits per week, with additional visits to be paid as an extra. If eventualities demand that the engineer be on site more often, the client may refuse to pay. As well, a client may not understand the extent of investigatory work needed to arrive at a satisfactory geophysical investigation. In some cases, the initial investigation must be supplemented with much more extensive and costly investigations. A client that has not budgeted for these extra costs may refuse to pay them, resulting in litigation.

However, many of these problems can be prevented with clear discussions about expectations and involvement before the project begins. One of the causes of poor understanding of these issues is poorly drafted retainer agreements that do not specify the scope of the consultant's services. For example, supervision differs from inspection, a difference which many designers take pains to ensure is documented in the contract.[5] An architect who is responsible for supervision will want to be on site every day; whereas inspection connotes a lesser degree of involvement. Although the architect may appreciate this distinction, the client may not.

Just as in all relationships, good communication is key to successful business relationships. Maintaining a good relationship with a client is vitally important, and doing so is the least expensive and most cost-effective way to avoid claims. Attention to simple communication matters, such as returning telephone calls, can mean the difference between a client who is prepared to forgive small errors and one who is looking for any excuse to sue.

15.5 Disclaimers

Some contracts contain exclusion, limitation, or waiver clauses[6] designed to preclude or limit the liability of one of the parties.[7] These clauses are also known as **disclaimers**. Examples of such clauses include consequential damages clauses, no-damage-for-delay clauses, and pay-when-paid clauses.

Disclaimers are not always written into contracts. They may be posted in conspicuous places or contained in letters. They may be a note or paragraph in a geotechnical report or at the top of drawings or specifications. For example, geoscientists often include disclaimers about a report's accuracy in the report introduction. Architects and engineers frequently place a stamp on shop drawings they have reviewed, advising that the designer's review is limited to general conformity with the intent of the design and is not a detailed review. Invitations for bid frequently include disclaimers about the accuracy of soils information and require the contractor to make an independent investigation.

Disclaimers and exclusion, limitation, and waiver clauses can be effective tools for shifting risk. But regardless of the type, disclaimers must be brought to the attention of the party against whom they are intended. In addition, they must be carefully drafted to be enforceable.[8]

Case Study 15.3

Zhu v. Merrill Lynch HSBC, 2002 BCPC 535

Facts

The plaintiff, Mr. Zhu, was a software engineer with a stock portfolio of $250 000 and considerable knowledge in the field of investment. Mr. Zhu traded stocks using a program known as NetTrader, run by the defendant through the Internet, with commissions payable to the defendant. Mr. Zhu

---

[5] Some contracts describe the extent of the architect's visits as "intervals appropriate to the stage of construction . . ." and explicitly state that continuous on-site inspection is not required.

[6] See Chapter 11.

[7] See Chapter 11.

[8] See Chapter 11, p. 126.

contended at trial that he used the defendant's NetTrader to sell 4000 shares from his RRSP account at the noted time of 14:47 on May 23, 2001, and then cancelled the trade immediately thereafter, also noted at 14:47 on May 23, 2001. He alleges that he received confirmation on his computer that 200 of the shares were sold by the time he made the cancellation but that the cancellation of the remaining 3800 was confirmed. He contended that he waited for five minutes and then placed another order to sell the remaining 3800 shares at the noted time of 14:52 on May 23, 2001. As it turned out, the cancellation of the shares had not been completed, resulting in duplicate trades of the same shares, thereby placing his account in a "short" position. When the defendant insisted that he make up his "short" position by buying back the 3800 shares, the price had increased to $5.26 (U.S) per share, up from the $3.70 per share he originally tried to sell them at. Zhu lost the sum of $9768.12 as a result of the buy-back necessitated by the duplicate sale orders.

When a trade is made on NetTrader, the following disclaimer appears at the bottom of the computer screen:

> Although we endeavour to maintain the accuracy of the above information, we cannot be held responsible for errors or omissions. All market data is delayed by at least 20 minutes unless otherwise noted. Click here for more information.[9]

A second disclaimer reads:

> Neither HIDC nor any information provider may be held liable to the customer or anyone else for: . . . Any inaccuracy, error, delay, interruption or omission of any information . . . or . . . any loss or injury caused in whole or in part by either negligence or contingencies beyond their control in procuring, interpreting, compiling, writing, editing, reporting or delivering any information or services through the service. . . . The customer agrees that the liability, if any, of HIDC and the information provider(s) arising out of any kind of legal claim (whether in contract, tort or otherwise) in any way connected with the service shall not exceed the amount the customer paid to HIDC for the use of the service.[10]

Question

1. Are these disclaimers enforceable?

Analysis

1. The Court found that they were not:

> It strikes me that the Defendant's Legal Disclaimer falls into the category of an agreement which "virtually eliminates liability for inaccuracy in the performance of the services contracted for by the customer" and can be construed as in fact "exonerating the broker from acts of . . . gross negligence" and in fact reserving "the right to be grossly negligent" to the broker. On this basis I find the Legal Disclaimer of the Defendant unenforceable.[11]

Thus, Zhu was awarded compensatory damages in the amount of $9768.12 plus interest, despite the disclaimer.

---

[9] *Zhu v. Merrill Lynch HSBC*, 2002 BCPC 535 at para. 9.

[10] See previous note at para. 10.

[11] See previous note at paras. 17 and 18.

15.6 Record Keeping

When disputes proceed to court, professionals can increase their likelihood of success by keeping accurate records. Even though oral evidence is admissible in court, written documentation provides the judge with substantive proof and provides the witness with a mechanism for refreshing memory. Without notes in a diary or some other written record, defendants may not have enough of a detailed memory to provide the court with a complete picture.

Accurate records can also prevent matters from ever reaching litigation. For example, if one party alleges an oral agreement and threatens to sue based on that alleged agreement, the likelihood of an action being commenced would be substantially diminished if the other party produced written documentation proving that no such agreement was reached.

It is therefore important to keep accurate diaries, to follow up meetings with minutes, to follow up oral agreements with written versions or at least with a letter confirming the agreement, and to generally provide sufficient documentation to support or defend a claim.

Recently, non-paper and digital documentation has become more commonly used as evidence in court cases. Photographs and videos are considered documentation. Film photographic evidence is routinely accepted in Canadian courts as proof of what is being shown in the picture. However, all photographs require authentication in order to be admissible as evidence. Digital photographs are less well received because of the ease with which they can be edited.

All documents presented as evidence must pass tests of authentication. Case law has developed the tests for paper documents; whereas federal and provincial *Evidence Acts* provide the tests for authentication of electronic documents, which assess the accuracy, genuineness, and authenticity of the evidence. Authenticity can be proven through a witness to the event, such as the person who took the image or an expert on the camera system. To authenticate an electronic document, the party adducing the evidence should prove the integrity of the system that created and stored the document. For this reason, organizations should have standard operating procedure in place for digital photographs. This procedure should specify how and when to download and store digital files, how and when to identify photographs, and how to document the chain of possession from the taking of the photograph to the presentation in court.

The following are suggestions for taking and maintaining digital photographs to be used as evidence. They should be included in standard operating procedure:

1. Start a paper audit trail as soon as possible, including the description, time and place of taking the photo, the downloading of data, and the creation of backups. Identify in the audit trail the persons taking each step.

2. Check the equipment routinely. When taking photographs, assess internal settings, time and date settings, space on the storage card, and battery strength.

3. Select image quality based on conditions, rather than on storage capacity.

4. Never delete an image, either intentionally or accidentally, since this action may raise questions.

5. Designate one computer for the storage of electronic photographs. Ensure that this computer does not have editing software.

6. Create a backup of the data immediately and identify it appropriately in the audit trail. Make the backup to a non-editable medium (such as to a Write Once Read Many (WORM) CD). For important files, maintain multiple copies, because CDs do not last forever. Maintain master and working copies separately, and limit and record access to the master.

7. Maintain printed and labelled versions of the photographs.

8. Maintain masters of the backup if there is any potential that it will be of evidential value.

The reason for maintaining a rigorous operating procedure is the protection of data. People within organizations come and go; and ensuring that you are able to prove the photographic record is crucial to being able to use it as evidence.

15.7 Problem Solving

When problems occur, professionals need to be proactive rather then reactive. Early meetings can sort problems out before they get out of hand. Refer to Chapter 14 for other methods for resolving disputes. The longer a problem festers, the more difficult it is solve and the more likely that it will disrupt the relationship and end up in court.

CHAPTER QUESTIONS

1. Which of the following is not a contractual risk allocation provision?
   a) An insurance clause
   b) An "unforeseen conditions" clause
   c) An indemnity clause
   d) A pay-when-paid clause
   e) None of the above

2. If an architect makes a mistake in a design, is that architect liable in negligence?

3. Do warranties always have to be expressly stated in writing in a contract?

4. Are disclaimers always enforceable? If not, under what circumstances might a disclaimer be unenforceable?

5. How may the authenticity of a document be proven in court?

ANSWERS

1. E

2. Not in every case. The mere fact that a mistake has been made does not automatically mean that there has been negligence. The plaintiff must still prove all of the elements of a negligence claim.

3. No. Implied warranties may exist. Courts often imply warranties of fitness for purpose in cases involving new products such as cars and houses.

4. No. If a disclaimer is too broad, a court may find it to be unenforceable.

5. The author of the document can take the witness stand and testify when the document was created. For other strategies, see p. 173.

CHAPTER 16

INSURANCE

Overview

Insurance policies are contracts of indemnity; and as such, they are governed by principles of contract law. However, insurance policies are a special form of contract and are thus interpreted according to specific statutes developed originally to protect insurers and policyholders, as well as a specialized body of case law.

The basic principles of contract law, including offer and acceptance, consideration, and mistake, all apply to insurance contracts. But other principles of contract law apply specifically to insurance contracts. Understanding the laws and principles that govern insurance contracts gives a professional greater control over negotiation of the terms of policies and helps prevent costly mistakes that prejudice the rights of the **insured**—the party covered by the insurance policy—and result in loss of coverage. This chapter explains basic principles of insurance and characteristics of insurance contracts, such as the duty of the insurer to defend a claim, subrogation, and insurable interest.

Insurance policies generally fall into two categories: liability insurance and property insurance. **Liability insurance** protects the insured against claims made by third parties, such as a claim against a professional for errors and omissions; whereas **property insurance** protects the insured against loss or damage to property as a result of certain causes, such as fire or theft. With property insurance, a claim by a third party against the insured is not a prerequisite for recovering under a policy, whereas it is for liability insurance.

Most insurance policies have the same basic elements. Most have a **deductible**, which is the portion of the costs of defending or paying for a claim that the insured must pay (also known as **self-retention**). All insurance policies have a **coverage period**, which is the length of time during which coverage is in place. Every policy of insurance must also state the **perils covered**, which are the events that trigger the obligation of the insurer to either indemnify, defend, or both. There is also a **limitation period**,[1] which is a contractual or statutory requirement that a notice of claim must be made or a legal action commenced within a certain time period.

---

[1] See Chapter 13, pp. 148–150.

Professionals need to be aware of their responsibilities regarding insurance. Many contracts require the parties to procure certain types of insurance. For example, the CCDC 2 construction contract stipulates that the contractor must obtain both liability coverage and property coverage for the duration of the project. In addition, professionals must be aware of the risk and possibility of loss of coverage, due to the financial implications. Loss of coverage may occur for a number of reasons, including expiry of a limitation period, lack of insurable interest, failure to provide timely notice, and material non-disclosure. In addition, insured professionals have responsibilities to the insurance company once a claim is made or comes to their attention. After providing notice of the claim to the insurer, the insured must cooperate with the insurer in defending the claim and can not, for instance, admit fault. Failure to give timely notice or to cooperate can relieve an insurer of its insurance obligations.

Other standard form contracts require the parties to obtain specialty coverage, such as boiler insurance, marine and transportation insurance, and delayed opening and business interruption insurance.[2] Some common perils are specifically excluded from typical builders' risk and commercial general liability (CGL) policies. To determine if there are any gaps in coverage, professionals must review all insurance policies with a knowledgeable broker or agent.

16.1 Operating Without Insurance

The basic purpose of insurance is to spread risk and to shift risk. Few architects, engineers, geoscientists, or contractors, with the exception of some very large and well-established firms, can afford to self-insure, which is another way of saying that they could absorb the full cost of any potential loss. Through the mechanism of insurance, professionals spread the risk of loss to many other insured people and organizations, some of whom receive payments from the insurer to cover losses, and others of whom realize no benefit from their premiums except peace of mind.

Operating without insurance creates the risk of bankruptcy as a result a large claim. Even so, some professionals choose to operate without insurance. Due to the market supply and demand for insurance, sometimes coverage is either prohibitively expensive or impossible to obtain. In addition, some individuals and corporations assume that lack of insurance makes them less desirable targets for litigation and are therefore less likely to be sued or pursued. There is some truth to that assumption; however, it is a great risk for anyone with assets they wish to preserve.

Operating without insurance is certainly not recommended for anyone with anything to lose. Even the most competent professionals get sued, sometimes for no valid reason. The cost of defending a frivolous claim can be enormous. For example, in Canada, successful litigants usually recover only a portion of legal costs, generally less than 50 percent. One advantage of carrying insurance is that the insurer is usually required to defend against lawsuits, although the insured may be required to pay the deductible first, depending upon the wording of the policy.

Professionals often try to make themselves *judgment-proof*, which means taking steps to put one's assets beyond the reach of creditors, such as putting assets in one's spouse's name or removing assets from a corporation. While there are perfectly legal means of protecting and transferring assets, there are risks and ethical considerations associated with these methods. For example, in most jurisdictions, transferring assets to a spouse after one knows that a lawsuit is imminent is considered a fraudulent transfer and would be reversed by a court in order to satisfy the creditor.

---

[2] Boiler insurance typically covers damage to property and other losses caused as a result of explosions of pressure vessels. Marine and transportation insurance covers shipments of goods. Delayed opening and business interruption insurance protects against loss of rents or profits.

Another method used to protect assets is to incorporate a company and keep a minimal level of assets in the company. There are two reasons why this may be ineffective. First, courts can pierce the corporate veil[3] in order to find the shareholders liable if the corporate shell is used for an improper purpose (although this is difficult to do). Second, and more importantly, an individual can be sued in his or her personal capacity for negligence, despite the fact that he or she was working as an employee or officer of a corporation at the time.

16.2 The Duty to Defend

Insurers under liability policies generally have a duty to indemnify and a duty to defend. The **duty to defend** is the obligation to pay for legal fees and other costs of defending a claim; whereas the **duty to indemnify** is the duty to pay claims. The duty to defend may exist even where there is no duty to indemnify. Sometimes the duty to defend is of more value, particularly where the claim is not likely to succeed. The insurer is obliged to appoint counsel to defend the insured at the expense of the insurer (subject in some cases to the deductible being exhausted first) and to pay for all other defence costs, such as hiring expert witnesses.[4]

If the claim alleges only conduct that is outside the scope of the policy and is therefore not insured, such as fraud, the insurer will likely have neither a duty to defend nor to indemnify. For this reason, an experienced plaintiff's lawyer usually tries to frame the claim in such a manner that at least a portion of the claim falls within the typical policy so that there will be insurance available to pay for the claim. While generally the duty to defend and the duty to indemnify go together, in some cases there may be a duty to defend even if there is no duty to indemnify. In addition, some insurance policies have only a duty to defend.

16.3 Subrogation

Subrogation is an inherent right of the insurer, which allows the insurer to step into the shoes of the insured and recover losses from third parties after the claim is settled. An insurer is subrogated to the rights of its insured as against third parties once it has paid out under the policy. In other words, the insurer is entitled to assume the legal position of the insured in order to recover from some other party the amount it has paid out on a claim. For example, suppose that an architect for a commercial building project retains a specification writer to draft the specifications. Later, the waterproofing section drafted by the specification writer proves to be inadequate, and the owner is forced to tear up the material already installed and redo it at great expense. The owner then sues the architect. The insurer defends the architect, and judgment is awarded against the architect, which the insurer pays. At this point, the insurer has the right to sue (or subrogate against) the specification writer, just as the architect could have if he or she had not been insured.

In order for the insurance company to make use of these subrogation rights, the insured must have legal rights that are capable of being assumed and enforced by the insurer. For this reason, the insured party must take care not to relinquish or impair its rights against the party who was responsible for the loss. But occasionally a waiver of subrogation clause is included within a contract. A **waiver of subrogation** clause eliminates the insurer's right to subrogation. Professionals should watch for these clauses, since unauthorized waiver of the insurer's subrogation rights may allow the insurer to escape its indemnity obligation.

---

[3] See Chapter 5.

[4] See Chapter 14, p. 162.

A simpler method of eliminating the right of subrogation is to have all of the project participants named as insureds in the same policy. It is a basic rule of insurance law that subrogation cannot be obtained against an insured. This principle was affirmed in *Commonwealth v. Imperial Oil*,[5] where the insurer tried to make a subrogated claim against an insured subcontractor.

16.4 Insurable Interest

Insurable interest means that the insured party benefits from the existence of or would be prejudiced by the loss of the insured property. In order for a person or corporation to be entitled to recover insurance proceeds following a loss, that person or corporation must have an insurable interest in the insured property. Insurable interest can become an issue when a party does not own the property involved but is simply performing a limited amount of work on it. A subcontractor would have an insurable interest in the entire value of a project, even though the subcontract value is only a fraction of the total. This is the second issue in the *Commonwealth* case in Case Study 16.1.

The issue of insurable interest also becomes relevant when a sole proprietor decides to incorporate. The assets of the proprietor are then transferred to the company. If the owner of the company forgets to transfer the insurance from the sole proprietorship to the company, then insurance coverage may be lost. The sole proprietor no longer has an insurable interest in the assets, even though he or she has an ownership interest in the company. The same analysis applies where assets are transferred from one company to another, or from a company to an individual.

Case Study 16.1

Commonwealth v. Imperial Oil, [1978] 1 S.C.R. 317

Facts

In March 1967, Imperial Oil Limited decided to proceed with the construction of a fertilizer plant at Redwater, near Edmonton. It hired a general contractor, Wellman-Lord (Alberta) Ltd. Wellman-Lord retained the subcontractor Commonwealth for the installation of process piping. In the course of Commonwealth's work, a fire broke out. The damage to Commonwealth's property was negligible; but damage to the rest of the project was $102 628.50. The total damage was covered under a multi-peril property insurance policy, described as a builders' risk policy. The "insured" under the policy was defined as "Imperial Oil Limited and its subsidiary companies and any subsidiaries thereof and any of their contractors and subcontractors." The damage in its entirety was claimed by and paid to Imperial, which had Wellman-Lord carry out the repairs. The insurer commenced a court action alleging subrogated rights obtained from the owner, Imperial, as well as the general contractor, Wellman-Lord. Commonwealth denied the right of the insurers to invoke any such rights.

---

[5] [1978] 1 S.C.R. 317. See also Case Study 16.1.

Questions

The principal issue was whether the insurer had the right to subrogate against Commonwealth. The Court subdivided this issue into two questions:

1. In addition to its obvious interest in its own work, did Commonwealth have an insurable interest in the entire project so that in principle the insurers were not entitled to subrogation against Commonwealth for the reason that it was insured with a pervasive interest in the whole of the works?

2. If Commonwealth was not such an insured, were the insurers entitled to take advantage of their basic right to subrogation considering:

    a) the wording of the subrogation clause and of the policy as a whole

    b) the contractual arrangements between Imperial, Wellman-Lord, and Commonwealth?

Analysis

The Court concluded that the insurer could not subrogate against Commonwealth. The reasons were based on "the basic principle that subrogation cannot be obtained against the insured himself." In other words, it would defeat the intent of the project policy to allow the insurer to pay out the loss to one insured, and then try to recover that same amount from another insured. The purpose of the policy is to protect the insured against loss, and if subrogation were allowed against an insured, that protection would be illusory.

A second issue raised in this case was whether a party with limited involvement in a project, such as a subcontractor, has an insurable interest in the entire project, or whether that insurable interest is limited to the work performed by the subcontractor. If the insurable interest were limited, then a subcontractor who has caused damage to the work of others would not be covered by the policy for the loss to the work of others. The Court decided that there was an insurable interest in the entire project, and required the insurer to pay for the loss:

> On any construction site, and especially when the building being erected is a complex chemical plant, there is ever present the possibility of damage by one tradesman to the property of another and to the construction as a whole. Should this possibility become reality, the question of negligence in the absence of complete property coverage would have to be debated in Court. By recognizing in all tradesmen an insurable interest based on that very real possibility, which itself has its source in the contractual arrangements opening the doors of the job site to the tradesmen, the Courts would apply to the construction field the principle expressed so long ago in the area of bailment. Thus all the parties whose joint efforts have one common goal, *e.g.* the completion of the construction, would be spared the necessity of fighting between themselves should an accident occur involving the possible responsibility of one of them.[6]

---

[6] *Commonwealth v. Imperial Oil*, [1978] 1 S.C.R. 317.

16.5 Claims-Made and Occurrence Policies

A **claims-made policy** covers claims that are made during the policy period and does not depend on when the work was done. **Occurrence policies** cover claims in which the insured event, such as a house burning down or damage occurring to a building, occurred during the policy period. Most professional liability policies are of the claims-made type. The distinction is best explained in Case Study 16.2.

Case Study 16.2
Claims-Made Policies

Suppose that a geoscientist was employed from 1990 to 2000 by Ace Geotechnical Ltd. During that period, the geoscientist was insured under a professional errors and omissions policy issued by ABC Insurance. In 2001, the geoscientist left the employer and moved to a new company, Deuce Geotechnical Corp. Deuce Geotechnical Corp. and its employees were insured by DEF Insurance under an errors and omissions policy. In 2002, while employed by Deuce, the geoscientist was sued over a project he had worked on in 1995, while he was working with Ace. A third party alleged that faulty research had led to loss of investment money.

Questions

1. Are the two errors and omissions policies mentioned claims-made or occurrence policies?

2. Would the claim mentioned be covered under a property or liability policy?

3. Is the geoscientist covered by the ABC policy, the DEF policy, neither policy, or both policies?

Analysis

1. Professional liability policies are almost always claims-made policies. Of course, one would have to check the wording of the specific policy to confirm that fact.

2. This case deals with a claim by a third party. Therefore, only a liability policy would apply.

3. A claims-made policy is one for which coverage exists only for claims that are made while the policy is in force. The claim was made in 2002. The only policy in force at that time was the DEF policy. Therefore, only the DEF policy could cover the claim. This is true even though the geoscientist's work was performed before his employment at DEF.

The insured also has obligations to the insurer. In *Reid Crowther and Partners Ltd. v. Simcoe & Erie General Insurance Co.*,[7] the Supreme Court of Canada ruled on whether an insurer was obligated to insure an engineer who had withheld information. Reid Crowther was an engineering firm that had done some municipal engineering work in Stonewall, Manitoba. The town of Stonewall had expressed concerns about the quality of the work. Reid Crowther was covered by a claims-made

---

7 [1993] 1 S.C.R. 252.

insurance policy but did not inform its insurer, Simcoe & Erie, of the town's concerns. Years later, the town sued Reid Crowther, which then informed Simcoe & Erie. Simcoe & Erie then learned of the earlier criticisms by the town and denied coverage to Reid Crowther on the basis that Reid Crowther had breached its obligation to provide prompt notice of claims. The Supreme Court held that Reid Crowther had an obligation to notify its insurer, because a reasonable engineering firm standing in Reid Crowther's shoes would have suspected that a claim was going to be brought, even if the magic words "we are going to sue you" had not been used. Ironically, because of some unique and complicated facts, the Court found that Reid Crowther was indeed insured. However, in general, insured parties should not necessarily expect the same results. They must be very careful to provide notice to their insurers as soon as they are aware that there is a reasonable potential for a claim; otherwise, they risk losing their insurance coverage.

As well, when changing insurers, the insured has an obligation to notify the prospective insurer of all potential claims as well as actual claims. Failure to do so will likely result in a lack of coverage for such claims. Professionals may be reluctant to disclose a possible claim that looks as if it might disappear; but if they have any doubt, they should disclose it. Moreover, if both the new and old policies are of the claims-made type, and a new claim or potential claim arises (for the first time) after the new policy came into force, the professional should be scrupulous about informing the new insurer of the claim or potential claim. If a claim that the insured was aware of is made after the new policy has come into force, the original insurance policy will no longer cover it, because the original insurer was not notified of the claim while its policy was in force.

In addition, professionals must be aware what type of policy they have when changing coverage; otherwise, a gap in coverage can result. Another way professionals leave themselves uninsured is through retirement. The prudent way to deal with this situation is to maintain coverage after retirement. Such coverage can usually be obtained at rates substantially less than those charged during periods of active practice. Another gap can be created if a professional's employer goes through a corporate change, which causes the named insured—the company—to change. Employees may have been insured under the former company's policy for present and past employees, but that former policy no longer covers them. The former employee is not a past employee of the new company. Moreover, the coverage of the newly formed company may not extend to work done under its previous name. This situation can leave a professional uninsured for past work.

Moreover, some policies are considered hybrids of the occurrence and claims-made types, which further limits the professional's protection. The policy period clause in a hybrid policy may state that the occurrence must have taken place during the policy period, and that the claim be made either during the policy period or a specified period following. The descriptions of the policy period may be subject to varying interpretations by different courts. For this reason, the prudent course of action is to report any potential claim as soon as the insured has notice of it.

16.6 Material Non-Disclosure and Prejudice to Third Parties

One might ask why it is necessary to protect insurers. Because of the nature of an insurance contract, the insurer is at a disadvantage when it comes to access to information. The insured must therefore disclose all material facts that might affect the premium on the insurer's decision to provide coverage.

Material non-disclosure refers to a breach of the duty of the insured to provide all relevant information to the insurer when purchasing an insurance policy. The duty to disclose, as applied to insurance contracts, has caused courts to impose a **duty of good faith** on the insured, which means that the insured must disclose to the insurer any and all facts that could influence either its decision to provide coverage or the amount of the premium to be charged. Even if the non-disclosure is unrelated to the loss, coverage may still be denied. For example, suppose that an engineering firm applies for errors and omissions insurance but fails to disclose that one of its employees had had a negligence claim against her in the past year. After the insurer issues a policy, a claim is made against the firm based on alleged negligence of a different employee. The insurance company could refuse to pay, arguing that had full disclosure been made, it would have charged a higher premium or would have chosen not to provide a policy.

In some business situations, frequently one party relies upon another party to procure insurance coverage. For example, the owner of a construction project may purchase a project policy to cover all participants, including the design team and all contractors. In this instance, the owner's failure to disclose a material fact could void the policy, resulting in a denial of coverage for all parties. Whenever one party relies upon another to arrange for insurance, such risks are present.

In addition, prejudice to the rights of others can also occur if the party procuring the insurance simply fails to pay the premiums. All insured parties would like to be notified of such circumstances in order to determine whether to pay the premium themselves, obtain separate coverage, or let the policy lapse. It is prudent to require the party obtaining the insurance to notify all the insured parties of any material change in coverage. Legislation in some jurisdictions requires the insurer to make such notification.

16.7 Cooperation and Conflict between Insurer and Insured

The **duty to cooperate** deals with the conduct of the insured after a claim has been made. Virtually all insurance policies stipulate that the insured has a duty to cooperate with the insurer in the defence of a claim. Cooperation may take many forms: it may mean supplying documents to the insurer, testifying at discovery and trial, or providing pre-trial assistance to defence counsel. It can also mean refraining from doing anything to prejudice the result, such as admitting liability or providing a release of liability to other parties. While cooperation may be costly in terms of lost time and expense, failure to cooperate can result in a denial of coverage and, as a result, may be even more costly.

In the defence of a claim, the interests of the insurer and insured may be identical. Both want to see the claim defeated. But situations may arise where the interests of the insurer and insurer diverge, such as the following:

- Denial of coverage by the insurer
- Claims alleging conduct that fall outside the scope of coverage, such as fraud
- Desire by the insured to defend, despite a reasonable offer to settle
- Claims that exceed the limits of coverage
- Offers to settle which require compromise of a counterclaim, such as a claim by a professional for fees

Where no conflict of interest exists, only one lawyer represents the interests of both the insurer and the insured. However, where there is a conflict or potential conflict, the insured must retain separate counsel to look after his or her own interests. In some jurisdictions, the insurance company may be required to pay not only the fees of the lawyer it selects but also the fees of the lawyer selected by the insured.

Retaining a separate lawyer may be useful to the professional, whether or not the insurance company pays for legal assistance. The insured should understand that the insurer's goal is to resolve the claim for the lowest cost possible. If the insured has any other objective, such as preservation of reputation or recovery of fees, he or she should consider separate counsel. Furthermore, if an insurer or employer raises any issue which could lead to denial of coverage, professionals should seek independent advice.

Be aware that claims that exceed the policy limits can expose the insured to personal liability. Claims that put the insurer in a difficult position are those for which liability is doubtful and potential damages are large. For example, a geoscientist who carries one million dollars in coverage for errors and omissions may end up being sued for two million dollars. Assume the claim has some chance of success, and the plaintiff offers to settle for the policy limits of one million dollars. In this situation, the insured would be well advised to put the insurance company on notice that the insured would like to have the claim settled within the limits. If the insurer then decides to defend rather than settle, the geoscientist may be able to make a claim against the insurer for bad faith for refusing to settle.[8]

16.8 Types of Policies and Their Common Exclusions

Insurance policies usually fall within one of two categories: liability policies and property policies. Some policies provide both types of coverage. Within these two categories, a few well-known types of policies are commonly used in the consulting and construction industries. These policies are builders' risk policies, CGL policies, and errors and omissions policies.

(a) Builders' Risk Policies

Most of the discussion thus far has focused on liability policies. **Builders' risk policies**, also known as **all-risk policies**, are property insurance. Unlike more common forms of property insurance, which cover only specifically listed perils, builders' risk policies are designed to cover all damage to the insured property except those perils specifically excluded. These policies specifically cover projects under construction. Once the project is complete and occupied, coverage typically ceases. Builders' risk policies have specific purposes, as described in the *Commonwealth* case:

> In England, it is usually called a "Contractors' all risks insurance" and in the United States it is referred to as "Builders' risk policy". Whatever its label, its function is to provide to the owner the promise that the contractors will have the funds to rebuild in case of loss and to the contractors the protection against the crippling cost of starting afresh in such an event, the whole without resort to litigation in case of negligence by anyone connected with the construction, a risk accepted by the insurers at the outset. This purpose recognizes the importance of keeping to a minimum the difficulties that are bound to be created by the large number of participants in a major construction project, the complexity of which needs no demonstration.[9]

Some all-risk policies exclude certain acts of negligence, and most exclude faulty design. Virtually all builders' risk policies exclude losses due to faulty design, materials, and workmanship (see Figure 16-1 on the following page). However, it would be unwise to assume that a policy excludes faulty construction methods, for example, without examining the particular wording of the policy.

---

[8] A claim for bad faith is not guaranteed to succeed in these circumstances. However, the law governing such claims is extremely complicated and is thus beyond the scope of this text.

[9] *Commonwealth v. Imperial Oil*, [1978] 1 S.C.R. 317.

FIGURE 16-1 Builders' Risk Coverage

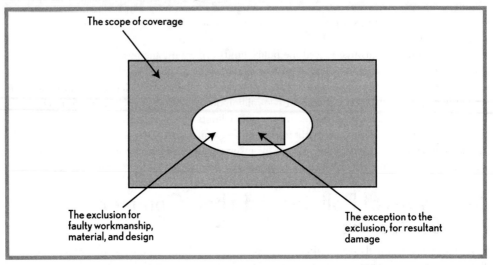

Note: shaded areas denote insured claims

However, courts interpreting the faulty design exclusion have not reached entirely consistent results. Plaintiffs in several cases have argued that construction methods are not included in the meaning of design, whereas insurers have argued that they are included.[10] In a case involving a wall that was blown down by high winds due to inadequate bracing, the Court held that design is wider in scope than the design of the finished product and is wide enough to encompass methods of construction.[11] But in a similar case, also involving collapse of a wall due to high winds, the Court came to the opposite conclusion: faulty design does not include methods of construction.[12] In general, American courts tend to construe the exclusion clauses more narrowly against the insurer.

Moreover, when interpreting exclusion clauses, courts have also considered whether negligence must be present in order for there to be faulty design, workmanship, or materials. Some courts have held that a design may be faulty even though there was no negligence on anyone's part. In the *Queensland* case,[13] a bridge was washed out by unforeseeably high floodwaters. The Court ruled that even though the designer of the bridge may not have been negligent, the mere fact that the bridge was not fit for its intended purpose renders the design faulty.

This reasoning has been followed in other cases as well. But with all due respect to the learned judges in those cases, this reasoning ignores the fact that no design is intended to withstand any and all perils. If legislation requires that designs be adequate to withstand winds, floods, or earthquakes up to a specified return period, that reflects a policy choice by the government. But it should not be translated into a "fault" in the design, thereby depriving an insured party of coverage.

[10] These two competing interpretations were discussed by the Court in the case of *Pentagon Construction (1969) Co. v. U. S. Fidelity and Guarantee Co.*, [1977] 4 W.W.R. 351 (B.C.C.A.).

[11] *Simcoe and Erie General Insurance Co. v. Willowbrook Homes (1964) Ltd.*, [1980] I.L.R. 1-1236 (Alta. C.A.).

[12] *Todd's Men and Boys' Wear Ltd. v. Diamond Masonry (Calgary) Ltd.* (1985), 12 C.C.L.I. 301 (Alta. Q.B.). The Court distinguished the *Willowbrook Homes* case on the basis that there was a clear element of design involved in *Willowbrook* from the outset, whereas there was no such element in *Todd's* case. But this rewards the contractor who makes no effort to calculate a safety factor over the contractor who makes imperfect calculations.

[13] *Queensland Gov't Railways and Electric Power Transmission Property Ltd. v. Manufacturers' Mutual Insurance Ltd.*, [1969] 1 Lloyd's Law Reports 219, 118 C.L.R. 314 (H.C. Aust.).

Even where a loss is excluded under the faulty design, materials, and workmanship exclusion, portions of the claim may in fact be covered. That is because these exclusion clauses typically contain an exception that provides for coverage to resultant damage. **Resultant damage** refers to damage to property other than the property containing the faulty workmanship, material, or design. For example, if two towers are being constructed side-by-side, one of which is under-designed and as a result collapses into the second tower, damage to the second tower may be covered even if damage to the first is excluded.

In some cases, the insured has tried to argue that one component of a construction project should be considered separate from the remainder of the project, so that damage to the remainder may be treated as resultant damage. Canadian courts have generally rejected this argument.

Case Study 16.3

B.C. Rail v. American Home Assurance Co. (1991), 54 B.C.L.R. (2d) 228 (C.A.)

Facts

B.C. Rail owned and operated a railway line consisting of approximately 1000 miles of main line track and 250 miles of branch lines. In 1985 a landslide occurred at Mile 307, which was located on a high, steeply sloping section of fill (the "embankment"). Earlier, in 1983, B.C. Rail had undertaken a capital improvement program from Miles 299 to 311. The purpose of the improvement program at Mile 307 had been to design a grade that would remedy ballast and alignment problems. A geotechnical engineer employed by B.C. Rail had conducted extensive pre-design investigations at Mile 307 and had found no signs of instability. One month after construction was completed, the shoulders of the new fill started to slump. Tension cracks in the new fill area had been observed on the east side of the track, and a 350-foot crack in the fill on the west side of the track had been observed. Shortly thereafter, the landslide occurred. It was determined that the landslide would not have occurred had the construction used a different fill design.

Question

1. The issue before the Court was whether any of the remedial costs were recoverable under the policy. Clearly, if there was found to be error in design, namely the fill design, the cost of replacing that fill was excluded from coverage under the policy. However, if there were remedial expenses incurred over and above the fill, as the geotechnical engineer tried to argue, those additional expenses constituted resultant damage and were therefore recoverable.

Analysis

1. The Court refused to draw a distinction between the fill and the remainder of the additional expenses. It ruled that none of the loss was insured. In reaching this decision, the Court relied on the *Sayers* case.[14] In *Sayers*, the Court explained its reasons as follows:

---

[14] *Sayers v. Insurance Corp. of Ireland Ltd.* (1981), 126 D.L.R. (3d) 681 (Ont. C.A.).

It would be taking too narrow a view of the case to isolate one part of the work from the total contractual obligation. The damage to the equipment was the product of the failure to take protective measures, and so that fault rendered the appellant's performance of its contractual obligations "faulty workmanship". The damage to the ducts and the switching gear was not, therefore, "damage resulting from such faulty . . . workmanship . . .," so as to come within the exception to the exclusion.[15]

The exclusion clause from the aforementioned *Sayers* case is typical of faulty workmanship, material, and design exclusions in builders' risk policies:

THIS POLICY DOES NOT COVER:

. . . cost of making good faulty or defective workmanship, material, construction or design, but this exclusion shall not apply to damage resulting from such faulty or defective workmanship, material, construction or design.

In addition, other perils are typically excluded from all-risk policies, such as fraud by employees and loss of revenue due to delay. If a loss occurs and there is a question as to whether coverage exists, a professional should seek legal advice. Even an insurer's denial of coverage may not be the last word on the matter, as shown by the many cases litigated over the interpretation of policies. Courts sometimes disagree with the interpretation given by the insurer, regarding not only all-risk policies but all policies of insurance.

(b) Commercial General Liability (CGL) Policies

Commercial general liability (CGL) policies cover claims by others for bodily injury and property damage. As such, they are liability policies, not property policies. Contractors usually purchase a CGL policy to provide coverage against claims made by third parties for loss arising out of physical injury or damage to persons or property. Some policies further restrict claims to loss or damage to tangible property. These policies typically exclude claims for damage to the insured's own product. Therefore, if a portion of a building collapses due to poor workmanship, the insurer would likely argue that the CGL policy does not cover the loss. This argument was advanced in the *Privest* case,[16] which involved the removal of asbestos. The owner of the building sued the contractor who had originally installed the asbestos many years earlier. One issue in the case was whether the contractor's CGL policy provided coverage for the cost of its removal. The presence of asbestos *per se* was found to constitute damage to property, but it was not damage to *tangible* property.

Where the damage is found in work performed by a subcontractor, the exclusion relating to the insured's product should only apply to the work performed by that specific subcontractor.[17]

Some CGL policies also exclude losses arising out of work performed by the general contractor and its subcontractors. Other policies only exclude work performed by the general contractor but not its subcontractors. In each case, the specific wording of the policy determines whether there is coverage.

---

[15] *B.C. Rail v. American Home Assurance Co.* (1991), 54 B.C.L.R. (2d) 228 (C.A.) at para. 8.

[16] *Privest v. Foundation Co. of Canada* (1991), 57 B.C.L.R. (2d) 88 (S.C.).

[17] *F.W. Hearn/Actes v. Commonwealth Insurance Co.* [2000] I.L.R. 1-3801 (B.C.S.C.).

(c) Wrap-up Policies

A **wrap-up policy** is a liability policy that insures all direct participants in a construction project, including the consultants, the owner, and all subcontractors. It typically runs from the end of a project until a specified time afterwards, such as the end of the contractor's warranty obligations.

(d) Errors and Omissions Insurance

Errors and omissions insurance, also referred to as *malpractice insurance*, is liability insurance for a professional, such as an architect, engineer, or geoscientist, that covers claims made by third parties for loss or damage caused by negligence of the insured. Such insurance is almost always written on a claims-made basis.

As with other types of insurance, errors and omissions insurance usually contains exclusions, such as the following:

- losses arising out of late delivery of drawings
- work done by the consultant as a contractor, including work done as a partner or joint venturer on a design-build team
- mould
- water ingress.[18]

Many provincial architecture, engineering, and geoscience statutes place obligations on these professionals either to carry insurance or to inform clients of the level of insurance that is carried. Both the client and the professional need to understand the coverage being provided, the exclusions to such coverage, the dollar-value limits on the coverage, the terms of the coverage, and the parties covered by the policy (*e.g.*, employees, independent contracts, subconsultants covered). Professionals must also be very aware of the insurance notice requirements.

CHAPTER QUESTIONS

1. A contractor procures a policy of property insurance covering the equipment of his competitor. Later, a loss occurs. Can the insurer deny coverage? If so, on what grounds?

2. Which of the following is not a potential defence available to an insurer to a claim under professional liability insurance policy?
   a) The professional failed to disclose an unrelated claim in the initial application for insurance.
   b) The professional failed to give timely notice of the claim.
   c) The professional agreed that he or she was liable for the claim.
   d) The professional agreed to a contract for services without obtaining specific approval of the insurer.

3. A claim is made against a geoscientist in negligence. The geoscientist has an errors and omissions policy in place. But part of the geoscientist's work involving mathematical modelling was subcontracted to an independent consultant. The negligence was in fact caused by an error of the subconsultant. The geoscientist decides to discuss the matter with the subconsultant, who has few assets and no insurance; and the geoscientist agrees to release the subconsultant from liability in exchange for $10 000. The geoscientist then reports the claim to the insurer. Can the insurer deny coverage? Explain.

---

[18] In light of the many leaky building cases on Canada's west coast, the recent decision of insurers to exclude claims relating to water ingress has caused much concern to architects and engineers.

4. "It is inappropriate for professionals to make themselves judgment-proof." Discuss this statement.

5. A contractor is constructing the concrete core of a high-rise tower. A blockout is left at each floor to allow electrical conduit to pass through in an enclosed duct. Due to design error, a fire erupts in the conduit, damaging not only the conduit and wiring but also the finishes in the room that the conduit passes through. The policy in place is a builders' risk policy. Discuss the coverage issues.

6. What is a waiver of subrogation?
   a) It permits the insurer to seek compensation from others as if the insurer were the insured.
   b) It eliminates an insurer's obligation to provide indemnities to the insured.
   c) It eliminates the insurer's right to seek compensation from others as if the insurer were the insured.

7. An individual purchases property insurance covering the perils of fire and theft. A theft occurs, and the insurer denies coverage, arguing that the insured failed to disclose the fact that he was a foreigner (*i.e.*, a landed immigrant), not a citizen, and that the insurer would not have accepted the risk of insuring an alien. Was this non-disclosure material? Why or why not?

8. What types of claims does a claims-made policy cover?
   a) Those made during the currency of the policy
   b) Those arising from facts that occurred during the currency of the policy
   c) Those for damage to property on the insured's office premises

9. A wall under construction collapses due to an error in design, destroying not only the wall itself, but also an electrical switchgear on the ground and a sports car parked next to the jobsite. The contractor and owner have used a standard form CCDC 2 contract and have procured the required insurance. Are the losses (the wall, the switchgear, the car) covered? Which policies cover what losses?

10. Which statement(s) is/are true of builders' risk insurance policies?
    a) They always cover all claims of any nature arising from construction.
    b) They often exclude coverage for repairing the contractor's own work product.
    c) They continue to cover a building for the life of the building.

ANSWERS

1. Yes. The insurer can deny coverage on the grounds that the contractor had no insurable interest.

2. D

3. Yes. The insurer can take the position that it has been prejudiced by the geoscientist's release of its subconsultant.

4. Professionals can legally arrange their affairs, including distribution of their assets, in any way they want before they receive notice of a potential claim. But after receipt of notice of a potential claim, the professional is prohibited from transferring his or her assets to place them out of the reach of creditors.

5. The facts of this problem are taken from the case of *Sayers & Associates Ltd. v. Insurance Corp. of Ireland Ltd.* (1981), 126 D.L.R. (3d) 681 (Ont. C.A.). In that case, the insurers argued that the appellant's loss was excluded under the terms of the insurance policy: "This policy does not cover: cost of making good faulty or defective workmanship, material, construction or design, but this exclusion shall not apply to damage resulting from such faulty or defective workmanship, material, construction or design."

As a result, the Court ruled:

> In the present case, the "fault" that underlay the "faulty workmanship" was the failure of the Appellant to take protective measures; by the terms of its contract its 'work' was to install electrical equipment and to keep it dry and clean until the contract was completed. It would be taking too narrow a view of the case to isolate one part of the work from the total contractual obligation. The damage to the equipment was the product of the failure to take protective measures, so that fault rendered the Appellant's performance of its contractual obligations "faulty workmanship." The damage to the ducts and the switch gear was not, therefore, "damage resulting from such faulty workmanship," so as to come within the exception to the exclusion.[19]

Even when a loss is excluded under the "faulty design, materials, and workmanship" exclusion, portions of the claim may be in fact covered. That is because the exclusion clauses typically contain an exception, which provides for coverage to other property that may be damaged. For example, if two towers are being constructed side by side and one of them is under-designed and as a result collapses into the second tower, damage to the second tower may be covered even if damage to the first is not.

The court in the *Sayers* case did not consider the surrounding work to be other property. It is sometimes difficult to predict whether a court will draw a distinction between the component of the building or project that contains the fault, and surrounding work which is damaged as a result. Recently, courts have tended to consider all of the work to be integrated so as to exclude coverage in total.

6. C

7. The denial was wrong because the immigration status of the individual was irrelevant to the policy that was purchased.

---

[19] *Sayers v. Insurance Corp. of Ireland Ltd.* (1981), 126 D.L.R. (3d) 681 (Ont. C.A.).

8. A

9. The contract requires the contractor to obtain liability insurance and the owner to purchase property insurance. Liability insurance, typically in a CGL form, does not cover work that is in the control of the insured. In other words, the project under construction is not covered by the liability policy. Therefore, the wall itself will not be covered under the liability policy; and it is uncertain whether the electrical switch gear will be covered by that policy either. The sports car, on the other hand, would not fall within that exclusion and, assuming that a claim is made by the owner of the car against the contractor or the owner, or both, the liability policy should afford coverage.

Meanwhile, the contract requires the property insurance purchased by the owner to be in an all-risk form. Virtually all policies of this form contain an exclusion clause with respect to faulty or defective workmanship, material, and design. The wall itself would also most certainly be excluded from coverage, given the facts stated in the problem description (*i.e.,* the collapse was due to an error in design). The switch gear, on the other hand, may fall within the exception to the exclusion, which provides for coverage to other property that may be damaged.

Therefore, in summary, damage to the wall is probably not covered by either policy; damage to the car would be covered under the liability policy; and damage to the switch gear would be covered under the all-risk property policy.

10. B

CHAPTER 17

BONDS

Overview

A **bond** is a special form of contract, whereby one party, the **surety**, guarantees the performance by another party, the **principal**, of certain obligations. The party to whom the obligations are owed is called the **obligee**. A bond is also called a *contract of suretyship*, and a surety is also called a *bonding company*. A *suretyship agreement* involves three parties: the principal, the surety, and the obligee.[1] Bonds have been described as three-party agreements.

    A bond is similar in some ways to a contract of indemnity, such as an insurance policy. Like contracts of indemnity, bonds are interpreted according to the rules of contract law. But they are subject as well to additional rules, such as the right of subrogation. Although a surety appears to be like an insurer, a bond is not insurance. Different types of bonds are used for different purposes. Three types of bonds are normally used on construction projects: bid bonds, performance bonds, and payment bonds. Outside of the construction context, fidelity bonds and performance bonds are also commonly used. Municipalities and other agencies sometimes require that a performance bond be posted to guarantee performance or completion of the work covered in the permit. Unless noted otherwise, in the examples used in this chapter, the principals (*i.e.*, the parties who are being bonded) are contractors, and the obligees are project owners. However, the basic principles explained in this chapter also apply where the principal is a party other than a contractor, and the obligee is a party other than an owner.

    A surety can secure release from its obligations under the bond as a result of certain acts by the obligee that cause prejudice to the surety, or because of other circumstances. This chapter presents the typical defences used by the surety in response to a claim by the obligee.

---

[1] The principal is also referred to as the obligor, because the principal owes the primary obligation to the obligee.

17.1 Roles and Responsibilities in Bonds

The role of the surety is to guarantee the performance of another party, often a contractor, for the benefit of a third party, usually the project owner. But a surety can also guarantee performance of a different party's obligations. For example, a subcontractor's performance may be guaranteed for the benefit of a general contractor. Similarly, a governmental agency may require a performance bond from a party performing environmental clean-up work. The purpose of a bond is to provide some comfort and protection to the obligee, due to the risks inherent in the industry in question. If the principal fails to perform, the obligee can look to the surety instead of just to the contractor, who may not have the money to satisfy a court judgment.

Note that two separate agreements are involved when a bond is issued, each of which contains separate obligations. The primary obligation is found in the contract between the principal and the obligee, and it exists separately and distinctly from any surety contract. The principal enters into a contract with the obligee; and that contract contains obligations owed by the principal to the obligee, such as performing the work in a timely manner. These obligations are referred to as the *primary obligations*. The bond itself contains the second set of obligations. A surety's obligations are considered secondary obligations, because they depend on the existence of primary obligations owed by the principal.

The principal pays the surety a premium, just like an insurance premium; and in exchange, the surety accepts the obligation to guarantee the performance of the principal. The premium paid by the principal varies, depending on the financial resources of the principal, its bonding capacity, billing level, and asset base. For an established contractor with a solid financial base, the premium will probably be approximately 1.5 percent of the contract value, subject to certain factors.

Typically, the surety requires the principal and its major shareholders to indemnify the surety against any loss. In fact, the indemnity given by the principal is one of the things that distinguishes a bond from an insurance contract. The indemnities given to the surety as well as the acts of the surety that can negate the indemnities are discussed on pp. 192–193.

One of the requirements of a contract of suretyship that differs from the requirements of contracts in general is that surety contracts must be in writing in order to be enforceable. Standard forms have been developed and are widely used in the industry.[2] A further difference is that in order to enforce a bond, the obligee must have the original bond. A photocopy or other facsimile is of no value. For that reason, it is critical that the obligee under a bond obtain the original and keep it in a secure location.

17.2 Indemnities and Other Surety Recourses

After a surety pays out money to satisfy its obligations under a bond, it is subrogated to the rights of the obligee. **Subrogation** allows the party who pays for a loss suffered by another party to assume the rights of that other party for the purpose of recovering the loss from a third party. In this context, subrogation allows the surety to assume the rights of the owner in order to sue the contractor to recover the monies paid out. This right of subrogation exists at common law as one of the characteristics of a suretyship agreement.

---

[2] In Canada, the Canadian Construction Document Committee (CCDC) publishes documents that are widely used. The bid bond is CCDC 220, the performance bond is CCDC 221, and the labour and material payment bond is CCDC 222.

In addition to the right of subrogation, sureties often demand indemnities from the contractor and its shareholders. The indemnity from the principal fulfills the same function as the right of subrogation: it allows the surety to recover any losses from the principal.

But an indemnity from a contractor and a right of subrogation may be useless remedies particularly if the contractor has become insolvent during performance of the contract. In the absence of any other form of security, the shareholders of the contractor might be willing to let the company go out of business in the face of a serious problem, rather than rescuing it with more money to complete the project. For this reason, sureties often demand indemnities or guarantees from the shareholders as well.[3] The managers of the contracting company are then forced to treat every claim by an owner as though it were made against the individuals behind the company and not against an empty corporate shell. Once the shareholders become indemnitors, they are personally liable for the losses incurred by the surety.

17.3 Bid Bonds

Bid bonds are used almost exclusively in the construction industry. A **bid bond** is a guarantee by a surety, in favour of a project owner, that if the owner accepts a bid by the contractor in question, and the contractor fails to enter into the contract that is the subject matter of the bid, the surety will pay the penalty specified in the bond.

Invitations for bid often require that the bid be accompanied by a bid bond in the amount of 10 percent of the amount of the bid. Where such requirement is included in the invitation, failure to include a bid bond with the submission often results in the bid being rejected out of hand. For this reason, the contractor must enter into an agreement with a surety to provide the bid bond. In some cases, an alternate form of security may be acceptable. Letters of credit have been used, as well as cash, as security in lieu of a bid bond.

What obligation of the contractor does the surety guarantee under a bid bond? The primary obligation is that the contractor is obligated to enter into a contract with the owner if the bid is accepted. If the contractor fails to do so, the surety agrees to pay the owner the difference between the amount of the contractor's bid and the amount of the bid accepted by the owner, up to a maximum of the value of the bond.[4]

However, in the absence of such a bond, a contractor who withdraws an irrevocable bid could be liable for the difference between its bid and the next lowest, even if that amount exceeded 10 percent of the bid amount. But the existence of a bid bond may limit the contractor's liability. In fact, the CCDC bond contains the statement: "The principal and the Surety shall not be liable for a greater sum than the specified penalty of this Bond." Thus, by requiring contractors to submit standard form bid bonds, owners may end up limiting the damages available from contractors to the amount of the bond.

Note that the positions of all the parties in a bid bond are not equal. The surety can never be in a worse position than the principal in its relationship to the obligee. In other words, the obligations of the surety to the obligee are dependent on the obligations of the principal to the obligee. Therefore, any defence that a principal has to a claim by an obligee is available to the surety as well.

---

[3] *Mortgage Insurance Co. of Canada v. Markwood Construction Ltd.* (1987), 31 C.L.R. 161 (B.C.C.A.).

[4] The owner cannot simply accept any bid. It must mitigate its loss. This usually means taking the next lowest bid from a qualified contractor. In the United States, the AIA Document A310, which is a commonly used bid bond form, explicitly states that the owner must "in good faith contract" with another bidder or risk its claim against the surety. The CCDC bid bond does not explicitly contain a "good faith" requirement; but the obligation to mitigate is implied in every contract.

Whenever a claim is made against a surety on a bid bond, the surety will first look to any defences available to the contractor.

One defence available to the contractor relates to revocability of the bid. If the bid is revocable, and the contractor revokes before it is accepted, that is a complete defence. Of course, revocability may depend on a number of factors, such as whether the bid was stated as being acceptable for a specified period, whether any consideration was paid to keep the bid open, whether there was detrimental reliance, and on custom and practice of the industry.[5] Other defences may be contained in the bond itself. For example, the CCDC bid bond provides that any suit under the bond must be made within six months from the date of the bond. In contrast, a suit by an owner against a contractor brought on a basis other than the bond, such as breach of a promise not to revoke, may not be subject to the same limitation period. Thus, the surety always remains in a less vulnerable position than the contractor.

A contractor company that becomes aware that it has made a mistake in its bid or that must revoke its bid for some other reason should obtain legal advice immediately. Failure to take immediate appropriate action could result in the bid being accepted; and if the contractor refuses to sign the contract, the bond could be forfeited, the cost of which would ultimately fall on the contractor and other indemnitors. A contractor should never treat a claim against a bid bond lightly. Sureties typically pursue all their indemnitors, possibly including shareholders, to recover monies paid out under a bid bond.

17.4 Performance Bonds

In the construction industry, a **performance bond** is a bond under which the surety guarantees a contractor's performance of a construction contract. Arguably, in a general sense, every bond is a performance bond, because under every bond, the surety guarantees performance of a primary obligation owed by a principal. Under a performance bond, the surety guarantees the obligation of the contractor to perform its contract with the owner. Like bid bonds, performance bonds can have any face value; but in the construction industry, they are usually written in an amount equal to 50 percent of the value of the construction contract. This amount, known as the *face value* of the bond, is the maximum potential liability of the surety in almost all cases. The contract between the obligee and principal is incorporated by reference into the bond (see Figure 17-1).

The roles that the parties play in performance bonds can differ. In some cases, where the principal is a subsidiary of a large, well-established parent company, the obligee may choose to purchase a corporate guarantee of performance from the parent company instead of a performance bond. On large projects, the prime contractor often requires the major subcontractors to provide performance bonds. If the subcontractor provides the bond, the prime contractor would be the obligee and the subcontractor would be the principal.

Claims under performance bonds follow the same sequence of events as for bid bonds. The surety responds only when the obligee notifies the surety that the principal has defaulted on its obligations. It is a condition precedent to any obligation on the part of the surety that the principal be in default.[6] Once the surety has been put on notice that its principal is in default, it generally has six options:

---

[5] Bid revocation is discussed in greater detail in Chapter 10.

[6] The CCDC performance bond contain the following language: "Whenever [the contractor] shall be, and declared by [the owner] to be in default under the Contract, [the owner] having performed [the owner's] obligations thereunder, . . ." This same language is also found in the AIA form of performance bond.

FIGURE 17-1 Performance Bond

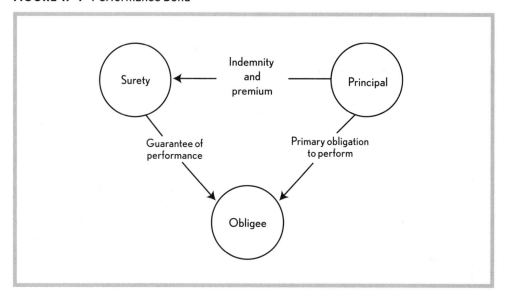

1. It can have the principal remedy the default by performing its obligations.

2. It can complete the contract itself in accordance with its terms and conditions.

3. It can solicit bids for completion of the work and pay the obligee the difference between the accepted bid and the remainder owing to the principal under the original contract, up to the face value of the bond.

4. It can pay the obligee the amount of the bond.

5. It can assert a defence and refuse to do anything.

6. If there is a genuine dispute between the obligee and the principal, it can take a "wait and see" approach to determine whether the principal was in default.

Once a surety has been notified of a default, it may obtain legal advice. It should conduct as thorough an investigation as possible before choosing a course of action. If the cost of completing the unfulfilled contractual obligations bears some relation to the amount owing to the contractor, the surety will likely solicit bids for completion. With this course of action, the cost to the surety could be relatively small. But if the cost to complete the project is well in excess of the amount owing, the owner may have inadvertently overpaid the contractor, which could provide a defence to the surety. Owners typically rely on architects or engineers to certify progress payments to the contractor; and if those professionals perform the certifications negligently, resulting in overpayment to the contractor, the surety or the owner may sue the consultant to recover the overpayment.

Case Study 17.1
Performance Bonds

Facts

Assume that a contract for a construction project has a value of $10 million. A performance bond issued for this project has a face value of $5 million. But the prime contractor company performs only 50 percent of the work before it defaults. Meanwhile, the architect certifies to the owner that 60 percent (rather than 50 percent) of the work is complete. As a result, the owner pays $6 million (60 percent of the contract value) to the defaulting contractor. (For this case, issues relating to construction lien holdback are not relevant: the owner has paid the entire $6 million.) The surety then tenders the remainder of the work, and the lowest bid for completion is $5.3 million. Of this amount, approximately $300 000 represents the inefficiency resulting from a second contractor taking over the work.

Questions

1. What is the exposure of the surety?

2. Does the surety have any recourse, and if so, against whom?

3. Is there a possible defence available to the surety?

Analysis

1. The surety's exposure is $1.3 million. But the owner still has to pay out the remaining amount of the contract to whoever is hired to complete the work. The owner has already paid $6 million on a $10 million contract; so there is still $4 million left to pay out. This money will now be paid to the completion contractor, who will receive a total of $5.3 million. The difference ($1.3 million), which will have to be paid by the surety, represents the surety's exposure. But this amount is far less than the face value of the bond. Had the face value been less than the difference between the amount left to be paid by the owner and the cost of completion, the surety would have been better off paying the face value of the bond, which it is entitled to do under the terms of the bond.

2. The surety now has recourse against the defaulting contractor, who would be an indemnitor. Moreover, if the surety had obtained indemnities from the shareholders or officers of the defaulting contractor, then the surety would likely try to recover from these indemnitors as well. Finally, the surety may have a claim against the architect because a portion of the surety's loss was caused in part by the over-certification by the architect. Had the architect certified the correct amount of $5 million (50 percent of the contract), the surety's loss would only have been $300 000. The surety's claim against the architect could be framed in negligence or perhaps as a subrogated claim based on the rights of the owner against the architect.

3. If the surety could prove that it had been prejudiced by an act of the owner, it might reduce or eliminate its own obligation. The owner's overpayment might support such a defence. However, in this example, if the architect certified the amount, and the owner paid the certified amount, such a defence is very unlikely to succeed.

As in Case Study 17.1, sureties must choose between soliciting bids and completing the work themselves, either with their own forces or by delegating the work to another contractor. If time is important, the surety may elect to complete the work. But sureties hesitate to do the work itself because then their obligations will have no financial limit. According to the wording of the bond, the option of soliciting bids is expressly limited to the face value of the bond; whereas completing the work puts the surety in the same position as the contractor, complete with warranty obligations.

Some sureties prefer to wait and see because they don't want to act as arbiter of disputes between the principal and obligee. A surety also does not want to become a **volunteer**, a party who pays money to another without any legal obligation to do so. Suppose the surety decided that the principal is in default and pays money to complete the contract. The principal then sues the obligee and proves that in fact the obligee was in default rather than the principal. The surety would then be unable to collect on its indemnity from the principal. The surety would be considered a volunteer at law and would have prejudiced its rights under the indemnity.

But if it turns out that the principal was in default, the surety may be liable for more than the face value of the bond, because its delay in responding aggravated the damages suffered by the owner. There is conflicting case law on the issue of whether a surety can be held liable for delay damages. Canadian case law seems to favour the view that a surety is not liable for delay. The *Lac La Ronge* case[7] contains a very thorough analysis of this issue, canvassing both Canadian and American authorities. In concluding that the surety should not be liable, the Court relied upon the following reasoning:

> A surety's obligation on this point can be no greater than if it had completed the contract or had found a responsible bidder to complete it. If either of those options had occurred, [the surety] would not be liable to the Band for the extra costs incurred by it.[8]

While some courts in Canada have held sureties liable for delay damages, the weight of authority seems to support the view held by the Court in *Lac La Ronge*.[9] This authority supports the sureties' position that liquidated damages for delay are not covered by a performance bond, and that a performance bond is only intended to cover the labour and material used in the construction, not other financial obligations of the contractor. In contrast, some courts in the United States have held the surety liable for failure to act, with damages in excess of the face value of the bond.[10]

17.5 Defences under a Performance Bond

As with bid bonds, the surety is entitled to raise against the owner any defences available to the principal, as well as any other defences available under the bond. For example, if the principal is able to raise a defence regarding the enforceability of the main contract, based on mistake, economic duress, misrepresentation, frustration, lack of consideration, or any other such defence, the surety can avail itself of that defence as well.

However, the surety's obligation under the bond is triggered only when the principal is in default. In some cases, even though the consultant has declared the principal to be in default, that finding may subsequently found to be erroneous. The surety may decide that no default has occurred and may defend the claim on that basis.

---

[7] *Lac La Ronge Indian Band v. Dallas Contracting Ltd.* (2004), 35 C.L.R. (3d) 35.

[8] See previous note at para. 104.

[9] See also Scott, Kenneth, and Bruce Reynolds. *Scott and Reynolds on Surety Bonds* (Toronto: Carswell, 2004).

[10] For example, see *Continental Realty Corp. v. Andrew J. Crevolin Co.*, 380 F. Supp. 246 (S.D. W. Va. 1974).

Several conditions govern claims under a bond. First, it is a condition precedent to any claim by the obligee, under a standard form CCDC performance bond, that the obligee must have performed all of its obligations under the contract. For example, if the obligee is in default of its payment obligations, the principal's refusal to perform further may be justified. When the obligee notifies the surety about the contractor's default under these circumstances, the surety would simply raise the obligee's default as a defence.

Another condition of a performance bond is that the obligee must give the surety timely notice of a claim. If notification is made after remedial work has already been performed, or in any manner that deprives the surety of its ability to either mitigate the loss or take other steps available to it, the lateness of the notice may allow the surety to escape liability.

A significant change in risk also affords the surety with a defence. When a surety agrees to provide a bond, it is assuming a risk, based on the terms of the contract between the principal and obligee, as well as on the financial strength of the principal and any other indemnitors. If the obligee does anything to significantly change the risk (unless the change is contemplated in the bond), that change may discharge the surety's obligations.

In addition, at common law, any material change in the contract time could be considered a material change in risk. A minor change would not have any effect on the surety's obligations; but a material change in the amount of work, type of work, or schedule of payment for work could significantly affect the risk and prejudice the surety. In the construction industry, changes are expected; and those that fall within the type generally contemplated do not materially alter the risk. However, an excessive number of changes or changes that are excessive in size could alter the risk. For this reason, when considering a significant change to the project, a prudent owner should contact the surety and obtain prior approval.

The surety has a few more defences. Substantial overpayment to the principal, as discussed above, also prejudices the surety to the extent of the overpayment, thereby providing a defence. In addition, the bond usually contains time limitation defences. The CCDC Document 222-2002 states that any suit under the bond must be made before two years from the date on which final payment under the bond is due.

17.6 Payment Bonds

A **payment bond** is a guarantee of the performance of a payment obligation. A party to a contract requests a payment bond whenever there is a concern that the other party may default on a payment obligation. While payment obligations in a contract usually flow in one direction, such as from owner to contractor, any party to a contract may have payment obligations, which, if breached, could cause problems.

For example, in a construction contract, the owner has an obligation to pay the contractor. As well, the contractor has an obligation to pay the subcontractors, suppliers, and workers. Failure by the contractor to meet these obligations could result in liens being filed and prejudice to the owner. Similarly, subcontractors have the same obligation. Failure by a subcontractor to pay those parties below it can result in prejudice to the general contractor. For this reason, owners and general contractors sometimes require a **labour and material payment bond** to protect against liens by unpaid subcontractors, suppliers, and those working below them in the contractual chain. A labour and material payment bond is structured somewhat differently than a performance bond. The owner is the obligee, but only as trustee for claimants as defined under the bond.

In general, payment bonds are typically required in two situations: first, an owner may require one from the contractor; and second, a contractor may require one from each of its major subcontractors. In situations where the financial stability or solvency of the owner is in question, the contractor may require a payment bond from the owner. However, use of payment bonds in such situations is relatively rare.

The payment bond creates a trust relationship. A **trust** is a legal doctrine that separates the legal ownership of a right or of a property from the beneficial interest or ownership. A **trustee** is a person appointed under the terms of a trust, who is given legal ownership of the trust property or rights but is required by law to act on behalf of and in the best interests of the trust beneficiaries. A **trust beneficiary** is a person for whose benefit the trust property or rights must be used. Because the obligee is named as trustee, it may be required to take steps to assist claimants under the bond to enforce their rights.

According to the CCDC labour and material payment bond (also known as an L&M bond), the claimants under the bond are the beneficiaries of the trust. A claimant is defined as anyone who has a direct contract with the contractor. This is known as a *one-tier bond*.[11] If a claimant has not been paid the money owed for work done on the contract, it may make a claim under the bond.

In a claim situation, the surety who has issued the payment bond is entitled to any defences available to the principal. For example, under a labour and material payment bond, if a claimant is not in fact owed the amount it has claimed, due to deficiencies or other set-offs, the surety is entitled to raise such defences against the claimant. Another defence is that the surety's obligation is limited to the face value of the bond. Furthermore, the surety is entitled to raise as a defence any limitation period contained in the bond that has not been complied with. These limitation periods vary from one bond to another, and claimants should be extremely careful when timing their claims.

One useful form of payment bond for the construction industry is the lien bond. A **lien bond** is a payment bond used as security to facilitate the discharge of a lien that has been filed against the land on which a project was constructed. The bond is used in substitution of the land as security for the lien, so that the lien may be removed from title to the land.[12] Under a lien bond, the surety is required to pay a lien claimant if the lien is found to be valid and the principal does not pay the amount ordered by the court. A lien bond is not put in place until after the lien has been filed, unlike a labour and material payment bond, which is put in place at the beginning of the contract.

CHAPTER QUESTIONS

1. A surety has provided a performance bond to the owner, guaranteeing performance of the contractor. The contractor in default has been paid $400 000 on a $1 million contract but has completed 50 percent of the work. The surety solicits bids, and the lowest acceptable bid is $550 000 to complete the project. How much money will the indemnitors have to reimburse the surety?

2. A surety has provided a labour and material payment bond. A subcontractor makes a claim against that bond in the amount of $10 000. The general contractor advises the surety that the subcontractor is owed only $5000. What should the surety do? Explain.

3. An owner receives irrevocable bids, each accompanied by a bid bond in an amount equal to 10 percent of the respective bid. The lowest bid of $90 000, is accepted. But the contractor refuses to enter into a contract for no excusable reason. The next lowest bid is $95 000. However, that bidder also refuses to enter into a contract. The third bid of $100 000 is

---

[11] In the standard form bond used in the United States, a claimant is defined as a person or corporation having a contract with the contractor *or one of its subcontractors*. This is known as a *two-tier bond*.

[12] For further explanation of liens and lien bonds, see Chapter 18.

accepted, and a contract is signed. What is the liability of the surety for the lowest bidder? Why? What is the liability of the surety for the second lowest bidder?

4. What is the difference between a bond and an insurance policy?
   a) Only a bond includes a right of subrogation.
   b) A person purchasing a bond must pay a premium.
   c) A surety, but not an insurer, obtains indemnities.
   d) All of the above

5. When a claim is made under a performance bond, which of the following is an option for a surety?
   a) Wait and see
   b) Pay the face value of the bond
   c) Defend the claim
   d) Only (a) and (b)
   e) All of the above

6. Can a sub-subcontractor qualify as a claimant under a standard form CCDC labour and material payment bond?

7. Why would a surety want to avoid being characterized as a volunteer?

8. If a surety against whom a claim has been made on a performance bond decides to wait and see if the principal is liable, is the surety liable for delay damages resulting from its decision to wait and see?

9. Which of the following parties cannot be an obligee under a payment bond?
   a) An owner
   b) A general contractor
   c) A subcontractor
   d) None of the above

10. Which of the following cannot qualify as a defence under a performance bond?
    a) The owner has failed to pay the contractor.
    b) The owner has overpaid the contractor.
    c) The owner has added a large number of changes to the work.
    d) None of the above

11. If a claim is asserted against a surety under a performance bond, which of the following statements is true?
    a) The surety cannot rely on the defences available to the principal, if those defences appear weak.
    b) The limitation periods available to the contractor may not be available to the surety.
    c) The limitation periods available to the surety may not be available to the principal.
    d) None of the above

ANSWERS

1. To get the surety to complete the work, the owner has to pay to the surety the remaining $600 000 owing on the contract. Of that amount, $550 000 goes to the completion contractor. The indemnitors do not have to reimburse the surety at all.

2. The surety should conduct its own investigation to determine whether the contractor is correct. If the surety simply takes its contractor's word as truth, it may overpay the claimant, in which case it would not have any recourse against the indemnitors for the overpayment.

3. The liability of the surety for the lowest bidder is at most the face value of the bond, which is $9000 (10 percent of the bid price). That amount is lower than the difference between the bid of the principal and the bid accepted by the owner. The same analysis applies to the second-lowest bidder. However, in the second case, the difference in bids is $5000, which is lower than the face value of the bond ($9500). Therefore, the second surety does not face exposure greater than $5000. Note that the courts would not allow the owner double recovery.

4. C

5. E

6. No, this form is a one-tier bond.

7. If a surety pays money as a volunteer, without a legal obligation to pay, it would be unable to recover this amount from its indemnitors.

8. While some courts in Canada have held sureties liable for delay damages, the weight of authority seems to be that delay damages are not covered by a performance bond.

9. D

10. D

11. C

CHAPTER 18

CONSTRUCTION LIENS

Overview

Every province in Canada has a construction lien statute.[1] These statutes vary in content but are similar in many respects. Their purpose is twofold: to provide security of payment for the suppliers of labour and materials and renters of equipment on construction projects, and to facilitate credit in the construction industry.

On almost every construction project, money flows from the lender to the owner, then through the general contractor to workers, suppliers, and subcontractors. In a typical project, each month the general contractor requests a progress draw, the consultant approves it in whole or in part, and the owner pays it. But the lender or owner does not release monies without evidence that work has been performed or materials have been delivered with value equal to the amount of money claimed. This means that workers, contractors, subcontractors, and material suppliers must do their work and supply their own materials before they get paid.

The fact that payment follows performance is not unusual or peculiar to the construction industry. Many businesses in the construction industry provide goods on credit. Large purchases such as houses and automobiles are made with little money paid in advance. What distinguishes the construction industry from others is that the security for payment used in other industries does not work in the construction industry. If the purchaser of a car or house fails to pay the vendor or mortgage lender, the vendor or lender can repossess the car or house. Conditional sales agreements, mortgages, and other security instruments are commonly used to create security and rights of repossession. But if drywall or painting subcontractors remain unpaid, repossession of the drywall or paint is impossible or impractical.

Thus, without some form of security, many contractors and other participants would be unwilling to supply labour and material without being paid in advance. Yet payment in advance would be unacceptable to most lenders and owners. Therefore, it is in the interest of all parties to have in place a mechanism such as a construction lien act to facilitate credit and secure payment.

---

[1] Historically, construction liens were called *mechanics liens.*

Lien acts use several methods to secure payment. The primary method is the lien itself. A **lien** is a charge or claim against property. A **construction lien** or **builders' lien** is a charge or claim against real property that has been improved by construction.[2] Construction lien legislation requires the lien to be registered in the appropriate land registry against title to the property. The effect of registering the lien is to make the property difficult to sell or borrow against until the lien is removed, giving the owner incentive to deal with the claim.[3] Ultimately, the lien claimant may force the sale of the property to satisfy the lien.

A construction lien is not the only remedy available to an unpaid party. The unpaid party also has a claim for breach of contract against the party who hired him or her.[4] A subcontractor can sue the general contractor in contract, a supplier can sue the subcontractor or general contractor with whom it contracted, and a labourer can sue his or her employer. Unfortunately, a claim in contract is a hollow remedy against a party who has no assets or who has left the jurisdiction. Lien legislation creates a **cause of action** (a right to sue) against the owner's property, even though the lien claimant may not have contracted directly with the owner. The rationale behind this remedy is that the owner's property value has increased due to the labour or material of the claimant.

Note that lien legislation is provincial and may have no application where federal interests or legislation take priority.[5] Because of the short limitation periods for filing a lien, a claimant should obtain immediate legal advice if there is any question regarding the applicability of legislation to the project in question. But even if the lien is barred due to lapse of time, legal action for breach of contract may remain possible, since the limitation period for actions for breach of contract is typically much longer than those of lien actions. This chapter reviews the procedural and substantive requirements of construction lien statutes regarding notice and filing requirements, limitation periods, and categories of claimants.

18.1 Making and Proving a Lien Claim

If a supplier of labour or material to a construction project has performed and has not been paid, then that supplier can file a claim of lien. This is usually done by filing one or more documents with the land registry office. The form and content of this notice is set out in the statute in each province. The legislation also outlines time limitations and procedural requirements. For example, the limitation periods may limit the amount of time between substantial completion and the date of filing, or between the last day of work performed by the claimant and the day of filing, or between notice to the owner and filing. The statute may also state what land registry and court registry must be used and how documents must be notarized. Failure to comply with procedural and substantive requirements can result in the claimant losing his or her lien rights. For that reason, it is prudent to obtain legal advice at this point in the process.

However, until a lien claim is proved in court, a claim of lien remains only a claim. Claimants must commence legal action and prove their claims. This legal action is called *perfecting the lien*.[6] To perfect the lien, claimants need to prove that the work was performed or material supplied to

---

[2] In some jurisdictions, the lien is described as a claim against the *interest* of the owner in the property. In those jurisdictions, if the "owner" has a leasehold interest or something less than a full ownership interest, the lien would attach only against that interest.

[3] This situation results because a purchaser would be unwilling to purchase subject to a lien, and a bank unwilling to finance the purchase.

[4] There may be other remedies as well, such as unjust enrichment.

[5] While filing a lien against title on a federally owned project is not permitted, some courts have held that the trust provisions of the statute still apply.

[6] In some jurisdictions in the United States, the action is called a *foreclosure action.*

the specific improvement and that they complied with all of the requirements of the construction lien act. Construction lien statutes also require that they commence the action within a specified period, such as one year after the date when claim of lien was filed or the work was performed. Because this limitation period for commencing action varies from one province to another, lien claimants should obtain legal advice. Furthermore, some construction lien statutes contain provisions that allow the limitation period to be abridged.[7] Professionals involved in lien claims also need legal advice because the procedural and technical requirements contained in lien legislation may be complex or may be difficult for a layperson to understand. For example, in some provinces a plaintiff must file a *lis pendens*, also known as a *certificate of pending litigation*, within a period of time after the lien was filed. The *lis pendens* is usually filed in the land registry to give notice to prospective purchasers and lenders that the property is the subject of a lawsuit. Failure to do so can result in the lien being dismissed. Moreover, even though a claim may be small, construction liens are sufficiently complex that without a lawyer, claimants frequently lose due to technical and procedural difficulties.

Once the claimant has commenced action, defendants may raise defences, such as deficiencies in the work, set-offs, proof of payment, and limitation defences. Then once the claimant has proved the claim of lien, he or she must enforce it. If security, such as a lien bond,[8] has been put up in lieu of the property, enforcement is relatively straightforward: the claimant is paid with the security. Otherwise, the claimant, who is now a judgment creditor, can either wait for the property to be sold or else commence proceedings to have the property sold. But property is rarely sold in order to satisfy a lien.

18.2 Who May Claim a Lien

Lien claimants typically include workers, contractors, subcontractors, and material suppliers. The claimant does not need to be in privity with the owner or general contractor. In fact, two or more tiers of contracts may be interposed between the claimant and the owner.

The issue of who is entitled to claim a lien has been litigated in many provinces, with differing results, depending in part on the specific wording of the lien statute in the province in question. Some statutes require the claimant to have either physically worked on site or to have provided materials to the site. The entitlement of an architect or engineer to claim a lien also varies from one province to another. In some provinces, an architect who has furnished plans to the owner is not entitled to the benefit of the lien legislation, but an architect who supervised work on the site is. British Columbia's *Builders' Lien Act*[9] provisions relating to architects and engineers provide an example of the intricacies of lien states. The starting point is the definition of "subcontractor" in the Act:

> "subcontractor" means a person engaged by a contractor or another subcontractor to do one or more of the following in relation to an improvement:
>
> (a) perform or provide work;
>
> (b) supply material;

---

[7] For example, in Alberta, an owner or other interested party may abridge the limitation period to 30 days, by delivering a notice to commence action.

[8] See p. 206, and Chapter 17, p. 199.

[9] S.B.C. 1997, c. 45.

but does not include a worker or a person engaged by an architect, an engineer or a material supplier.[10]

This definition excludes subconsultants, who under this legislation would not be entitled to a claim of lien. But would a prime consultant be entitled to claim a lien under this legislation? Note first the statute's definition of "work":

"work" means work, labour or services, skilled or unskilled, on an improvement.[11]

Then consider the definition of "services":

"services" includes

(a) services as an architect or engineer whether provided before or after the construction of an improvement has begun, and

(b) the rental of equipment[12]

Thus, according to the Act, an architect or engineer who merely provides drawings but does not set foot on site is still providing "services" and is entitled to a lien, as long as that architect or engineer is not a subconsultant. However, the services must be in relation to an "improvement" as defined by the Act. If the architect provides drawings for a project that is never built, the architect would be unlikely to be entitled to claim a lien.

Material suppliers often have difficulty proving their claims, because in some provinces, material supplier claimants have to prove that the material was delivered to the site. Material suppliers frequently supply to several jobsites for the same contractor and are unable to say with certainty what material went to what site. For example, a lumber supplier may prepare a load of lumber for its client, the general contractor, who picks it up and makes the rounds from one site to another. The supplier has no way of knowing how much was delivered to each site. Without being able to prove delivery, the supplier may be unable to prove a lien. To prevent this problem, the supplier must designate separate orders for each jobsite and keep accurate records of deliveries.

18.3 Substitute Lien Security

Anyone who purchases or lends money against a property with notice of a lien does so subject to the lien claimant's interest. When a claim of lien is registered against title to a property, such registration serves as notice to the public that the property is subject to the lien. Therefore, if the lien was registered first, then the interest a purchaser has in the property or the interest of a lender of funds secured against the property both rank lower in priority than the lien. If the property is sold, the lien claimant must be paid out of the proceeds first in accordance with the lien's priority.

But priority may depend on factors other than the timing of registration and other charges against title to the property, such as mortgages, easements, and judgments. In some jurisdictions, once a lien is registered, its priority relates back to the last date work was performed. Priority is important because it determines the order in which parties get paid out of the proceeds of sale of the property. If there is not enough value in the property to pay all of the registered charges against title, those at the bottom of the priority list remain unpaid. For example, Alberta's *Builders' Lien Act* contains this priority provision:

---

[10] See previous note, s.1.

[11] See previous note, s.1.

[12] See previous note, s.1.

11. (1) A lien has priority over all judgments, executions, assignments, attachments, garnishments and receiving orders recovered, issued or made after the lien arises.[13]

In addition, the order of priority between the mortgage lender and lien claimants varies from one jurisdiction to another. In some provinces, the construction lender ranks ahead of lien claimants only to the extent of funds advanced before the lien was filed. That means that funds advanced after a lien was filed rank below the lien. Under those circumstances, a lender should always conduct a search of the title to the property to confirm that no liens have been filed before releasing funds each month to pay the progress draw. If a lien has been filed, the lender should refuse to advance funds until the lien is removed.

There have been cases involving priority claims between unpaid subcontractors and Canada Revenue Agency (CRA), which raises an interesting issue. When a contractor becomes insolvent, in many cases it not only defaults in its payment obligations to subcontractors, but also fails to pay CRA for unpaid income taxes or GST. CRA may issue to the owner a "Requirement to Pay," which is akin to a garnishee notice. Such notice is a demand that the owner pay CRA any amounts that are due and owing to the contractor. Of course, the unpaid subcontractors in such circumstances will assert priority claims, usually in the form of trust claims under the applicable lien statute.

CRA has, in the past, taken the position that the *Income Tax Act* gives CRA priority over the unpaid claimants. However, this argument was put to the test recently in a British Columbia case known as *PCL Constructors Westcoast Inc. v. Norex Civil Contractors Inc.*[14] In that case, the Requirement to Pay was issued to the prime contractor, with respect to a defaulting subcontractor, who had failed to pay its suppliers as well as CRA. The Court held that as long as the contractor has valid right of set-off, meaning that it has the right to deduct from the subcontractor the costs of paying the unpaid suppliers, those suppliers' claims take priority over CRA. The important lesson to learn from this case is that contracts should be drafted to allow a right of set-off, otherwise the contractor (or the owner, as the case may be) may end up having to pay the holdback twice.

Despite these priority limitations, a lien can still be a powerful tool for encouraging payment, because once the contractor and others learn that payment is not forthcoming, they stop working, and parties can start quickly filing other liens. In order to avoid that scenario, lien statutes often allow the owner or contractor to remove a lien by providing alternate security for the claim. This allows the flow of funds to continue.

When an owner provides substitute security, the land and building under construction cease to be security for the lien; instead, the lien is secured by the alternate form of security. The most common forms of substitute security are cash, letters of credit, and lien bonds. The value of that security is usually required to be equal to the amount of the lien claim, plus an amount for security for costs. The Ontario statute specifies that the security for costs be equal to the lesser of $50 000 or 25 percent of the lien.[15] In other provinces, the amount of security for costs varies depending upon the statute and standard practice. In one province, a recent court decision has removed the requirement to post security for costs.[16] In some jurisdictions, the amount of security may be less than the amount of the lien itself. This may occur if the court is convinced that the lien is exaggerated or inflated.[17]

After substitute security is in place, it is then held in trust or in court until the lien claim is resolved or abandoned. If the claimants prove their claim, they are paid out of the cash, letter of credit, or lien bond. If their claims are defeated or abandoned, the security is returned to the party that put it into place.

---

[13] R.S.A. 2000, c. B-7, s. 11(1).

[14] 2009 BCSC 95.

[15] *Construction Lien Act*, R.S.O. 1990, c. C.30, s. 44(1).

[16] *Tylon Steepe Homes Ltd. v. Charles Eli Pont*, 2009 BCSC 253.

[17] For example, see *Q West Homes Inc. v. Fran-Car Aluminium Inc.*, 2007 BCSC 823.

18.4 Trust Provisions

A *trust*[18] is a legal mechanism for separating the legal interest in property from the equitable interest. The property need not be land. For the purposes of construction lien legislation, the property is money. In practical terms, it means that while one party, the trustee, has legal ownership and perhaps possession of the property for the time being, the trustee must use the property for the use and benefit of another party, the beneficiary. The **trust provision** of a construction lien statute creates a trust in the funds paid by the owner to the general contractor. As soon as the funds are paid to the contractor, they are impressed with a trust. The contractor is deemed a trustee, and all workers, subcontractors, and suppliers directly below the contractor in the contractual chain are deemed beneficiaries. The terms of the trust require the contractor to pay all the beneficiaries before using any of the funds for any other purpose. A similar trust is then created as soon as the contractor pays a subcontractor; the subcontractor becomes a trustee of the funds received by that subcontractor and must pay its workers and suppliers first.

For example, suppose a general contractor receives $100 000 as a progress payment, of which $90 000 is owed to subcontractors and labourers. If the contractor used $15 000 to pay down a bank loan or to buy equipment and was left with insufficient funds to pay its subcontractors, the contractor would be in breach of the trust provisions of the construction lien statute.

One of the consequences of a breach of trust is that it may give rise to *quasi-criminal penalties*. A quasi-criminal penalty is a fine or jail term specified in a statute other than the *Criminal Code*. Construction lien statutes contain such penalties for breach of trust. Furthermore, if the party committing the breach of trust is a corporation, its directors and officers may be personally liable to repay the trust funds and may face criminal sanctions together with the company. Anyone else who knowingly participates in the breach of trust may also be liable.

Breaches of trust sometimes occur because a contractor experiencing cash flow problems uses funds from one project to make payments for another project, honestly intending to make up the shortfall. But this is a dangerous practice. Instead, a contractor should always maintain separate bank accounts for each project and refuse to co-mingle funds. The temptation may be great if the money is sitting in another account or if a new lucrative contract has just been awarded; but the consequences are so severe that acting on this temptation is simply not worth the risk.

Case Study 18.1
Mackenzie Redi-Mix Co. v. Miller Contracting Ltd.[19]

Facts

The Ministry of Transportation and Highways was the owner of a bridge project. Miller Contracting Ltd. was the general contractor, and Procon Builders Ltd. was the concrete-placing subcontractor. The plaintiff, Mackenzie Redi-Mix Company, supplied concrete to Procon Builders (see Figure 18-1).

In the course of the project, Miller paid Procon less than the full subcontract price they had agreed upon. In fact, if Procon had completed its work, Miller would still have owed $46 000 to Procon. However, Procon failed to complete, and Miller hired a replacement subcontractor to complete the Procon contract. Miller then paid the replacement subcontractor an amount exceeding

---

[18] Trusts are defined in Chapter 17, p. 199.

[19] (1987), 20 B.C.L.R. 283 (C.A.).

FIGURE 18-1 Contractual Relationships in the Mackenzie Case

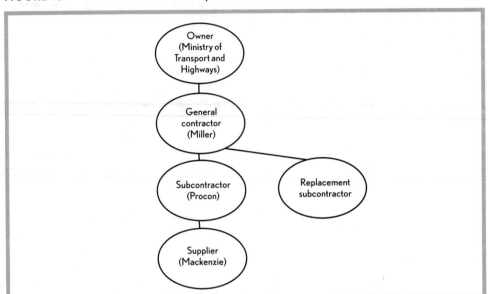

$46 000 to complete the Procon subcontract. Meanwhile, Mackenzie had not been fully paid for the concrete it had supplied to Procon.

Mackenzie's argument at trial and on appeal was that as a trust beneficiary, it was entitled to be paid before Miller used any of its contract funds to pay itself. If accepted, such a finding would make Miller a guarantor of its subcontractors' financial obligations to those working under the subcontractor. The courts in at least one other province have held for plaintiffs in the position of Mackenzie; but the British Columbia Court of Appeal did not agree:

> If a contractor who has paid a subcontractor in full may also be liable for that subcontractor's debts to materialmen, he will be unwilling to pay the subcontractor without assurances that those below in the construction chain have already been paid. The subcontractor will have to pay those materialmen before receiving contract moneys from the contractor, but from what funds is the subcontractor to make such a payment? The logical flow of money is from the contractor to the subcontractor and then to the materialmen. This payment flow will be hindered if each trustee must ensure that all those below have been paid before he pays those with whom he contracts.
>
> There is another problem if the appellants' interpretation is adopted. None of the contractors or subcontractors could appropriate any of the money to expenses or overhead, nor take any of the proceeds of the project as profits, until the whole job was complete and everyone fully paid. I think it rather unlikely that the legislature intended such an interpretation . . .
>
> In my view, the answer to the question raised by this appeal must be that a head contractor who has paid to a subcontractor the amount properly payable to that subcontractor has discharged his trust obligation with respect to those who would have had a claim against the funds before they were paid to the subcontractor. In this case, Miller held money in trust for Procon and those below, including Mackenzie. When Miller paid all that was payable to Procon and Procon's replacement, it discharged its trust obligation to Procon and those further down that line, including Mackenzie.[20]

---

[20] See previous note.

Questions

1. Would Mackenzie have had a lien remedy as well as a trust remedy?

2. If the answer to Question 1 were yes, under what circumstances would Procon want to pursue the trust remedy as well as the lien remedy?

3. Is there a cause of action against Procon? If so, why would it be necessary to pursue Miller and the owner?

Analysis

1. Normally yes. However, in British Columbia, there is no lien remedy against a highway. If this had been a private owner on privately owned land, there would probably have been a lien claim.

2. In this situation, because of the holdback provisions of the statute (discussed in the following section) MacKenzie's lien remedy in British Columbia would have been limited to 10 percent of the value of the contract between Miller and Procon at the time that contract was terminated. This 10 percent value may be less than the claim. As well, the statute provides that the 10 percent holdback must be shared with other lien claimants. Therefore, if there are multiple lien claimants, MacKenzie's lien remedy may be less than the amount of its claim. The trust remedy is not limited to 10 percent of contract value; in some cases, it can provide a remedy where the lien is inadequate.

3. There is certainly a cause of action against Procon, as well as its directors and officers. Procon appears to have received more money on the contract than it paid out to those below. However, if the company is insolvent or its principals cannot be found, Mackenzie may not find it worthwhile to pursue that action.

18.5 Holdback

Holdback, sometimes called **retainage**, is a percentage of contract value that must be withheld from each progress draw, not to be paid out until certain time periods have passed and certain conditions have been met. Holdbacks are a portion of the contract price kept out of the hands of the contractor and subcontractors until the project is complete, in order to determine whether other parties have been paid. All construction lien statutes contain holdback requirements. Holdbacks in construction are typically 10 percent and are kept as a reserve to deal with liens at the end of the job.

Ontario, Saskatchewan, and British Columbia use what is known as a *multiple holdback system*, which is a legislative requirement for the owner, general contractor, and all subcontractors to retain the 10 percent holdback from others below them on a construction project. In contrast, Alberta uses a *single holdback system*, in which the owner retains the holdback from the general contractor.[21]

Even under the single holdback system, contracts with subcontractors frequently contain provisions that create a trickle-down effect. The general contractor does not want to finance the holdback and so usually requires its subcontractors to agree to the same holdback percentage.

Release of holdback funds is typically triggered by substantial completion of the contract. Lien statutes specify that release of funds take place after a fixed period (*e.g.*, 45 or 55 days) following substantial completion. A consultant's certificate is normally used to determine the substantial completion date. The consultant must act impartially and fairly in making that determination.

---

[21] British Columbia used a single holdback system until 1997, at which time the *Builders' Lien Act* was amended, and the holdback system changed.

(a) Holdback Calculation under a Multiple Holdback System

While the retention of holdback is an obligation, it is also a defence available to the party holding back the funds. Lien statutes typically provide that the maximum liability to lien claimants of a party who has retained the holdback is the holdback amount. In other words, an owner faced with multiple lien claims may discharge those lien claims by paying the holdback, which is then distributed to the lien claimants who have proven their claims. The same is true for contractors who must discharge liens that have been filed by parties below them in the contractual chain. This principle is illustrated by the following two hypothetical cases.

Case Study 18.2
Holdbacks 1

Facts

Assume that the owner, ABC Development, hires a contractor, DEF Contracting Ltd., to construct a building for the contract price of $10 million, in a province with a 10 percent holdback requirement. DEF subcontracts the mechanical work to a subcontractor, S1 Ltd., for the subcontract price of $1 million. DEF also subcontracts the electrical work to a subcontractor, S2 Ltd., for the price of $2 million (see Figure 18-2). However, DEF defaults in its payments to both S1 Ltd. and S2 Ltd. Both subcontractors file liens. The S1 lien is in the amount of $800 000, and the S2 lien is in the amount of $1.2 million. Assume both liens are valid and are proven in the amounts claimed.

FIGURE 18-2 Contracts, Liens, and Holdback

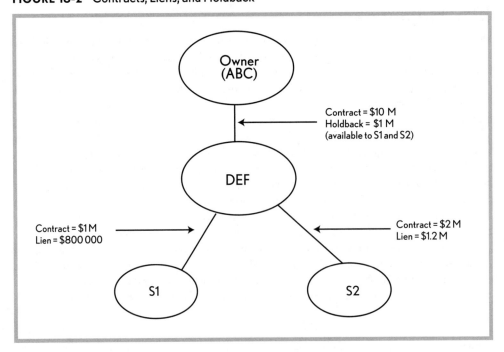

Questions

1. What is the holdback that the owner should have retained?

2. Assuming the owner retains the correct amount of holdback, what is the maximum exposure of the owner to the lien claimants?

3. How much money will each of the claimants receive?

Analysis

1. The owner should have retained 10 percent of the prime contract value. Ten percent of $10 million is $1 million.

2. The owner can pay the holdback into court and have the liens discharged. Even though the liens total $2 million, the owner is only liable for $1 million.

3. Typically, each claimant will recover a *pro rata* share of the holdback fund. The holdback ($1 million) divided by the total value of all lien claims ($2 million) is 50 percent. Therefore, each claimant should recover 50 percent of the amount if its lien. S1 has a lien of $800 000 and therefore should recover $400 000. Similarly, S2 should recover 50 percent of $1.2 million, or $600 000.

Case Study 18.3
Holdbacks 2

Facts

Assume an owner, ABC Development, hires a contractor, DEF Contracting Ltd., to construct a building for the contract price of $10 million, in a province with a 10 percent holdback requirement and a multiple holdback system. DEF subcontracts the mechanical work to a subcontractor, S1 Ltd., for the subcontract price of $1 million. DEF also subcontracts the electrical work to a subcontractor, S2 Ltd., for the price of $2 million. However, S1 Ltd. fails to pay its suppliers, M1 and M2, and each of those suppliers files a lien in the amount of $250 000 (see Figure 18-3).

Questions

1. Who has the responsibility of removing the liens from title to the property?

2. What is the holdback available to the two suppliers M1 and M2?

3. How much will each of those suppliers recover on account of their lien?

4. Would the result differ under a single holdback system?

Analysis

1. The general contractor is responsible to the owner for discharging the liens. The clause governing this responsibility is usually in the prime contract, either as an express term or as an implied obligation. S1 also owes a duty to the general contractor to discharge the liens.

FIGURE 18-3 Contracts, Liens, and Holdback

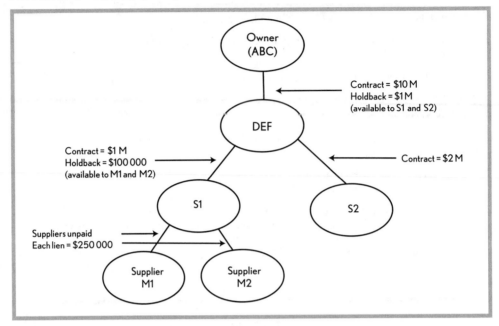

2. The holdback is based on the contract between DEF and S1. Ten percent of that contract is $100 000.

3. The ratio of holdback ($100 000) to total value of the liens ($500 000) is 20 percent. Therefore, each claimant will receive 20 percent of its lien, or $50 000.

4. Under a single holdback system, the holdback fund would be $1 million. Each lien claimant would recover the full amount of its lien.

Bear in mind that these examples are based on simplified facts. The following complications could change the result:

- The amount held back from the claimants is greater than the 10 percent statutory holdback.
- The lien claimants' claims were based upon unapproved extras that would have the effect of increasing the various contract values.
- The lien claimants were responsible for deficiencies in the work.

(b) Lien against the Holdback

In a recent development, the Court in a British Columbia case known as *Shimco*,[22] reinterpreted the province's *Builders' Lien Act*[23] such that it now creates two separate liens: one against the lands, and the other against the holdback fund. In other provinces, lien acts specifically permit a lien against the holdback. The practical effect of this ruling is that if an unpaid claimant misses the limitation period for filing a lien against the lands, that claimant may still obtain a remedy by asserting a lien against the holdback. This lien against the holdback must be asserted by commencing legal proceedings before the holdback has been paid out.

[22] *Shimco Metal Erectors Ltd. v. North Vancouver (District)*, 2003 BCCA 193.
[23] S.B.C. 1997, c. 45.

18.6 Risk to the Contractor and Owner

A general contractor may take all appropriate precautions and disburse all funds received from the owner and yet still be faced with liens at the end of the project. This may occur because a subcontractor has breached the trust provisions by failing to pay its employees or suppliers, or because a subcontractor has become insolvent.

Under the terms of its contract, a general contractor is usually required to remove all liens at its own expense. If it has already paid all of its subcontractors, workers and suppliers 100 percent of the amounts owed to them, and a second-tier subcontractor files a lien, then the general contractor may end up paying twice for the subcontractor's work.

To avoid having to pay for the same work twice, the general contractor must take certain precautions.[24] One of the more effective precautions is to require subcontractors to provide performance bonds and labour and material payment bonds. If a subcontractor then defaults by failing to pay its workers or suppliers, the subcontractor's surety will be responsible. [25] Another precaution is to require each subcontractor to submit each month a statutory declaration as a precondition to payment. A **statutory declaration** is a sworn statement declaring that to the best of the knowledge of the declarant, all the financial obligations of the subcontractor have been met, including payments to workers, subcontractors, suppliers, and assessments of all authorities having jurisdiction, such as workers' compensation premiums and taxes. If a person knowingly swears a false statutory declaration, that person may be held criminally liable for perjury.

CHAPTER QUESTIONS

1. A general contractor contracts with a subcontract for final cleanup. But the cleaning subcontractor does not begin work on site until substantial completion of the project. Meanwhile, the statute requires liens to be filed within 30 days of completion of the prime contract. "Completion" in this jurisdiction is defined as substantial rather than total completion. What effect, if any, might this definition have on the cleaning subcontractor's lien rights?

2. Which of the following may a lien claimant recover?
   a) More than the holdback amount
   b) Less than the full amount of the holdback
   c) Only the exact amount of the holdback

3. Which statement is true about the amount of money available to a claimant under a breach of trust claim?
   a) It is always less than under a lien claim.
   b) It is never less than under a lien claim.
   c) It is sometimes less than under a lien claim.

4. An owner, ABC Development, hires a contractor, DEF Contracting Ltd., to construct a building for the contract price of $10 million in a province with a 10 percent holdback requirement and a multiple holdback system. DEF subcontracts the mechanical work to a subcontractor, S1 Ltd., for the subcontract price of $1.5 million. DEF also subcontracts the electrical work to a subcontractor, S2 Ltd., for the price of $800 000. However, S1 Ltd.

---

[24] Many of the following precautions are also available to an owner facing possible default by the general contractor.

[25] For further discussion, see Chapter 15.

fails to pay its suppliers, M1 and M2; and each of those suppliers files a lien. M1's lien is in the amount of $200 000, and M2's lien is for $200 000. Moreover, S2 fails to pay two of its suppliers, E1 and E2; and each of those suppliers files a lien. E1's lien is in the amount of $50 000, and E2's lien is for $150 000. What is the holdback available to the lien claimants?

5. Based on the case in Question 4, how much should each claimant recover on account of their lien?

6. Based on the case in Question 4, what remedies might the claimants pursue, in addition to their liens against the land?

7. What is the purpose of a lien bond?
   a) To guarantee the performance of the contractor
   b) To ensure that no liens are filed
   c) To replace the land as security for a claim of lien

8. Which parties may be potentially liable for a breach of trust under a *Construction Lien Act?*
   a) A contractor
   b) A worker
   c) A material supplier
   d) A subcontractor
   e) A and D only

9. What is the difference between a multiple holdback system and a single holdback system?
   a) The amount of holdback available to a claimant
   b) The number of parties who are required to retain the holdback
   c) The percentage of money that must be held back
   d) A and B only

10. If a general contractor pays a subcontractor in full, and that subcontractor fails to pay one of its suppliers, is the general contractor potentially liable to that supplier for breach of trust?

11. What is the difference between security posted for a lien and security for costs?

12. What is one reason why it is important to include a right of set-off (the right to deduct) in a contract between the owner and its prime contractor?

ANSWERS

1. The cleaning subcontractor may not have any lien rights. As unfair as it may seem, the subcontractor is not entitled to a lien unless it is filed within 30 days of substantial completion of the prime contract. If the subcontractor files late, the lien would be invalid.

2. B

3. C

4. M1 and M2 will jointly pursue (and share *pro rata*) the holdback of $150 000, which is 10 percent of the contract between DEF and M1. E1 and E2 will share in the holdback of $80 000.

5. M1 and M2 have liens of equal value, so they will each receive one half of $150 000 ($75 000). E1's lien is one quarter of the value of the total liens filed under this chain of subcontracts, so E1 will recover $20 000, and E2 will recover $60 000.

6. Potential remedies include breach of trust, a claim against the holdback, breach of contract, and a claim against a labour and material bond (if one had been put in place).

7. C

8. E

9. D

10. In those provinces that follow the *MacKenzie Redi-Mix* [26] case, the general contractor would not be liable.

11. Security posted for a lien will be cash, or a letter of credit, or a bond, in an amount that is either equal to the amount of the lien or, in some jurisdictions, an amount that the court believes is sufficient to protect the lien claimant. Security for costs, on the other hand, is an amount that is meant to cover some of the legal costs that may be incurred in commencing and pursuing legal proceedings to prove the claim of lien, and is typically a small fraction of the lien security.

12. Without the right of set-off, if the prime contractor defaults in its payment obligations to its subcontractors as well as its obligations to CRA, the owner might have to pay the holdback amount twice: once to CRA and once to the unpaid claimants.

---

[26] *Mackenzie Redi-Mix Co. v. Miller Contracting Ltd.* (1987), 20 B.C.L.R. 283 (C.A.). See also pp. 207–209 in this chapter.

CHAPTER 19

DELAY AND IMPACT CLAIMS

Overview

The effect of delays on a project can be significant. When delays or impacts occur, they adversely affect most or all of the participants. The client loses revenue or sales, the consultants end up providing additional services, and the contractors incur both increased direct costs and extended overhead charges. When significant delays or impacts occur, the project participants try to isolate the cause and identify the responsible party in order to recover damages from that party.

A **delay** is an event that causes extended time to complete all or part of a project. An **impact** or **disruption** is the effect of an event that detrimentally affects a project but which may or may not extend the completion time of the project. An **acceleration** is an increase in labour or equipment which shortens the completion time or mitigates the effects of an impact or a delay.

The law relating to delay and impact claims is applicable to contracts and projects outside the construction industry. For example, a software engineer engaged to develop a software system for a client who requests numerous changes will face delays and impacts. However, because most of the case law on delay claims has been decided through cases involving construction projects, and much of the terminology used in these cases has come from construction cases, this chapter uses construction-related terminology, such as *critical path activities, concurrent delays,* and *no-damages-for-delay* clauses.

Claims consultants specialize in evaluating, presenting, and negotiating delay and impact claims. Claims consultants are often engineers who are skilled in schedule interpretation and evaluation. The emergence of this branch of the consulting industry is evidence of the substantial sums that are often at stake in such claims.

19.1 Scheduling Principles

A schedule is a tool used by all parties on a project to plan the sequence of activities. Schedules can be simple, hand-drawn efforts or complex, computer-generated works of art. Scheduling may seem like a straightforward, scientific exercise, but in the real world, it is more of an art. A well-planned schedule provides all of the parties with an opportunity to complete the work on time and on budget, while a poorly planned schedule creates significant risks of delay and overspending.

The most common type of schedule used to be the *bar chart*. A bar chart schedule gives each activity a start date and an end date. The bar chart's simplicity is both its greatest advantage and its greatest weakness. However, most bar charts do not show an inter-relationship between the activities. Without this information, activities cannot be sequenced in such a way as to achieve maximum efficiency.

FIGURE 19-1   Bar Chart

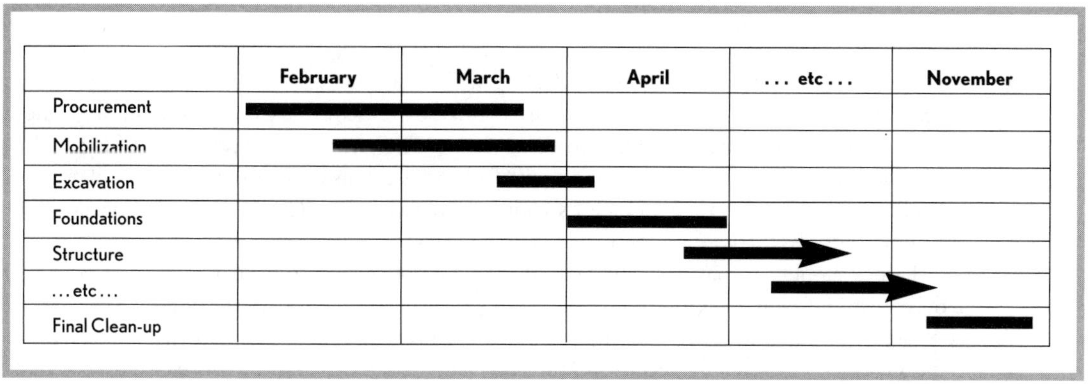

	February	March	April	... etc ...	November
Procurement					
Mobilization					
Excavation					
Foundations					
Structure					
...etc...					
Final Clean-up					

Network diagrams, on the other hand, require activities to be linked to one another such that the dependency of one activity on another is visible. For example, before a pile foundation can be poured, a hole must be dug. A bar chart is not able to show this relationship or the dependency of the start of concrete supply on the finishing of excavation work. But a network diagram clearly identifies the relationship between these two activities. One of the principal types of network diagram is the *critical path method* (CPM). This method identifies the activities that must be achieved and in what order to achieve the most efficient schedule.

Float is the time contingency in a given activity. If an individual activity can be delayed without delaying the final completion date, then that activity is said to have float. For example, if an activity is scheduled to start on June 1 but could start as late as June 15 without impacting the overall schedule, that activity has 14 days of float. Activities with no float are said to be on the critical path. The *critical path* is the series of activities for which any delay will cause an equal delay to the project completion.

19.2 Compensable and Excusable Delays

A **compensable delay** is a delay for which one party is contractually entitled to recover damages from the delaying party. An **excusable delay** is a delay that is not attributable to the fault of any party and which may or may not be compensable; but it entitles the parties to extend the time for completion. A **contractor-caused delay** is a delay caused by the contractor or by a party for whose actions of the contractor is responsible; it is almost always non-excusable and non-compensable.

Whether a delay is compensable, excusable, neither, or both depends on the wording of the contract. Most construction contracts make some effort to distinguish between compensable and excusable delays. The CCDC 2 contract contains the following compensable delay provision:

> If the Contractor is delayed in the performance of the Work by an act or omission of the Owner, Consultant, or anyone employed or engaged by them directly or indirectly, contrary to the provisions of the Contract Documents, then the Contract Time shall be extended for such reasonable time as the Consultant may recommend in consultation with the Contractor. The Contractor shall be reimbursed by the Owner for reasonable costs incurred by the Contractor as the result of such delay.[1]

Contracts also provide means for dealing with excusable and *force majeure* delays. **Force majeure** generally means events beyond the control of the contracting parties, although these events are usually specifically defined in the contract. The *force majeure* provision of the CCDC 2 contract provides that the contractor is entitled to an extension of time for excusable delays, including causes beyond the contractor's control. However, the CCDC 2 delay section does not specifically address the issue of contractor-caused delays.

Parties should also pay particular attention to the notice provisions in a contract. The CCDC 2 delay section requires parties to give notice of a delay claim within 10 working days so that the contractor can claim an extension.[2] But this clause "only bars a contractor's right to an extension of the contract"; it does not affect the owner's obligation "to reimburse the contractor for any costs incurred by it as a result of the owner caused delay."[3] Hence, the entitlement to an extension of time and to reimbursement of costs are separate; and reimbursement is possible even when a time extension is barred. Parties should note that there is a requirement to give timely notice of monetary claims under clause GC 6.6.1, but it is unclear what is defined as "timely notice."

Many contracts state that a delay must affect a critical path activity in order to qualify for an extension. While the CCDC 2 contract does not contain such a clause, parties must review this type of provision carefully. Moreover, the law usually presumes that in the absence of any provision to the contrary, there is an implied term in the agreement that each party will not impede or hinder performance by the other.

---

[1] Clause GC 6.5.1. See also GC 6.5.2, which contains a compensable delay provision.

[2] Clause GC 6.5.4.

[3] *Pacific Coast Construction Co. v. Greater Vancouver Regional Hospital District* (1986), 23 C.L.R. 35 (B.C.S.C.). Adopted in *Foundation Co. of Canada v. United Grain Growers* (1997), 23 C.L.R. (2d) 23 (B.C.C.A.).

Case Study 19.1
Compensable and Excusable Delays

Facts

Assume that an office tower construction project uses a standard form CCDC 2 as the contract between owner and contractor. The schedule shows that the elevator installation has two weeks of float and therefore is not on the critical path. A temporary hoist provided by the contractor for its own use is being used during elevator installation. The hoist is scheduled to be dismantled as soon as the elevators are operational. But the owner causes a two-week delay in the elevator work. For the contractor, a two-week delay will result in an increased rental charge for the hoist, but will not delay completion.

Questions

1. Should the contractor be entitled to compensation and an extension of time?

Analysis

1. Because the delay is an owner-caused delay, CCDC 2 provides clauses for both extension of time and compensation. Even though the delay is not on the critical path, CCDC 2 does not distinguish between critical path and non-critical path delays. The contractor would certainly be entitled to compensation for the delay, including the additional rental costs for the hoist. It is less clear whether the contractor would be entitled to a time extension, since CCDC 2 calls for an extension for a reasonable time. It is also arguable whether an extension is reasonable, since the owner would argue that the float for that activity exceeds the delay period; whereas the contractor would argue that it is entitled to retain and benefit from the float so that a subsequent contractor-caused event would have the benefit of the entire float.

19.3 Concurrent Delays

In many cases, compensable, excusable, and contractor-caused delays, or two of those three types of delay, occur at the same time and delay the same activities. For example, an architect may be slow in resolving design discrepancies, a labour disruption may occur, and the contractor's productivity may decline due to poor supervision. In case law, these are referred to as **concurrent delays**, in which two or more independent delays coincide. Where concurrent delays occur, apportioning responsibility becomes either difficult or impossible, and therefore neither party can recover damages from the other for the period of concurrency. Courts have thus been reluctant to award compensation to contractors or to enforce liquidated damages provisions in favour of owners for concurrent delays.

19.4 No-Damages-for-Delay Clauses

A **no-damages-for-delay** clause is a contractual limitation that protects the owner from claims for damages associated with delay and acceleration; but it does not stop the contractor from insisting on an extension of time. These clauses are considered limitation clauses. As a result, courts construe them very strictly *against* the party they are intended to protect. In extreme cases, courts may refuse to enforce them at all because of the harshness of the results. Only where the language is clear will a court deprive a contractor of the right to damages where the delay is caused by the owner.[4]

Different no-damages-for-delay clauses have different degrees of enforceability. Some, like the CCDC 2 exclusion of damages for *force majeure* events, are reasonable in scope and represent a considered allocation of risk.[5] They are generally enforceable. Others are wider in scope and preclude claims for owner-caused delays where the delay was unintentional. Still others try to exclude all delays, including those caused intentionally by the owner. Some courts will enforce these clauses, while others interpret the contract according to the following rules of interpretation:

- Words intended to exempt a party from liability because of its own fault are to be construed strictly against it.
- A party for whom another is doing work under a contract has a duty to facilitate the work. The no-damages-for-delay clause cannot be worded in such a way that it becomes an obstruction to another party's work.
- A contract will not be so construed so as to put one party at the mercy of another. Allowing the owner to delay the contractor at will, without recourse, puts the contractor at the mercy of the owner and makes the clause difficult to enforce.[6]

19.5 Acceleration

Parties can deal with delays in one of two ways: they can extend the time for completion, or they can accelerate the work to make up the period of delay. An **acceleration claim** is a claim for the cost of making up lost time. Acceleration usually costs less than delay. But in some cases they can be greater than or equal to delay costs, since acceleration is generally achieved through the use of overtime and additional labour, which can result in a decrease in efficiency. Courts treat acceleration claims in much the same way as they treat delay claims, including allocation of responsibility or fault.

Directed acceleration claims, in which a client (usually the owner) acknowledges that a delay has occurred and instructs the contractor to take all steps necessary to meet the contractual completion date, are relatively easy to recognize. However, constructive acceleration claims are more difficult to recognize. **Constructive acceleration** occurs when the contractor has been forced to accelerate the work, without any acknowledgement by the owner that the contractor is being asked to accelerate. If the contractor has been delayed by a cause that entitles the contractor to an extension of the contract time, and the contractor's request for an extension is refused, the owner's insistence upon the original completion date creates a condition of constructive acceleration and may entitle the contractor to additional compensation for costs incurred in accelerating the work.

---

[4] *Perini Pacific Ltd. v. Greater Vancouver Sewerage and Drainage District,* [1967] S.C.R. 189.

[5] Clause GC 6.5.3.

[6] Summarized from *Wilson and English Constr. Co. v. N.Y. Central Railway Co.,* 269 N.Y.S. 874 (N.Y.C.A.).

19.6 Impact Claims

An *impact claim* refers to a claim for costs that arise from inefficiency created by delays, interference, and changes of the sequence of the work. For instance, an owner that delays a summer project into winter weather may be faced with a claim for increased winter construction costs.

However, a claimant has responsibilities to disclose impact costs at the time of the delay or change. In the *Doyle Construction* case,[7] a contractor claimed for impact costs that had resulted from multiple change orders. But those change orders had not identified any impact costs. The owner defended the claim on the basis that the change orders were all-inclusive, and that the contractor had not given timely notice of his impact claims. The Court agreed, on the ground that the owner could have elected not to proceed with some of the changes had it been notified in advance of the impact costs.

Thus, when submitting a price for extra work, a contractor should include all impact costs. If that is not possible, the contractor should state in its quotation for the extra work and include in the change order itself, that it reserves the right to claim for impact costs at a later date. If the owner does not accept that reservation of rights but still wants to proceed with the extra work, then the owner may issue a *change directive*. Under a change directive, all costs associated with the extra work, including impact costs, should be recoverable.

19.7 Proving a Delay, Impact, or Acceleration Claim

As with any claim for breach of contract or negligence, the claimant must prove both liability (also known as *entitlement*) and damages. Entitlement for a time extension is often easier to prove than damages. A claimant must prove that another party breached a duty owed to the claimant, contractual or otherwise, which caused the claimant delay and entitled the claimant to a time extension.

But practical difficulties often arise in proving damages for delay, impact, or acceleration because of lack of adequate documentation. Thus, claimants need to understand what kinds of damages are possible, what kinds of documents to keep, and what types of delay can be claimed.

(a) Heads of Damage

Heads of damage are the recoverable categories of damages. Some of the different heads of damage claimed by contractors resulting from delay were listed in the *Kelson* case:[8]

- additional labour costs of both supervisors and field labourers
- travel and related expenses
- costs associated with lost productivity and loss of profit
- additional site and head office overhead
- costs associated with extending warranties and obtaining additional bonding
- loss of interest on holdback
- costs associated with paying subcontractors.

---

[7] *Doyle Construction Co. v. Carling O'Keefe Breweries of Canada Ltd.* (1988), 27 B.C.L.R. (2d) 89 (C.A.).

[8] *Geo. A. Kelson Co. v. Ellis-Don Construction Ltd.* (1998), 38 C.L.R. (2d) 215 (Ont. H.C.).

Of these costs, extended overhead is one of the more difficult to prove. Overhead is comprised of two components: field overhead and head office overhead. Head office overhead claims are usually difficult to quantify and require various formulas to calculate. Delays may also cause increases in material and labour prices. Some material, such as cast-in-place concrete, cannot be purchased in advance. However, these costs are relatively easy to document once entitlement has been proven.

Finally, contractors may also want to claim loss of profit for projects other than the one that was delayed. But these are considered consequential damages and are likely to be considered too remote to be recoverable. However, if the contractor can prove that the loss of profits is not too speculative and was reasonably foreseeable by the owner at the time the breach occurred, they may be recoverable.

(b) Required Documentation

The first documents to assemble in preparation for a delay claim are the project schedules. The contractor should have a bid schedule, hopefully in critical path form. Schedules should be updated on a regular basis as significant events impact the schedule. In this way, the final update should represent the as-built schedule. A lack of updated schedules will cause difficulty for the lawyer attempting to reconstruct the delays and their effects for the court. If an as-built schedule were not available, this information would have to come from other sources, such as project diaries, correspondence, and photographs.

Techniques for schedule interpretation and analysis all have the same purpose: to demonstrate the effect of delays on the completion date and other activities. The claimant or its delay expert incorporates each compensable delay into the schedule in order to demonstrate what other activities were affected and to what extent. He or she also adjusts each update of the schedule accordingly and removes the effects of non-compensable delays. Unless the schedule is computer-generated, this task may prove prohibitively expensive. Even using a computer program to calculate the effects, proving a delay claim is frequently a very costly exercise.

A claimant can also hire a delay claims consultant to help prepare the documentation. Delay claims consultants can offer advice about the effect of delays and about the compensability of delays. However, different topics require different types of expertise, and one expert may not have expertise in all required areas. For example, if the claim relates to an earth-moving operation, the consultant must have expertise in earth moving in order for his or her opinion to carry any weight or be admissible in court. Some contractors prefer to hire a claims consultant rather than lawyers to negotiate a resolution of a claim and avoid the expense of litigation; but if the negotiation fails, the consultant's report may be useless or even damaging in a lawsuit. Legal advice should be obtained before a consultant prepares a report, even if it is just to point the consultant in the right direction.

Furthermore, the claimant must maintain adequate project records as a matter of course, and the best time to start compiling documents to support or defend a claim is at the first indication that a delay may occur. Relevant documents include transmittals showing deliveries, labour records, correspondence about the delay, diaries, minutes of meetings, photographs, telephone records, and invoices. If a contractor believes that the owner is delaying the project, it should send letters to put the owner on notice. Many contracts contain strict notice requirements for claims; and failure to comply with the contract requirements may result in the claim being barred.[9] If the

---

[9] For example, CCDC 2 clause GC 6.5.4 requires the written notice of claim to be made not later than 10 working days after commencement of the delay.

claim is large, a claimant should seek legal advice for the drafting of the notice. Such advice is relatively inexpensive and should preserve the claimant's rights if done in a timely manner. Similarly, an owner should obtain legal advice as soon as notice of a large claim has been made.

Productivity and job cost reports may also be necessary for a delay claim. If the claim includes loss of productivity, the claimant needs to show what productivity was achieved on portions of the job that were not delayed in order to provide a comparison. Contractors track productivity for many purposes, including maintaining a database for estimating future jobs. But for these records to be useful in a delay claim, they need to be completed at frequent and regular intervals. Some superintendents may view record keeping and paperwork as a waste of time; but lack of adequate documentation has led to the defeat of many valid and valuable claims. Inexpensive software packages that record productivity data during the progress of a construction project are now commercially available.

(c) Total Cost and Measured Mile Approaches

The **total cost approach** values a delay claim by calculating the difference between the expected and actual cost of a project. It involves taking two snapshots of the project cost: one at the beginning and one at the end. The beginning cost is the contractor's estimate; and the end cost is the sum total of all of the contractor's costs to perform the work. The argument used in a total cost claim is that the difference in cost is due to breaches of contract by the owner, such as delay. The claimant must prove that such breaches occurred and that the total increase in the project cost is recoverable as damages.

However, this approach is fraught with difficulties. Claimants might be tempted to use it because it avoids costly schedule interpretation and analysis and compilation of documents required for a detailed individual analysis of delays. But it can cause the rejection of the entire claim. Problems arise when some of the delays and increases in cost were not caused by the owner. The owner has only to prove that some of the increase in cost is due to other causes, or that the costs would have occurred in the absence of the breaches, and those costs will be deducted from the claim. Owners can almost always claim that contractor error and excusable, but non-compensable, causes contributed to the delay.

Yet the total cost approach will not be rejected every time. Some cases are complicated and do not lend themselves to isolating all of the causes and tying each one to an increase in cost. Where such impracticability exists, and where the claimant was not substantially responsible for the increase in cost, this approach has a chance of success.

The **measured mile approach** involves using a portion of the contract unaffected by delays as a yardstick to compare productivity for the rest of the contract. The contractor is thus able to demonstrate what could have been achieved in the absence of the owner's breaches. But the contractor must still prove the breaches and tie them to the loss in productivity. Moreover, the owner can still try to prove that other causes were at work. The measured mile approach can also be used defensively by the owner, if there is a portion of the contract for which there was no breach, and for which the contractor provided poor productivity. The measured mile approach has been used with success in Canadian courts.[10]

---

[10] For example, see *W.A. Stephenson Construction (Western) Ltd. v. Metro Canada Ltd.* (1975), 27 C.L.R. 113 (B.C.S.C.).

CHAPTER QUESTIONS

1. The network diagram for construction of an office tower shows two critical paths: one involving completion of the mechanical room, and the other the completion of the elevators. During the project, the elevator subcontractor causes a delay of two weeks. Meanwhile, the mechanical subcontractor causes a concurrent delay of one week. The owner sues the general contractor for the two-week delay, and the general contractor names the two subcontractors as third parties. The mechanical subcontractor's defence is that its delay did not cause any loss because the delay to the elevators took the mechanical room off the critical path. Should that argument succeed? Explain.

2. An owner and contractor enter into a contract that states: "The contractor is entitled to an extension but no compensation for delays caused by the owner." Later, the owner suspends work on the site for a period of two months because the delay is advantageous to the owner in its dealings with prospective tenants. The contractor claims compensation. Should the clause be enforceable? Explain.

3. What is the measured mile approach for?
    a) To measure head office overhead
    b) To calculate filed office overhead
    c) To calculate loss of productivity

4. Which statement is true of constructive acceleration?
    a) It only applies to construction projects.
    b) It occurs when the owner acknowledges a delay and directs the contractor to make up lost time.
    c) It occurs when the owner does not acknowledge any delay but directs the contractor to meet the original completion date.

5. In the *Doyle Construction* case (see p. 221), why did the Court refuse to allow the contractor's claims for impact costs?

6. Which statement is true of a delay to a non-critical path activity?
    a) It never results in additional compensation.
    b) It never results in a schedule extension.
    c) It typically results in a schedule extension.
    d) It sometimes results in additional compensation.

7. Explain why neither party is usually able to recover damages for a concurrent delay.

8. Which of the following statements is true in a standard CCDC 2 contract?
    a) A compensable delay is not excusable.
    b) A contractor-caused delay is compensable.
    c) An excusable delay is always compensable.
    d) A contractor-caused delay is not excusable.

9. Under a CCDC 2 contract, what are *force majeure* delays?
   a) Contractor-caused delays
   b) Compensable delays
   c) Excusable delays

10. What term describes activities that have no float?

ANSWERS

1. The argument would likely not succeed. Both parties contributed to the delay. The delays were concurrent. Both subcontractors could offer the same argument; and if one were to succeed, then both would succeed, and neither party would be liable, which is obviously an incorrect result.

2. This is an example of an intentional delay, caused by an owner. Keep in mind the principle that one party should not be put at the mercy of another, as well as the principle that a party should not intentionally hinder the progress of the other party. A court would likely interpret the clause so that it would not apply to deliberate delays.

3. C

4. C

5. The contractor did not give notice of the claim in advance. The owner believed that when it approved each of the extras, the quoted costs were all-inclusive.

6. D

7. In a concurrent delay, each party can successfully argue lack of causation. The delay would have taken place "but for" their own fault; therefore, they did not cause it. This situation is *not* identical to question 1, wherein two subcontractors have each caused delay to another party who is claiming against both. The critical difference between question 1 and this question is that in question 1, the delay was caused by two parties who are not asserting a delay claim of their own; whereas in a concurrent delay, each party would be asserting its own claim against each other.

8. D

9. C

10. They are critical path activities.

CHAPTER 20

LABOUR LAW

Overview

Labour law generally refers to the law that governs union-management relationships as well as employee-union relationships. In contrast, *employment law* refers to the law that governs employee-employer relationships where there is no union representation. In Canada, labour law is mostly governed by provincial legislation.[1] For this reason, laws vary from one jurisdiction to another. However, the same basic principles underlie all provincial labour legislation.

    Labour law is not a subset of the law relating to architecture, engineering, and geoscience, but rather, a well-established area of law in its own right. However, architects, engineers, geoscientists, and other professionals need a basic understanding of labour law principles so that they can avoid common mistakes and know when to seek legal advice.

    One basic purpose of labour legislation is to promote industrial peace. Another purpose is to allow employees to organize and bargain together as a group. These purposes are stated explicitly in the labour legislation of most provinces. Chief Justice Hughes of the U.S. Supreme Court authored a particularly succinct explanation, which is equally applicable in Canada:

> Employees have as clear a right to organize and select their representatives for lawful purposes as the respondent has to organize its business and select its own officers and agents. Discrimination and coercion to prevent free exercise of the right of employees to self-organization and representation is a proper subject for condemnation by competent legislative authority. Long ago we stated the reason for labour organizations. We said that they were organized out of the necessities of the situation; that a single employee was helpless in dealing with an employer; that he was dependent ordinarily on his daily wage for the maintenance of

---

[1] Federal labour legislation covers a relatively small percentage of the workforce. That sector is governed by the *Canada Labour Code*, R.S.C. 1985, c. L-2.

himself and his family; that if the employer refused to pay him the wages that he thought fair, he was nevertheless unable to leave the employ and resist arbitrary and unfair treatment; that union was essential to give labourers opportunity to deal on an equality with their employer.[2]

Unions developed in response to oppressive work conditions. In the decades before workers began to organize, extreme inequality of bargaining power marked the employer-employee relationship. That inequality is evident today in countries that either prohibit worker unions or make organization of workers difficult. Such inequality is characterized by subsistence-level wages and unsafe and inhumane working conditions. While such conditions still exist in Canada, they are not as prevalent or extreme as in other countries or as in the nineteenth and early twentieth centuries. Inequality in Canada persists during periods of high unemployment, because an employer can usually survive without an individual worker longer than the worker can survive without wages. But by bargaining as a group and threatening to withhold labour as a group, a union can bring real pressure on the employer during negotiations.

The right to form a union is now recognized as a protected right under s. 2(d) of the *Charter*,[3] which guarantees freedom of association. The application of the *Charter* to collective bargaining was first recognized by the Supreme Court of Canada in 2007.[4] However, freedom of association does not necessarily extend to all union activities. As explained by the Court in that case,

> ... s. 2(d) of the *Charter* protects the capacity of members of labour unions to engage, in association, in collective bargaining on fundamental workplace issues. This protection does not cover all aspects of "collective bargaining," as that term is understood in the statutory labour relations regimes that are in place across the country. Nor does it ensure a particular outcome in a labour dispute, or guarantee access to any particular statutory regime. What is protected is simply the right of employees to associate in a process of collective action to achieve workplace goals.[5]

Labour conflict has obliged governments to develop laws to deal with various labour issues, including the appropriateness of the employer's conduct in resisting a union organizing campaign, limitations on the right to picket, and ways of resolving jurisdictional disputes. Some of these issues are of particular relevance to architects, engineers, and geoscientists.

Labour legislation designates unions as the sole agent for the employees they represent for the purpose of negotiating a collective agreement. Thus, the rules of agency[6] all apply. As agent, a union is authorized to enter into an employment contract with the employer (the third party) on behalf of its principals (the employees). However, unlike a typical agency relationship, an employer cannot negotiate directly with a represented employee once union representation is established.

Labour disputes can also be very costly, not only in terms of lost production or lost wages, but also in terms of the future of the company. For example, for a company, union certification can significantly reduce competitiveness. But at the same time, laws restrict and control how much management can interfere with the unionization process. For this reason, companies should obtain legal advice from an experienced labour lawyer before taking any action regarding labour conflict or unionization, such as meeting with employees to discuss the situation. Even minor actions can be interpreted as interference.

---

[2] *NLRB v. Jones and Laughlin Steel Corp.*, 301 U.S. 1. (1937).

[3] See Chapter 1, s. 1.1.

[4] *Health Services & Support-Facilities Subsector Bargaining Assn. v. British Columbia*, 2007 SCC 7.

[5] See previous note at para. 19.

[6] See Chapter 9, pp. 68–70.

20.1 Establishing Union Representation

Employees of a company may decide that they need a union to represent them and may approach a union agent to assist them. But more frequently, a union decides that employees at a particular company could benefit from representation; and that union initiates an organizing drive. In either case, once the union has obtained the support of a majority of employees, it approaches the employer and demands that the employer recognize the union as the bargaining agent for the employees. The employer can either agree or demand an election. An election is a democratic vote of employees to determine whether a particular union should be certified as their representative. In some provinces, a union may be certified without a vote, as long as a large enough majority of employees (for example, 55 percent) has joined the union.

(a) Definition of "Employee"

Although the law distinguishes between employees and independent contractors, this distinction is not always obvious. Most provincial case law has defined tests that consider a number of factors, including the level of control that the employee has over the methods of work. An independent contractor usually determines the means and methods used to achieve results. In contrast, an employee does not usually determine the means and methods. The test under labour law is similar to tests used to distinguish between an employee and an independent contractor under tax laws.[7]

Labour legislation often creates a third category, known as "dependent contractor." The British Columbia Labour Relations Board defines a dependent contractor as follows:

> A dependent contractor is one who "performs work or services for another person". He is in the labour market, not the product market. He does this work "for compensation or reward" not gratuitously. He may furnish "his own tools, vehicles, equipment . . ." (including the dump trucks used here). He need not be "employed by a contract of employment." However, he does have to be in "a position of economic dependence" which leads the Board to judge that his relationship resembles more closely that of "an employee than that of an independent contractor.[8]

A second category of employee that causes some difficulty under labour law is the *supervisory employee*. Although a supervisor is an employee in the usual sense of the word, his or her duty to the employer is more likely to come into conflict with the union's interest. For that reason, some labour relations statutes exclude supervisors from the definition of employee. For example, in Ontario, managerial employees and others are excluded under section 3 of its legislation:

> Subject to section 97, for the purposes of this Act, no person shall be deemed to be an employee,
>
> (a) who is a member of the architectural, dental, land surveying, legal or medical profession entitled to practise in Ontario and employed in a professional capacity; or
>
> (b) who, in the opinion of the Board, exercises managerial functions or is employed in a confidential capacity in matters relating to labour relations.[9]

Note that the first part of this definition excludes architects from the definition of employee and thereby excludes them from inclusion in a bargaining unit. But this exclusion is not typical of the definitions in other provinces. An engineer or other professional in some provinces may find him- or herself included in a union bargaining unit and bound by the collective agreement, as well as by union membership and the rules that go along with such membership. In such cases, the professional should remember that professional responsibilities, such as in the codes of ethics in

---

[7] See Chapter 21, pp. 242–243.

[8] *West Fraser Mills Ltd. and IWA-Canada, Local 1-245*, B.C.L.R.B., No. B97/93, Case No. 13822.

[9] *Labour Relations Act 1995*, S.O. 1995, c. 1, Sch. A, s. 3.

force in each province, prevail over union rules. For example, in the event of a strike, an engineer may decide that refusing to come to work will endanger public safety. In such a situation, the duty to public safety must prevail. For more detailed discussion on conflicts of interest, see Chapter 3.

Organization of employees in the construction industry may be more difficult than in other industries because employees are often hired on a project basis rather than on a permanent basis. When a project is complete, employment is often terminated. Thus, a union is unlikely to have obtained a majority of support before one project is complete.

(b) Employer Resistance to Union Representation

Many employers faced with the prospect of union representation vigorously oppose it. In rare cases, such opposition may take the form of threats, coercion, increased pay and benefits, transfers, firings, meetings, and posting of notices expressing the employer's view. Most of these methods are considered unfair labour practices and may result in either invalidation of an election or, in extreme cases, an order requiring the employer to bargain with the union.

As an example of legislation defining unfair labour practices, consider the following prohibition in the Manitoba statute:

> Every employer, and every person acting on behalf of an employer who, at a time when a union is seeking to be certified as the bargaining agent of a unit of employees of the employer or is attempting to enlist members from among employees of the employer, discharges or refuses to continue to employ, or refuses to re-employ, or lays off, or transfers, or suspends, or alters the status of, an employee who is a member of the union or who has applied for membership in the union, unless he satisfies the board that the decision to discharge, to refuse to continue to employ, to refuse to re-employ, to lay off, to transfer, to suspend or to alter the status of, the employee was not in any way affected by the employee's membership in the union or application for membership in the union, as the case may be, commits an unfair labour practice.[10]

However, this fundamental principle conflicts with the constitutional right to freedom of speech in situations where the employer expresses animosity toward the union. Courts have struggled to determine which principle should prevail, and the answer may depend upon to what degree the employees have been coerced by the owner's communication.

For example, in 2005, in Jonquière, Quebec, just when the United Food and Trade Workers' Union (TUAC) was trying to organize workers at the local Wal-Mart, the corporation decided to close the store. The corporation denied that its reason for closing operations had anything to do with the union drive, stating that the operation was being closed because it was unprofitable. The union believed that the closure was in fact related to the organizing drive. Regardless of who was correct, the employer was successful in closing the store; and the closure had a chilling effect on other unions in other jurisdictions that were trying to unionize Wal-Mart stores.

20.2 Trade Unions and Jurisdictional Disputes

In many industrial settings, one union represents all eligible employees. This arrangement is typical of manufacturing plants, mines, refineries, and offices. However, in the construction industry, each trade is usually represented by a different union.[11] This multi-union situation presents a unique set of problems.

---

[10] *Labour Relations Act*, C.C.S.M, c. L10, s. 9.

[11] There have been some attempts to establish a common union for all trades in some jurisdictions, but the individual trade union situation still prevails.

Over many years, each trade (formerly known as a *craft*) has carved out a portion of the work it considers to be its own. For example, a union electrician does not allow a union carpenter to lay conduit or pull wire. To a degree, specialization of trades is necessary to allow workers to develop greater skill in their trades. But tremendous coordination between the various trades is required to avoid inefficiencies associated with many trades working in one area.

Moreover, because of the interdependence between trades, it is possible for one trade to shut down an entire construction project. For example, if the electricians or operating engineers refuse to work, a project could easily come to a halt. A very small minority of workers could cause many others to lose work. In order to avoid such a scenario, procedures have been put in place to resolve disputes between trades.

A **jurisdictional dispute** is one in which two or more unions each claim that the work in question should properly be performed by its members. For example, the carpenters and plumbers might both claim as their own work the installation of block-outs in formwork for piping runs. Rather than shutting down the project, the two trades would attempt to settle the dispute and, if necessary, submit it to the labour board for decision. It may be of little interest to the employer which trade performs the work, as long as there is no work stoppage. In making its determination, the labour board considers the skills and work involved, company and industry practice, and agreements between unions and between employers and unions. Labour legislation typically prohibits work stoppages on account of jurisdictional disputes.

20.3 Union Security and Right-to-Work

Several terms and phrases have been used to describe union security arrangements, including closed shop, union shop, open shop, and Rand formula. A **closed shop** is one in which membership in the union is a precondition of hiring. A **union shop** is one in which employees must become union members after they are hired. **Open shop** means that there is no union membership requirement; and in some places, it is synonymous with *non-union shop*.

The **Rand formula** is a term in a collective agreement that requires employees to pay union dues but not necessarily join the union. The rationale behind the Rand formula is that all employees, whether union members or not, benefit from the presence of a union because the union bargaining power raises the level of wages and benefits. Thus, an individual might see no personal value in joining the union, because that individual will realize the benefit without joining. However, those who contribute to the union do not wish to subsidize those who don't. As a compromise, non-members are not forced to join, but they are not given a free ride. The Supreme Court of Canada held the Rand formula permissible under the *Charter of Rights*.[12]

Closed shop provisions still exist in Canada, although in the United States they have been found to be an infringement of an individual's right to freedom of association. Freedom of association is one of the rights guaranteed by the *Canadian Charter*, but Canadian courts have not interpreted that right to preclude closed shops.

20.4 Work Stoppages

Strikes and lockouts are the ultimate weapons in a labour dispute. But they are costly to both employer and employee. Lost production may be recovered through costly use of overtime, but lost market share may never be recovered. Lost wages for the employee are usually unrecoverable.

---

[12] *Lavigne v. O.P.S.E.U.*, [1991] 2 S.C.R. 211.

Most lockouts and strikes occur when an impasse in negotiations occurs; and in such circumstances, a lockout or strike is a legitimate tool. But as long as a collective agreement is in force, strikes and lockouts are illegal.

In some provinces, the law prevents employers from hiring replacement workers during a work stoppage. For example, the British Columbia *Labour Relations Code* contains the following restriction:

> During a lockout or strike authorized by this Code an employer shall not use the services of a person, whether paid or not, who is hired or engaged after the earlier of the date on which the notice to commence collective bargaining is given and the date on which bargaining begins, . . . to perform the work of an employee in the bargaining unit that is on strike or locked out.[13]

However, even in the provinces without such a law, employers face practical impediments to continuing operations during a work stoppage, such as persuading replacement workers or delivery companies to cross a hostile picket line. Workers face no legal impediment to obtaining alternate employment during a work stoppage.

Most employers expect picketing during work stoppages. Picketing is protected free speech as long as it is done peacefully and without intimidation. However, just as workers have the right to picket, the public has the right to either respect or ignore a picket line. If the number of pickets is so great as to prevent access to a site, a court injunction may be obtained to limit the number or location of pickets.

Picketing of construction sites creates special problems. Other employees may be part of the project, but have no relationship with the union that is picketing. They have the right to work if they so choose. Thus, construction strikes involve two competing rights: the right of the union and its members to picket, and the right of disinterested parties not to be interfered with.

20.5 Secondary Activity

In order to pressure an employer to bow to their demands, unions sometimes bring pressure on neutral parties doing business with that employer. For example, a union may picket a site with the intent to cause others to cease doing business with that employer. Courts have struggled to balance the competing interests of the union and disinterested parties.

20.6 Successor Employers and Common Employers

Where an individual or corporation purchases a business or the substantial assets of a business, the purchaser may be found to be a *successor* for the purposes of union bargaining obligations. Ontario has enacted the following provision governing related employers:

> Where, in the opinion of the Board, associated or related activities or businesses are carried on, whether or not simultaneously, by or through more than one corporation, individual, firm, syndicate or association or any combination thereof, under common control or direction, the Board may, upon the application of any person, trade union or council of trade unions concerned, treat the corporations, individuals, firms, syndicates or associations or any combination thereof as constituting one employer for the purposes of this Act and grant such relief, by way of declaration or otherwise, as it may deem appropriate.[14]

---

[13] R.S.B.C. 1996, c. 244, s. 68.

[14] Ontario *Labour Relations Act, 1995*, S.O. 1995, c. 1, Sch. A, s. 1.

Case Study 20.1

Vulcan Packaging Ltd., [1997] O.L.R.D. No. 2662

Facts

Vulcan Packaging Ltd. was declared bankrupt. Under the bankruptcy, the Court ordered and declared that "all Vulcan's employees have been terminated at the time of Vulcan's assignment in bankruptcy and that the Trustee [the person responsible for administering the winding up of the bankrupt party's assets] shall not be nor be deemed to be a successor employer of Vulcan under the *Labour Relations Act, 1995* (Ontario)." Immediately following the bankruptcy order, the Trustee sold the assets of Vulcan Packaging Ltd. to a purchaser (called the "Purchasing Group"). The Purchasing Group sold the assets of Vulcan Packaging to two newly created corporate entities, who continued to operate the business, without interruption, using the same employees.

Questions

1. Should the new companies should be bound by the collective agreements that had been signed by Vulcan Packaging Ltd. prior to its bankruptcy?

Analysis

1. Had there been no intervening bankruptcy, a mere sale by one company to another of its assets could not eliminate the obligation of the new owner to be bound by the union agreement. But in considering this problem, the Ontario Labour Relations Board made the following decision:

    > Here, given the fact that many of the key people are the same for Packaging and the Purchasing Group, (Containers and Ontario), both as operating officers of the three companies and as directors, given that the 100% shareholder of Packaging (Telfer) is a key member of the Purchasing Group and given that the business has not changed in any material respect, it is apparent that these corporations have acted together as one employer. While the legal ownership of Packaging has changed, in the sense that the majority shareholder has changed from Telfer to Bird, the key people structured the sale through a bankruptcy, and the same group of people still operate what in essence is the same business. Although Packaging was the legal employer prior to the bankruptcy and sale, the Bank was effectively dictating how it operated, with the meaningful input of the Purchasing Group. It was the Purchasing Group that was negotiating concessions with the unions, not Packaging, at a time when the Purchasing Group was not the employer. For practical purposes, it was the Purchasing Group (as prompted by the Bank) that would have initiated the termination of the Petrolia employees by Packaging on May 14, 1997. The three named responding parties are intertwined in their labour relations activities and in the manner in which they conducted their businesses. These businesses were one and the same, and the business of Packaging and now that of Containers and Ontario, were under the common control and direction of Bird, Telfer and the rest of the Purchasing Group, for some considerable time prior to the assignment in bankruptcy and sale of May 15, 1997, and after the sale was completed.

The Board therefore finds that Packaging, Containers, and Ontario are a single employer within the meaning of section 1(4) of the Act.[15]

Thus, the new owners were bound by the union agreement. The rationale behind this decision, and in fact the rationale behind the legislation, was to

> prevent the subversion of bargaining rights by transactions which are designed to get rid of the union. We have encountered situations where there are transactions between various corporate entities which are in effect "paper transactions," and are a form of corporate charade engaged in for the purpose of eliminating the trade union.[16]

As Case Study 20.1 shows, the implications for a company considering the purchase of a business or its assets are obvious: there is a risk that along with the business or its assets will be attached an obligation to bargain with a union. Thus, an employer cannot rid itself of its union obligations by engaging in corporate sleight-of-hand: a mere change in ownership designed to remove a union will not succeed.

Some employers have taken more drastic measures—closing the operation down or moving it to a different location. But a labour board can rule that companies with common ownership and/or control constitute a single employer and therefore find that the related employer is either liable for unfair labour practices or bound by the collective agreement. Such rulings occur even if the related employer has ceased operating.[17]

20.7 Enforcement of Collective Agreements

Disputes over the interpretation of collective agreements are not uncommon. Most collective agreements contain provisions that require such disputes to be resolved through arbitration. This use of arbitration is not to be confused with the use of arbitration to arrive at a collective agreement, which is known as *interest arbitration. Grievance arbitration* is concerned with determining rights under an existing collective agreement.

A **grievance** is a complaint made either by the union on behalf of an employee or by the employer, stating that the other party has breached the collective agreement. Because collective agreements may run for several years, the parties usually anticipate that grievances will occur during the term of the agreement and insert a provision requiring that all grievances be finally resolved through arbitration. An independent party or tribunal acts as judge and jury. The arbitrator may be named in the collective agreement or may be appointed after a grievance has been filed.

---

[15] *Vulcan Packaging Ltd.,* [1997] O.L.R.D. No. 2662 at paras. 75–76.

[16] See previous note at para. 47.

[17] In some provinces, as evidenced by the section of the Ontario *Labour Relations Act* reproduced above, companies with common direction (common officers, directors, employees, or headquarters) *or* control (shareholders) may be found to be sufficiently related so as to bind both to the collective agreement. This could have serious implications for anyone who wants to operate both a union and a non-union company simultaneously. The same issues arise in the United States: this argument was advanced indirectly in the *Garland Coal & Mining Co.* case, 121 L.R.R.M. 2029 (U.S.C.A. 8th Cir. 1985), where the union asked the NLRB to find that one company was the "alter ego" of another.

20.8 Layoffs and Seniority

A **layoff** is the temporary suspension of employment. Most collective agreements contain a clause that allows the employer to increase or decrease the size of the workforce, depending on business needs. For example, a pulp and paper mill may need to increase production or inventory levels and so hires additional workers. Alternatively, if the market for paper diminishes, it may decide to reduce the size of its workforce.

The standard arrangement in most collective agreements provides that employees may be laid off, which means that their employment is suspended. This is not the same as termination, which is a permanent end of the employment relationship. Typically, employers must lay off employees in order of seniority, so that workers with the most years of employment with the employer are the last to be laid off. When employers decide to increase the size of their workforce, they must hire back the employees who have been laid off in order of seniority. As a result, employers are somewhat restricted in their choice of employees: they cannot lay off a senior employee instead of a more junior one.

But the concept of layoff has no place in a common law employment relationship (see Chapter 21). In the absence of a union agreement, the common law allows employers to terminate the employment of any employee if there is just cause, or, in the absence of just cause, if they provide reasonable notice of termination. The right to terminate for just cause still exists in most collective agreements; but the right to terminate without just cause does not. In many industrial settings, collective agreements have been written so that employees cannot be fired except in extreme cases. Typically, an employer is required to give multiple warnings before terminating an employee, unless the conduct under complaint is of such an extreme nature that immediate termination is warranted.

When an employer disciplines an employee by giving a warning or a suspension, the union may file a grievance. An arbitrator hearing the grievance often has a wide choice of remedies, including reinstating a terminated employee, ordering the employer to pay suspended wages, or changing working conditions. In contrast, the power of a court to order these remedies is more limited. For example, ordering the reinstatement of a terminated employee is not permissible under common law.

CHAPTER QUESTIONS

1. Which of the following is *not* an unfair labour practice?
   a) A threat to fire someone who is organizing the employees
   b) An offer to promote someone who is organizing the employees
   c) A threat to close the business if the union is certified
   d) Closing the business

2. Why is it necessary for several trade unions in the construction industry to bargain at the same time?

3. What is the difference between successor employers and common employers?

4. A group of dump truck drivers have decided to organize in order to bargain for increased compensation. They each own their own truck and have been hired by the same excavation contractor. They are currently being paid $45 per hour, which is intended to cover

their truck maintenance and operating costs as well as their wages or profit. They receive direction from the contractor's dispatcher each morning and at each project site by the contractor's superintendents. How would you classify these people? Are they employees or independent contractors?

5. Which statement is true of an employer under a typical collective agreement?
   a) The employer cannot terminate an employee under any circumstances.
   b) The employer can only terminate an employee if he gives reasonable notice.
   c) The employer cannot lay off an employee because of a downturn in business.
   d) None of the above

6. Which of the following does not require all employees to pay union dues?
   a) Open shop
   b) Closed shop
   c) Rand formula
   d) Union shop

7. Assume that there is a collective agreement in place with one month until expiry. Is the employer allowed to lock out its employees in order to put more pressure on them during the negotiations?

8. Under a typical collective agreement, if a union employee is late for work, can the employer lay off or terminate that employee?

9. Which statement is true of crossing a picket line?
   a) It is illegal for a member of a union.
   b) It is only legal if the picket line itself is unlawful.
   c) It is legally permissible for a member of the public.

10. What is a jurisdictional dispute?
   a) A dispute over which union should represent the employees
   b) A dispute over which union's members should perform work
   c) A dispute over whether a dispute should be heard by a court or by an arbitrator
   d) None of the above

ANSWERS

1. D

2. If each union were to bargain separately, then each union would have the ability to shut down all construction sites. It is not in the interests of either tradespeople or employers to have a small number of employees going on strike one after the other. For example, on a building construction site, there may be only five or six elevator installation workers, but the project cannot proceed without the elevators. If those workers were to shut down a site (or the entire industry) until they obtained the wages they were looking for, the next union whose agreement is up for renegotiation (*e.g.,* the electricians) would do the same. Under such circumstances, construction projects would shut down frequently.

3. A successor employer is one who buys a business or a substantial part of a business. A common employer is one who carries on business at the same time as another business, with common direction or control over both businesses. In some jurisdictions, both common direction and control may be necessary to establish common employer status.

4. In most provinces, case law distinguishes employees from independent contractors based on several factors, including the level of control that the employee has over the methods of work. An independent contractor usually determines the means and methods to be used to achieve a result. In contrast, an employee does not usually determine the means and methods. If these were the only two options, the drivers would most likely be considered employees. However, if the category of "dependent contractor" is available, that is the one that most closely fits.

5. D

6. A

7. No. As long as the collective agreement is in force, strikes and lockouts are illegal.

8. No. A warning would be required before serious disciplinary action can be taken. Moreover, after repeated offences, employers may have other options available, including termination for cause. But layoff is not usually one of those options.

9. C

10. B

CHAPTER 21

EMPLOYMENT LAW

Overview

Canadian employment law is governed by the basic principles of contract law outlined in Chapter 6. Requirements and principles of contract law, including consideration, offer, and acceptance, legal purpose, and mitigation all apply to employment contracts. However, some aspects of employment contracts are unusual or specialized.

Because the common law employment relationship is a private contractual one, it is governed by provincial law. While the basic principles are quite consistent from one province to another, there may be subtle differences either in the governing legislation, such as the employment standards statute governing minimum wages and working conditions, or in the case law.

The relationship between employer and employee is governed by a contract of employment, which may be either written or oral, or partly written and partly oral. In rare instances where there are no express terms, the contract comprises only implied terms. Salary and length of vacation are usually express terms, while other terms, such as length of notice for termination, are often implied.

21.1 Implied Terms in the Common Law Employment Contract

Certain obligations are implied into every employment contract unless there are express terms to the contrary. One of these terms is the obligation to mitigate where a breach has occurred. The most frequent application of this principle is situations where the employer has improperly terminated the employment relationship. Other terms that are implied into employment contracts include the following:

- The employee's obligation of fidelity (loyalty)
- The employee's duty of competence
- The employer's duty to give adequate notice of termination or to pay severance in an amount equal to the salary and benefits that would accrue during the notice period.

(a) Duty of Fidelity

Employees are considered agents of their employer. One of the attributes of an agency relationship is that the agent owes a duty of fidelity, or loyalty, to the employer. One way in which this obligation is applied is that the employee is required not to disclose or make use of confidential information.

(i) **TRADE SECRETS AND CONFIDENTIAL INFORMATION** **Confidential information** in the employment context means data or industrial secrets or information obtained or learned through the employment relationship. Confidential information generally includes trade secrets. A **trade secret** is information usually contained in a document, product, formula, or patent that has at least some of the following attributes:

- It is or may be used in a particular industry.
- It is not widely known in that industry.
- It has value because it is not widely known.
- The company tries to maintain it as secret through reasonable efforts.

Note that trade secrets do not have to be mysteries and secret formulas, such as how to get the creamy filling into the chocolate bar; rather, they can be day-to-day operational information, such as a corporation's system for purchasing goods or managing its employees.

Trade secrets allow companies to compete in the marketplace. For example, most companies have a set profit margin that they apply to their cost of goods or services; and if that profit percentage were disclosed to competitors, that company would find it difficult to compete successfully on future work. Another example is client lists. Client lists are often considered confidential information. But if a client list could be generated through public documents, such as a telephone book, it would not likely be found to be confidential. On the other hand, if it would be difficult to create the list, a court might find it to be confidential. As a general rule, if disclosure of the information is valuable to a competitor, it will usually be impressed with a duty of confidentiality.

But despite the seriousness of breaches involving confidential information, precisely what circumstances give rise to a duty of confidentiality are not well defined. In the 1989 *Lac Minerals* case, the Court ruled that information is confidential based on the following criteria:

> It was transmitted with the mutual understanding that the parties were working towards a joint venture or some other business arrangement and, in my opinion, was communicated in circumstances giving rise to an obligation of confidence.[1]

It is not always clear what information is to be considered as confidential. The case of *Monarch Messenger Services Ltd. v. Houlding*[2] contains a good discussion of the relevant principles and the limits of what should be considered confidential information:

> …an employee will not be permitted, following termination of his employment, to use for his own benefit confidential information acquired in the course of his employment or information which is "special or peculiar" to his ex-employer. On the other hand, it is equally clear that following termination of the relationship, an employee is free to use for his own benefit or for the benefit of third parties, any skill and general knowledge which he acquires during the course of his employment…

For this reason, employers take steps to protect their confidential information. Those who highly value their trade secrets may have employees and third parties sign a *confidentiality agreement*. However,

---

[1] *Lac Minerals Ltd. v. International Corona Resources Ltd.,* [1989] 2 S.C.R. 574.

[2] 5 C.C.E.L. 219, 56 A.R. 147, 2 C.P.R. (3d) 235 (1984 Alta. Q.B.)

a written agreement is not needed for an employee to be bound by an obligation of confidentiality. The obligation of confidentiality begins on the first day of employment and extends beyond termination. An employee cannot quit his or her job and then use confidential information for any reason, including competing against his or her former employer. Additionally, senior or key employees are usually restricted from competing against the former employer for a period of time after leaving the company through restrictive covenants (discussed below); but less senior employees are not restricted, except that they may not use confidential information obtained through the former employer.

Employees cannot take with them or use in any form confidential information of a previous employer. But recreating the confidential information from memory is still actionable. However, there is a fine line between confidential information and experience. An employee is entitled to take his or her experience to a new position, as long as he or she does not disclose confidential information directly or indirectly to the new employer.

For example, assume that Mary is fired from her job with PQR Resources, a mining exploration firm. Mary has been working on a highly secretive mining plan in Nunavut. Mary can take her new skills regarding the latest geological mapping software to Pirate Co., but she cannot disclose to Pirate Co. that it should search for diamonds in Nunavut. Mary's obligation to PQR is to forget the information that she learned at PQR.

If a former employee breaches the duty of confidentiality and uses confidential information to successfully compete with the former employer or to otherwise profit from the information, a court may find that the profits earned through use of the confidential information really belong to the former employer on the basis of what is known as a *constructive trust*. Thus, the court may require the former employee to disgorge those profits.

(ii)   RESTRICTIVE COVENANTS A departing employee can cause serious damage to an employer by taking his or her expertise to a competitor. For this reason, companies with specialized skill often require new employees to sign an agreement with a restrictive covenant. A **restrictive covenant** forbids the employee from working for a competitor, usually for a specified period of time after termination of employment, in a specified geographic area, and in a specified area of business.

A restrictive covenant can be particularly onerous for the departing employee. It is difficult to ask someone to restrict his or her ability to earn a living. Therefore, courts that have been asked to interpret these restrictive covenants have often found that they are unenforceable. The test for determining whether such an agreement is enforceable is the reasonableness of the restrictions. For example, a requirement that an engineer not work for anyone in the same field, in the same province, for a period of 15 years, may be found too restrictive. Conversely, an agreement that restricts an engineer for a period of one year, within the same city, and within a narrow area of practice, is more likely to be enforceable. These cases depend on their specific facts, and there is no "bright line" test for determining what is reasonable. If an engineer was hired to develop proprietary software and spends 10 years working on it, restricting that engineer's competition is more reasonable than restricting that for an engineer brought in to work as a project manager for a construction company for a short time.

Clarity is as important as reasonableness in determining whether a restrictive covenant is enforceable. In a recent decision of the Supreme Court of Canada,[3] it was held that because the geographic restriction contained in a restrictive covenant was ambiguous, the covenant itself was unenforceable. In that case, a Toronto lawyer drafted the document using the term "Metropolitan City of Vancouver," an ambiguous term with no legal definition. The parties intended it to cover more than just the city limits of Vancouver, such as suburbs, but did not define those limits. While the ambiguity in this case dealt with geographic limits, it is possible that ambiguity in any other aspect of the restrictive covenant (*e.g.*, scope of practice) could have similar consequences of unenforceability.

---

[3] *Shafron v. KRG Insurance Brokers (Western) Inc.*, 2009 SCC 6.

(b) Duty of Competence

An employee is expected to be competent in the area of work for which he or she has been hired. Theoretically, if an employee makes a mistake, as everyone does on occasion, and that mistake causes the employer to suffer a loss, the employer may have a right to recover the amount of the loss from the employee.

This rule may seem counter-intuitive because employers rarely sue their employees or former employees for negligence. If an employee makes a mistake, and that mistake causes injury of loss to a third party, the third party often has the right to recover the loss from the employer. The liability of a party, such as an employer, for the acts or omissions of another party, such as an employee, is known as **vicarious liability**.[4] However, in practice, lawsuits by employers against employees almost never occur, for two reasons: first, there is often insurance to cover the loss, and such insurance usually covers both the employer and employee; second, any employer who pursues an employee for negligence would find it difficult or impossible to retain or find employees in the future. Claims by employers against former employees do occur, but generally only in cases where a former employee sues for wrongful dismissal.

The duty of competence does have one practical application: it may give the employer the right in extreme circumstances to terminate the employment for cause. But termination of employment can be an expensive proposition for the employer, especially if it is done without just cause, and without adequate notice.

(c) Termination, Notice, and Severance

In a union/collective bargaining situation, there is a concept known as layoff.[5] Almost all collective agreements contain a clause allowing the employer to lay off employees, usually in order of reverse seniority, to suit the amount of work available, and the laid-off employees have the right to be called back, usually in order of seniority, when there is more work. In a common law employment relationship, there is no right of layoff. The corollary is that a terminated employee has no right to be called back.

But lack of a union does not mean that an employer can terminate an employment contract at any time, or for any reason. Termination falls into one of two categories. Either it is for cause, also known as just cause or sufficient cause, or it is without cause. **Just cause** occurs when the employee has committed a fundamental breach of the employment contract, justifying termination. The question of what constitutes just cause for termination has no simple answer. The legal landscape is constantly changing, and what was considered just cause a few years ago might not be considered sufficient today. It is beyond the scope of this text to describe in detail what is considered as just cause for termination. However, here are a few basic guidelines:

1. If a breach of the employment contract by the employee is serious enough to destroy the trust that must exist between employer and employee, a court may consider that to be sufficient cause.

2. Dishonesty and theft are usually considered just cause.

3. Minor breaches, such as isolated incidents of tardiness or absence, are not just cause. However, if the employer gives a warning each time this occurs, and enough warnings are given, eventually the cumulative effect will be considered to be just cause.

4. In some circumstances, gross incompetence or insubordination may constitute just cause.

5. Isolated incidents of intoxication will not usually justify dismissal, unless there have been a sufficient number of warnings.

6. Lack of business or a drop in the employer's business is not just cause.

---

[4] See Chapter 13, p. 151.

[5] See Chapter 20, p. 234.

Case Study 21.1

Richardson v. Davis Wire Industries Ltd. (1997), 33 B.C.L.R. (3d) 224 (S.C.)

Facts

A shift manager at an industrial plant was sleeping on night shift. The employer obtained evidence using a hidden camera that this employee had been sleeping on the job for at least four days in a row. The employer confronted the employee and asked how long this conduct had been going on. The employee lied and said that it had only happened on two nights. The employer fired the employee immediately. The employee sued for wrongful dismissal.

Questions

1. Is sleeping on the job just cause for dismissal?

2. Does lying about sleeping on job constitute just cause for dismissal?

Analysis

1. The Court decided that sleeping on the job would not constitute just cause. It is arguable that sleeping during working hours is like theft from the company, because the employer is paying for the employee to work. However, the Court rejected that argument. Only if such conduct had occurred frequently, and warnings had been given, would the cumulative effect have amounted to just cause.

2. The Court agreed with the employer about the second question. Lying about sleeping on the job was found to be just cause. Therefore, the plaintiff lost his case.

In contrast, where no just cause exists, the employer must provide reasonable notice of termination. If the employer fails to give notice, the employee can sue for wrongful dismissal and recover wages and benefits that would have accrued during the period of reasonable notice. The difficulty in such an analysis is to determine what constitutes reasonable notice.

Reasonable notice depends on a number of factors. These factors include the age of the employee, the length of service, the type of position held, and external factors, such as enticement from another job. For example, a junior engineer who has been working for one year as an estimator might be entitled to one month of notice or severance, whereas a senior executive who has been with the company for 20 years might be entitled to 18 to 24 months. Reasonable notice has no precise formula, but the courts have come up with a general range based on similar cases. For this reason, specific legal advice should be sought in most cases involving termination or constructive dismissal.

If an employee is fired without notice, he or she must look for suitable alternate employment. This is the employee's duty to mitigate. If the employee finds a suitable job with the same level of pay after looking for only one week, then the employer's liability is then capped at one week—whether or not the terminated employee accepts the new job. Thus, an employee's failure to mitigate reduces the employer's liability.

An employer has duties as well. Termination of an employee must be done in the least traumatic manner possible. Failure to do so may increase the liability of the employer for damages. This was the finding in the *Wallace* case:

The point at which the employment relationship ruptures is the time when the employee is most vulnerable and hence, most in need of protection. In recognition of this need, the law ought to encourage conduct that minimizes the damage and dislocation (both economic and personal) that result from dismissal. In *Machtinger, supra*, it was noted that the manner in which employment can be terminated is equally important to an individual's identity as the work itself (at p. 1002). By way of expanding upon this statement, I note that the loss of one's job is always a traumatic event. However, when termination is accompanied by acts of bad faith in the manner of discharge, the results can be especially devastating. In my opinion, to ensure that employees receive adequate protection, employers ought to be held to an obligation of good faith and fair dealing in the manner of dismissal, the breach of which will be compensated for by adding to the length of the notice period.[6]

In addition, insubordination is one of the grounds that courts have relied on as just cause for termination. However, insubordination is not always considered just cause. An employee may properly refuse to perform work that is demeaning, unsafe, or beyond the scope of the employment agreement. For example, a professional engineer can properly refuse the request to compromise a safety factor or to wash the boss's car.

Finally, employers should be aware of the requirement to act fairly in deciding whether an employee should be fired. Failure to act fairly may entitle the employee to relief. In the case of *Pelletier v. Canada*,[7] involving a senior bureaucrat terminated by the government, the Court stated, "there is no doubt that there must be a duty to act fairly when a person's employment is at stake . . . and in those cases . . . a high standard of justice is required."[8]

Employers sometimes try to get around employment law by using constructive dismissal. In a **constructive dismissal**, an employer unilaterally changes the employment contract to the detriment of an employee, and the employee refuses to accept the changes; thus, the employment contract is terminated. For example, an employer might say, "We can't afford to continue paying you $5000 per month, so we are reducing your salary to $4500 per month." The employee is entitled to refuse this change and treat it as a repudiation of the employment contract by the employer without notice. In other words, the employee can consider him- or herself as fired and sue for severance.

21.2 Independent Contractor or Employee

It is not always clear whether a person hired to perform services is an employee or an independent contractor. Proper characterization of the relationship may determine the rights and liabilities of the parties under a number of circumstances, including the following:

- taxation treatment
- vicarious liability
- builders' lien holdback
- insurance
- workers' compensation coverage
- contractual limitation clauses
- duty of loyalty.

---

[6] *Wallace v. United Grain Growers Ltd.*, [1997] 3 S.C.R. 701.

[7] [2005] F.C. 1545.

[8] See previous note at para. 45.

For example, an independent contractor is entitled to deduct expenses for income tax purposes, whereas an employee may not be entitled to claim those deductions. An employer must withhold taxes from the paycheque of an employee, but not from an independent contractor. Conversely, builders' lien holdback may have to be deducted from an independent contractor, but not from an employee. For these reasons, employers must structure the relationship in the manner that benefits both parties.

When courts are asked to determine the nature of the relationship where one person has been hired to perform services for another, they consider the following factors:

> The most that can be said is that control [over means and methods of work] will no doubt always have to be considered, although it can no longer be regarded as the sole determining factor; and that factors, which may be of importance, are such matters as whether the man performing the services provides his own equipment, whether he hires his own helpers, what degree of financial risk he takes, what degree of responsibility for investment and management he has, and whether and how far he has an opportunity of profiting from sound management in the performance of his task.[9]

As a rule of thumb, if a person is providing similar services for more than one client or employer, the relationship more likely to be characterized as an independent contract relationship, rather than as employer-employee relationship. The reverse is true in situations where the person is providing services to only one party.

21.3 Employment Standards Legislation

Employment standards legislation exists in every province. These statutes address the issues of wages, safety, and working conditions.

The best-known requirement of employment standards legislation is the minimum wage provision. Minimum wages vary from one province to another, but they exist in every province. However, they may not apply to all categories of employees. For example, the regulations in force under the British Columbia legislation exclude most professionals, including architects, engineers, and geoscientists. In other provinces, students of particular professions are excluded as well. In fact, employment standards legislation often excludes all professionals. For example, the regulations enacted under the New Brunswick and Ontario statutes exclude most professionals, including architects and engineers. Similar exemptions are found in most provinces.

Employment standards statutes also contain requirements for hours of work. For example, they may require that if employees work more than seven and a half hours in a day or 38 hours in a week, then they are entitled to be paid overtime at a specified rate. Minimum vacation pay is also mandated in such statutes.

These statutes also contain minimum standards for notice of termination. Generally, the notice provisions are much lower than those than required under common law. Whereas an employee might be entitled to six months' notice at common law, the applicable statute might only provide for three weeks' notice. This raises the question why anyone would choose to pursue a remedy under the statute where a greater remedy exists at common law. The answer is that to enforce one's rights at common law, one must sue in court. To enforce one's remedy under the statute, one may file a complaint with the employment standards branch of the provincial government, and the government will pursue the employer on behalf of the employee.

---

[9] *Lake v. Callison Outfitters Ltd.* (1991), 58 B.C.L.R. (2d) 99 (S.C.).

Other requirements of employment standards legislation include minimum holiday pay, maternity leave, and parental leave.

21.4 Human Rights

Every province has enacted human rights legislation. One common feature of such legislation is the specialized tribunal, separate from the courts, which deals with human rights complaints. Human rights legislation covers more than the employment relationship; it also covers such issues as discrimination in housing. However, dealing with discrimination issues other than those in an employment context is beyond the scope of this text.

Not every form of discrimination is prohibited. In order for discrimination to give rise to a remedy, it must be prohibited under the statute. Prohibited grounds generally include religion, ethnic origin, skin colour, age, disability, and gender, but not employability skills such as education. As a result, employers are allowed to discriminate based on level of education. If an employer hires an employee because that person has a higher level of education than other candidates, the unsuccessful candidates has no grounds for complaint under the human rights statute. Similarly, paying one person more than another based on education or experience is permissible as well, because neither education nor experience is a prohibited ground of discrimination.

One ground of discrimination that causes controversy is disability. An employer should arguably be entitled to discriminate based on disability only where that disability affects the ability of the individual to perform the work. Human rights tribunals and courts have generally found that if the disability can be dealt with through reasonable accommodation, such as modifying equipment or furniture to allow the employee to work properly, then employers must make the accommodation. However, some jobs require a high level of physical ability, such as a firefighter who must be able to carry a person out of a building. In situations where the physical ability is a genuine job requirement, then employers are allowed to discriminate based on disability.

Provincial human rights statutes also maintain that everyone has the right to work in an environment that is free from harassment, which includes physical, sexual, and psychological harassment. Sexual harassment claims are often brought before human rights tribunals, and these tribunals often take a wide view of what constitutes harassment. If an employer becomes aware that one employee is harassing another, then that employer must take immediate steps to end the harassment. The employer may be vicariously liable for the acts of the harassing employee. Even minor incidents of harassment should not be tolerated.

Note that all breaches of human rights and other employment laws can create liability for employers. Damages may include lost wages, reinstatement, and punitive damages; therefore, these costs can be very high.

21.5 Employees Facing Termination

Employees facing termination have access to several remedies, depending whether they choose to make their claim at common law or through employment standards legislation or human rights legislation. The advantages and disadvantages of these remedies are summarized in Table 21-1.

21.6 The Charter of Rights and Freedoms

In Canada's Constitution, the *Canadian Charter of Rights and Freedoms* codifies basic human rights, such as freedom of expression, freedom of religion, and freedom of association. But the Constitution applies to government action, not to the action of private individuals. For example,

TABLE 21-1 Comparison of Remedies for a Terminated Employee

	Advantages	**Disadvantages**
Common Law	• high levels of damages for breach	• expensive to enforce • no reinstatement remedy
Employment Standards Legislation	• inexpensive to enforce	• low levels of damages available • no reinstatement remedy
Human Rights Legislation	• inexpensive to enforce • reinstatement is an available remedy	• must prove discrimination based on a prohibited ground

if a provincial government enacted a law that infringed on a person's *Charter* rights, a court could declare the law invalid.[10] Cases involving the *Charter* are frequently decided by the Supreme Court of Canada, because they affect a large segment of society.

CHAPTER QUESTIONS

1. Which of the following is not a factor in determining the appropriate notice period for termination of employment?
   a) Age
   b) Skin colour
   c) Length of service

2. Which of the following is not a prohibited ground of discrimination under most provincial human rights statutes?
   a) Religion
   b) Skin colour
   c) Length of service

3. A geologist is hired by a mining exploration company. As a condition of employment, the employer demands that the geologist sign a contract of employment that contains a clause stipulating that the geologist, upon leaving the company, will not work for another mining company in the same province for a period of 20 years. Is such a clause enforceable?

4. If an architect is found guilty of driving while intoxicated, is the architect's employer justified in dismissing the architect for cause?
   a) Only if the architect was arrested during working hours
   b) No, under any circumstances
   c) Yes

5. Can an employer refuse to hire a site inspector on the grounds that the candidate is hard of hearing? What if the site inspector uses a wheelchair?

---

[10] However, there are exceptions. If it can be demonstrated that the law is justified "in a free and democratic society," the court may allow the infringement of rights.

6. A geoscientist works for only one company, 40 hours per week, under a written agreement that says the geoscientist is an independent contractor. Is the company required to withhold income tax deductions?

7. A company has a mandatory retirement policy. Is that an infringement of the employees' rights under the *Charter of Rights*?

8. A software engineer has worked as an employee for five years while developing a product. The company patents and successfully markets the product. The software engineer then leaves the company to set up her own company. The software engineer uses her knowledge to write new software for a competing product. Is this a breach of confidentiality?

9. A firm of architects runs out of work and can no longer afford to employ its draftsman. The firm tells the draftsman, "We are laying you off. As soon as we have more work, we'll call you back." The draftsman sues for wrongful dismissal. What is the likely result of the lawsuit?

10. An employee engineer makes a mistake in her calculations, causing loss to the client. Who may be liable to whom?
    a) The employer may be liable to the client under the doctrine of vicarious liability.
    b) The employee may be liable to her employer.
    c) All of the above
    d) None of the above

ANSWERS

1. B

2. C

3. The clause is extremely broad, in terms of length of time and scope of work; therefore, a court will likely find it unenforceable.

4. A

5. The employer could discriminate on either account, but only if the employer can prove that the impairment or disability would make performing the work impossible for the candidate. Moreover, there would also have to be no reasonable possibility of making accommodation for the impairment or disability.

6. Yes. This is really a contract of employment, even though it purports to be a contract with an independent contractor.

7. No. The *Charter* does not apply to private companies.

8. This situation would be breach if the employee is using confidential information. Nothing prevents an employee from using his or her skills, regardless where they were obtained. But an employee cannot use confidential knowledge of the former employer's product to compete with the former employer.

9. The draftsman would likely succeed. Lack of work is not just cause for dismissal.

10. C

CHAPTER 22

HEALTH AND SAFETY LAW

Overview

Health and safety law is a mix of common law and federal and provincial legislation. The common law principles of contract and tort previously discussed in Chapters 6 and 12 apply to health and safety issues. But both the federal and the provincial governments have also implemented legislation relating specifically to health and safety issues. These include legislation concerning occupational health and safety (OH&S) that is designed to protect workers from injury by setting minimum health and safety standards that employers must provide in the workplace, and legislation that provides no-fault compensation for injured workers through a workers' compensation board (WCB) in exchange for workers giving up their right to sue employers for workplace injuries.

Canadian OH&S legislation is primarily based on three fundamental rights of workers:

1. The worker has the right to be informed of known or foreseeable safety or health hazards in the workplace.

2. The worker has the right to participate in the prevention of occupational accidents and diseases either as a member of a joint health and safety committee or as a health and safety representative.

3. The worker has the right to refuse dangerous work and be protected against dismissal or disciplinary action following a legitimate refusal.[1]

Both federal and provincial OH&S legislation generally follow these principles. Employers must comply with regulations specifying technical requirements, meet set standards, and follow prescribed procedures to reduce the risk of occupational accidents and diseases.

OH&S legislation is enforced through site visits. Officials appointed by federal and provincial governments have the power to inspect workplaces and give orders to employers and workers relating to any violations of the legislation. Employees can request these inspections, and the law protects these employees against dismissal or disciplinary action for seeking the enforcement of the occupational health and safety legislation.

---

[1] Human Resources and Skills Development Canada at **http://www.hrsdc.gc.ca**.

Note that the emphasis of this chapter is workplace safety issues, as opposed to design safety issues. The phrase "design safety" means building code compliance and choice of safety factors in the design of a building or a piece of equipment. Issues relating to codes and standards of design are discussed in Chapter 13.

22.1 Occupational Health and Safety

(a) The Importance of Safety

There has been an increased focus on workplace health and safety as an issue in Canada. Today, health and safety issues are usually considered fundamental concerns, and an important legal issue for all companies doing business in Canada and for their employees. While this is the case in many countries with similar legislation, many health and safety regulators are becoming increasingly vigilant about enforcement, and Canadian companies have been forced to understand the importance of health and safety, and the potential costs from a human, economic, and reputational perspective. Simply put, most Canadian companies pay significant attention to occupational health and safety, not only because they are required to do so at law, but also because they face significant reputational and economic consequences if they do not.

All employers and employees must be familiar with the occupational health and safety obligations of their employer, and with their own obligations as individuals, as they can be personally fined for failing to take appropriate safety measures and prevent accidents, should they occur.

(b) Federal and Provincial Law

Federal and provincial legislation protects workers from unsafe work conditions and punishes employers, supervisors, and employees for failing to follow safety requirements.

(i) **THE CRIMINAL CODE** In Canada, the importance of health and safety law was clearly demonstrated when the federal government amended the *Criminal Code* in 2004 to create a legal duty for all persons directing work to take reasonable steps to ensure the safety of workers and the public. Anyone who undertakes, or has the authority to direct, how another person does their work is under a legal duty to prevent bodily harm to that person and any other person arising from that work. Section 217.1 states:

> Every one who undertakes, or has the authority, to direct how another person does work or performs a task, is under a legal duty to take reasonable steps to prevent bodily harm to that person, or any other person, arising from that work or task.

Because of this new legal duty in the *Criminal Code*, any person or organization who breaches this duty may be found guilty of the offence of criminal negligence causing bodily harm or death. A conviction for criminal negligence causing bodily harm can result in imprisonment. Significantly, under these provisions, a company can be found negligent not only through the conduct of one of its representatives, but also through the combined acts or omissions by several representatives.

In 2008, a Quebec paving stone company, Transpavé Inc., became the first organization to be convicted of criminal negligence under these provisions, in a case involving a young worker who was fatally crushed when he attempted to clear a jam in one of the company's machines.[2] After an investigation by the provincial health and safety regulator and the provincial police, the company

---

[2] *R. v. Transpavé Inc.*, 2008 QCCQ 1598.

was charged with criminal negligence for having allowed the worker to operate a machine with its motion detector safety mechanism deactivated, and because the worker had not been properly trained. The company pled guilty, and was levied a fine of $110 000. As the company was a relatively small company, employing only 100 workers, and as the company had subsequently spent more than $500 000 on safety improvements, the Court believed that the amount of the fine was sufficient to have a meaningful economic impact.

Individuals can also be charged under s. 217.1 of the Code. In an earlier case, a construction supervisor, Domenico Fantini, in Ontario was originally charged with criminal negligence causing death on the basis that he had failed to adequately supervise two workers digging a ditch. The supervisor had advised the workers about the dimensions of the trench, but not given any directions concerning sloping or shoring, and then left the site moments before the trench collapsed, killing one of the workers. In that case, the Crown withdrew the criminal charges in exchange for the supervisor pleading guilty to several violations of Ontario's OH&S legislation, and being fined $50 000.

(ii) **THE CANADA LABOUR CODE** The federal *Canada Labour Code*[3] (CLC) sets out OH&S requirements for federally regulated workplaces in Part II, and includes regulations titled *Canada Health and Safety Regulations* (CHSR). The CLC and CHSR apply to work performed on federal government property, including Aboriginal reserves and federal government buildings.

(iii) **PROVINCIAL OH&S STATUTES** Each province and territory has its own OH&S legislation. Most provinces maintain separate OH&S legislation, but British Columbia combines OH&S with WCB legislation. Enforcement of OH&S varies among provinces. In some provinces, a designated body, such as Worksafe New Brunswick, conducts inspections and administers the applicable OH&S legislation, while in others, the provincial Crown or Ministry of Labour is responsible for enforcement.

(iv) **OTHER SAFETY STATUTES** In addition to the OH&S regulations, other statutes throughout Canada contain safety regulations, many of which are industry-specific. For example, regulations have been created pursuant to the *Canada Transportation Act*.[4] Other regulations govern safety in the maritime industry and the airline industry. In addition, some provinces have enacted legislation that creates OH&S regulations specifically aimed at mining. However, covering all Canadian safety regulations is beyond the scope of this text. The reader should simply be aware that regulations are not restricted to just one statute in each province. Those interested in finding the OH&S regulations in a given province should visit that province's OH&S website.

(c) Responsibility for Health and Safety

In general, OH&S legislation improves working conditions through prescribed safety requirements that impose duties and responsibilities on all owners, employers (including their officers and directors[5]), supervisors, and workers. These statutes usually stipulate fines and other penalties for failure to comply with these requirements. In most instances, fines increase dramatically with repeat offences.

Understanding and following OH&S legislation is important in keeping workers safe and avoiding prosecution. However, simply following the legislation may be insufficient because OH&S duties and responsibilities are minimum requirements rather than maximum ones. Other codes and standards, such as the National Building Code and other industry-specific publications, must

---

[3] R.S.C. 1985, c. L-2.

[4] S.C. 1996, c. 10.

[5] See Chapter 20.

be followed as well. If these codes and standards fail to address a specific situation, professionals are obliged to use a standard appropriate to the circumstances. For instance, if the OH&S code, if followed, would still create an unsafe condition, a professional must use their judgment to apply a higher standard.

All employers and employees should be actively involved in preventing accidents. Accident prevention includes the following:

- Developing, implementing, and enforcing organizational and project-specific safety plans, and auditing thereof
- Providing proper training and supervision of all employees
- Maintaining clear records of all safety-related activities, including training
- Enforcing appropriate discipline for employees who violate OH&S requirements and policies
- Clearly delineating authority and responsibility, including identification of the prime contractor or constructor (or similar designation, as this varies from province to province) and of appropriate supervisory roles
- Implementing proper communication procedures and protocols
- Providing appropriate resources and proper goal setting for organizations and projects
- Taking appropriate steps when an unsafe condition is identified to prevent an accident from occurring

In most jurisdictions, there is an increasing focus on the role of supervisors in preventing accidents, and an expanding sense of what parties may play a supervisory role on a project. While it is not always clear as to which parties in a multiple-employer workplace have supervisory responsibilities, it should be assumed that the law will place such responsibilities on anyone who has authority over others on a work site, or who, by nature of their role, is in a position to identify safety hazards or concerns.

This was demonstrated by a recent case in British Columbia, where the provincial regulator imposed penalties against a consulting engineering company following a fatality in which a worker fell into an improperly shored excavation and died. While the regulator imposed the larger penalties on the owner, general contractor, and developer, it also levied a small penalty on the engineering company because the engineering inspector who conducted regular inspections of the site should have been aware that the shoring cages at the site were inadequate.[6]

(d) When an Accident Occurs

Unfortunately, workplace accidents are inevitable. When a workplace accident occurs, workers and employers must follow prescribed OH&S procedures for reporting and investigating accidents, and cooperate with the OH&S regulator conducting the investigation. These procedures usually include preparing and submitting a report to the appropriate OH&S regulator concerning the causes and contributing factors that may have led to the accident. The form and content of these reports is prescribed in many jurisdictions' legislation.

Companies should be aware of their duties to cooperate with regulators conducting investigations, and of any procedural rights that representatives may have with respect to responding to the regulator's investigation. For instance, in some jurisdictions, people being interviewed about an accident may have the right to have another person present during the interview. Often, people involved in an accident may volunteer information about their or the company's role in

---

[6] See http://www.worksafebc.com/news_room/news_releases/2006/new_06_12_01.asp.

an accident that may be influenced by guilt or fear. As the information provided by witnesses immediately following an accident is usually fundamental to any findings by the regulator, it is important for companies to ensure that, where possible, its employees are aware of their rights and responsibilities in terms of participating in such investigations. To the extent possible while still fulfilling statutory and contractual obligations, statements and evidence should be provided only after receiving appropriate advice from a lawyer or other designated person within a company.

If an accident occurs, the following should be considered:

- Taking all reasonable steps to minimize the effects of the accident, arranging for appropriate medical attention, and eliminating any hazards
- Fulfilling all statutory and contractual obligations for reporting the accident to the appropriate parties in accordance with the appropriate safety plan
- Ensuring the preservation of evidence, including the taking of photographs
- Seeking appropriate legal and other advice and doing so, where possible, in advance of giving statements to investigators
- Conducting internal investigations to determine the cause, perhaps with the assistance of legal counsel, and taking steps to prevent similar accidents.

(e) The OH&S Regulators

Given the importance of OH&S in Canada, it is not surprising that regulators are given significant regulatory and investigatory powers. Companies and individuals who are new to Canada are often surprised by the sweeping and significant role played by OH&S regulators.

On the prevention side, there are very strict and detailed requirements for various construction techniques. Failure to understand these requirements can lead to significant increases in costs, and a failure to follow these requirements can lead to fines and delays. These problems can be prevented by ensuring that everyone is aware of the obligations that are imposed by OH&S legislation.

OH&S legislation also provide regulators with significant tools to enforce OH&S regulations. These can include stop-work orders, compliance orders requiring immediate remedial action, and significant fines and penalties. In addition, the OH&S record of a corporation is usually taken into account in considering future fines and penalties, meaning that such fine or penalty not only has immediate effect, but also can have severe long-term consequences.

In many jurisdictions, OH&S regulators have imposed increasingly significant fines and penalties in recent years. For example, two companies in Alberta were issued fines of $350 000 in 2007, while in British Columbia a company was assessed a fine of more than $297 000 in that same year. Individuals can also face significant fines, as noted above in the Domenico Fantini case.

(f) Common Interests

It is in everyone's interest to promote occupational health and safety. Despite the significant focus on OH&S in Canada, accidents still occur daily. Employers and employees who promote OH&S and who actively participate in both education and functional implementation of safety measures are acting in the common interest of all. It is crucial for everyone, including professionals, to take a leadership role in such education and implementation.

22.2 Contracts

Contracts sometimes contain specific obligations with respect to health and safety. These obligations may be specific requirements for a project, references to provincial regulations, the owner's health and safety policies, or a combination of the above. Contracts also generally describe obligations regarding public safety and the worksite, including requirements to fence off the project. Finally, contracts often contain an indemnity[7] requiring the contractor to hold the owner harmless from claims of personal injury on the project site.

In addition, contracts may determine who is responsible for OH&S regulations on a work site. OH&S legislation generally requires one party to take overall responsibility for OH&S on a project site. In some provinces, this contractor is called the prime contractor, and in others, the constructor. Owners must designate one (and only one) prime contractor or constructor by contract, but this designation may change at specific milestone events, such as substantial completion. While owners cannot contract out of their own OH&S obligations, they can have another contractor take over the prime contractor or constructor role. If the owner fails to designate a prime contractor or constructor, or if the owner designates more than one at any time, the prime contractor or constructor obligations default to the owner.

22.3 Torts and Workers' Compensation Legislation

Workers' compensation board (WCB) legislation provides workers with a no-fault compensation scheme for injuries incurred on the job. This legislation applies to all workers, from those in an office to those on a shop floor to those on a construction site. All employers are required to make payments to WCB to fund the compensation scheme. In all provinces, the payments are calculated on a per-employee basis and vary depending on the risk of the employer's business. In some provinces, the employer's safety record is taken into account in setting that employer's payments.

In exchange for this compensation scheme, employers are protected from worker lawsuits for job-related injuries. Protection from such lawsuits applies not only to the worker's direct employer, but also to all registered employers and employees of registered employers. Under the WCB scheme, claims in contract or tort are not generally available for injured workers. However, tort claims may be made by people injured while not working. Torts are discussed in detail in Chapter 12.

---

[7] See Chapter 9, pp. 70–71.

A comprehensive explanation of the WCB scheme is found in the *Pasiechnyk*[8] case:

Workers' compensation is a system of compulsory no-fault mutual insurance administered by the state. Its origins go back to 19th century Germany, whence it spread to many other countries, including the United Kingdom and the United States. In Canada, the history of workers' compensation begins with the report of the Honourable Sir William Ralph Meredith, one-time Chief Justice of Ontario, who in 1910 was appointed to study systems of workers' compensation around the world and recommend a scheme for Ontario. He proposed compensating injured workers through an accident fund collected from industry and under the management of the state. His proposal was adopted by Ontario in 1914. The other provinces soon followed suit. Saskatchewan enacted *The Workmen's Compensation Act,* 1929, S.S. 1928-29, c. 73, in 1929.

Sir William Meredith also proposed what has since become known as the "historic trade-off" by which workers lost their cause of action against their employers but gained compensation that depends neither on the fault of the employer nor its ability to pay. Similarly, employers were forced to contribute to a mandatory insurance scheme, but gained freedom from potentially crippling liability. Initially in Ontario, only the employer of the worker who was injured was granted immunity from suit. The Act was amended one year after its passage to provide that injured Schedule 1 workers could not sue any Schedule 1 employer. This amendment was likely designed to account for the multi-employer workplace, where employees of several employers work together.[9]

22.4 Ethical Considerations

Architects, engineers, and geoscientists often become aware of safety issues in the course of their professional duties. In some cases, professionals may be asked by their employer or by a client to ignore safety concerns or to relax the rules. For example, under some provincial OH&S regulations, all formwork and falsework must be inspected by a professional engineer prior to the pouring of concrete. Because delaying a concrete pour can be very costly, clients are known to be displeased about delays due to inspections. Thus, if during the inspection process, an engineer finds inadequate cross-bracing in the formwork, that engineer may be faced with a difficult choice of relaxing the safety standards and allowing the pour to proceed or delaying the concrete pour until the problem is remedied. In these circumstances, the duty of the engineer is to ensure that safety is not compromised. The codes of ethics require that the dilemma be resolved in favour of protecting the safety of others. These issues are discussed more fully in Chapter 3.

---

[8] *Pasiechnyk v. Saskatchewan (Workers' Compensation Board),* [1997] 2 S.C.R. 890.

[9] See previous note at paras. 24–25.

CHAPTER QUESTIONS

1. Zoe is a software engineer and is driving from her office to a client's office. Zoe's car is hit from behind by a police car. Who can Zoe successfully sue?
    a)   Her employer
    b)   The police officer's employer
    c)   The police officer
    d)   None of the above

2. If an owner designates two contractors as the prime contractor or constructor simultaneously, what is the potential consequence to the owner?

3. Can an owner contractually require safety measures that are more costly than OH&S requirements?

4. An employer asks a mining engineer to inspect a shaft. But the engineer believes the shaft is unsafe and does not want to go down. The employer threatens to fire the engineer for insubordination if the engineer refuses to follow the employer's orders. Which of the following statements is true?
    a)   The engineer can refuse to carry out the order.
    b)   The employer has no right to fire the engineer for refusing the order.
    c)   The engineer can report the safety concern to the OH&S authorities.
    d)   Only A and B
    e)   A, B, and C

5. If an employer is in full compliance with OH&S requirements, is it still necessary to comply with other legislation that contains health or safety requirements?
    a)   No
    b)   Yes
    c)   Only if the other legislation is inconsistent with the OH&S requirements

6. What is the "historic trade-off" that underlies workers' compensation schemes?

ANSWERS

1. D. Because Zoe is employed and is injured while she is employed, she is covered by the *Workers Compensation Act* and is therefore unable to sue her employer, another employer, or another employee. She will be compensated by the Workers' Compensation Board rather than through a private lawsuit.

2. The owner could become responsible for any breaches of the OH&S requirements, since by designating two prime contractors, the owner ends up being considered the prime contractor.

3. Yes.

4. E

5. B

6. The trade-off is between the entitlement of employees to share compensation from the workers' compensation fund in exchange for the protection of owners against lawsuits by employees injured during work.

CHAPTER 23

ENVIRONMENTAL LAW

Overview

Environmental regulation is not new. According to historians, European cities have had problems with pollution for centuries:

> By the last decades of the thirteenth century, London had the sad privilege of becoming the first city in the world to suffer man-made atmospheric pollution. In 1285 and 1288 complaints were recorded concerning the infection and corruption of the city's air by coal fumes from the limekilns. Commissioners of Inquiry were appointed, and in 1307 a royal proclamation was made in Southwark, Wapping, and East Smithfield forbidding the use of sea coal in kilns under pain of heavy forfeiture.[1]

Environmental regulation in England in 1307 had relatively harsh penalties: the first two offences under the royal proclamation resulted in heavy fines, while the third was punishable by death.

Modern environmental law is an area of specialization within the legal system, a mix of common law and statutory regulation. The common law uses existing contract and tort concepts to impose liability on certain parties for environmental contamination. In addition, the tort concepts[2] of trespass, nuisance, negligence, misrepresentation, and strict liability are all used against certain parties in claims for compensation for environmental contamination. At the same time, federal and provincial statutes and municipal bylaws protect the environment and are used to obtain funds to clean up environmental contamination. Governments have created a broad range of statutes that include criminal and quasi-criminal penalties, obligations to remediate, and to provide enhanced rights for individuals seeking compensation. However, listing all statutes and regulations enacted by each province or by the federal government is beyond the scope of this text. The reader should consult local regulatory bodies for more information.

---

[1] Gimpel, Jean. *The Medieval Machine* (New York: Holt, Rinehart & Winston, 1976) at p. 82.

[2] See Chapter 12.

Environmental contamination is usually discovered and then remediated through a three-stage **environmental site assessment (ESA)**. Environmental audits, which are often confused with environmental site assessments, investigate a corporation's potential environmental liabilities, including its real property.

In addition to punishable offences and compensation provisions, environmental legislation contains proactive measures designed to prevent future environmental contamination. The principal preventative tools are federal and provincial environmental impact assessment requirements. Many architectural, engineering, and geoscience projects have a permanent impact on the environment. For this reason, new projects that meet certain statutory criteria must undergo environmental impact assessments. This assessment attempts to balance the desire to protect the environment with the need for economic development.

23.1 Environmental Site Assessments and Audits

Environmental site assessments (ESAs) and environmental audits are usually conducted in accordance with Canadian Standards Association (CSA) guidelines. The document Z768 is a guideline for environmental site assessments, and a separate document, Z773, is a guideline for environmental auditing. While not legally mandatory, following these guidelines is considered good professional practice. They contain legally accepted codes and standards and, as such, represent a minimum practice standard against which courts will usually hold a professional.[3]

Confidentiality is often an issue in ESAs and environmental audits. Assessors work for their clients under consulting contracts, which usually contain confidentiality provisions. Confidentiality clauses are important in every consulting agreement but are particularly important in a consulting contract for an ESA or an environmental audit. One of the legislative requirements common to all statutes is the obligation of all parties, including assessors, to report environmental contamination. Furthermore, provincial codes of ethics[4] governing the practices of the professions often contain an overriding obligation to safeguard the environment. Thus, the legislative and ethical obligations to prevent and to report contamination may trump a contractual confidentiality requirement. However, if a consulting contract contains a tightly worded confidentiality provision, or if the environmental assessment or audit was prepared at the request of a lawyer, the client may have a legal argument that the report should be exempt from an obligation to report. In such circumstances, a contracting professional should seek specific legal advice before reporting.

(a) Environmental Site Assessments

An ESA assesses a property for possible contamination; and if contamination is discovered, it recommends a remediation protocol. ESAs are generally conducted by a group of professionals, including engineers and geoscientists. ESAs are often requested by potential sellers or purchasers of property, by lenders, and by parties wanting to develop property.

ESAs are divided into phases. The first phase determines the likelihood that a property is contaminated. A **Phase I ESA** is defined by the CSA as "the systematic process . . . by which an Assessor seeks to determine whether a particular property is or may be subject to actual or potential contamination."[5] A Phase I ESA involves little or no drilling or physical sampling and consists mainly of gathering information about the property from parties such as regulatory authorities,

---

[3] See Chapter 13, pp. 151–153.

[4] See Chapter 3, pp. 16–17.

[5] CSA Guideline Z768.

present and former owners, and managers of the property. The degree and type of information to be gathered is at the discretion of the assessor but may include historical title searches, aerial photographs, regulatory databases, and fire insurance maps. Surrounding properties are generally included in the inquiry to determine whether contamination may have migrated from a neighbouring property. The Phase I ESA report should clearly describe the findings and conclusions of the assessor following a systematic process and an appropriate level of inquiry.

A Phase II ESA is generally conducted if recommended in the Phase I ESA. A **Phase II ESA** is defined by the CSA as "the systematic process . . . by which an Assessor seeks to characterize and/or delineate the concentrations or quantities of substances of concern related to a site and compare those levels to criteria."[6] The Phase II ESA examines the most probable locations of contamination by sampling groundwater and soil and then comparing the results with relevant regulatory and industry standards.

If a Phase II ESA recommends remediation, then a Phase III ESA is necessary. A **Phase III ESA** provides a detailed description of the contaminants, a recommended remediation process, confirmatory sampling throughout the remediation process, and a measurement of the success of the remediation program. A Phase III ESA is often an integral part of an environmental remediation.

(b) Environmental Audits

Many organizations now undertake corporate environmental audits, sometimes called *compliance audits*. These audits fulfill statutory obligations or provide defences against environmental claims or prosecutions. They also assist an organization in understanding its environmental risks.

Corporate environmental audits have many goals, including the following:

- protecting the health and welfare of employees and others
- protecting the organization from liability claims by tenants and customers, and preventing health risks associated with environmental contaminants
- assisting in establishing defences to statutory offences and claims
- reducing remediation expenses
- ensuring the marketability of a property.

According to CSA Guideline Z773, an **environmental audit** is "a systematic process of objectively obtaining and evaluating evidence regarding a verifiable assertion about an environmental matter, to ascertain the degree of correspondence between the assertion and established criteria, and then communicating the results to the client. A verifiable assertion is a declaration or statement about a specific subject matter which is supported by documented factual data." CSA guideline Z773 is the applicable standard for environmental audits, although other guidelines have been developed for environmental management systems, such as ISO 14000. Some audits also include ESAs.

23.2 Remedies For Private Landowners

A private landowner can be affected by environmental contamination that existed on the property before it was purchased or by contamination introduced by another party after the property was purchased. In either case, the landowner will want to seek recovery for the cost of remediating the property or for the diminution, or reduction, of value of the property.

---

[6] CSA Guideline Z768.

For this reason, whenever there is any concern about the environmental condition of a property, both buyer and seller should obtain at least a Phase I ESA. Phase I ESAs can sometimes be requested jointly. However, if the seller commissions an ESA and that ESA contains concerns about the environmental condition of the property, the consultant generally makes the report available to both the seller and the buyer, both of whom are clients of the consultant. That does not mean that a vendor who obtains a report without involvement of the buyer is free to withhold such information. In the sale of real property, the law implies an obligation to disclose that information to the buyer.[7] The vendor has an obligation to the purchaser to disclose all relevant facts about the property, including the results of an ESA, unless the facts are patently obvious.

If the Phase I ESA reports possible contamination, then a Phase II ESA is usually necessary. This step identifies the source of the contamination, since different remedies are available, depending whether the contamination is wholly contained in the purchased property or is emanating from a neighbouring property.

(a) Contamination Wholly on the Property

For contamination identified as contained solely upon the landowner's property, the owner may be able to make a claim against the prior landowner for breach of the purchase and sale contract, or innocent, negligent, or fraudulent misrepresentation. The owner can sometimes also make a claim against the realtors involved in the sale. The likelihood of success of these actions depends on how much the seller and the realtors knew about the contamination before the sale, and on the representations in the purchase and sale contract.

(b) Contamination from a Neighbouring Property

Most case law about contamination from neighbouring properties relates to contamination that migrates through soil or groundwater. All property owners are responsible for contaminants that flow from their property.

Common law actions for such contamination include trespass,[8] negligence,[9] misrepresentation,[10] and strict liability.[11] In addition, an owner can sue for **nuisance**—undue interference with the use and enjoyment of rights to land. With the tort of nuisance, liability is strict liability. This means that the injured party does not have to prove that the other party was negligent.

For example, in *Rylands v. Fletcher*,[12] the defendants' underground water reservoir on a neighbouring property caused an old mine shaft owned by the plaintiff to collapse. Although the Court found that the defendants were not negligent, they were still strictly liable for damages. Canadian courts establish strict liability where the plaintiff has proven the following:

1. The defendant made a non-natural use of his or her land.

2. The defendant brought onto his or her land something which was likely to do mischief if it escaped.

---

[7] See Chapter 12, pp. 132–135.

[8] See Chapter 12, pp. 142–143.

[9] See Chapter 12, pp. 131–132.

[10] See Chapter 12, pp. 139–140.

[11] See Chapter 12, p. 142.

[12] [1868] L.R. 3 H.L. 330.

3. The substance in question escaped.

4. Damage was caused to the plaintiff's property (or person) as a result.[13]

In addition to common law remedies, a property owner may be able to make a claim based on environmental statutes. For example, in British Columbia, CN Railway used the *Environmental Management Act* to obtain compensation for remediation costs associated with environmental contamination from a neighbouring property.[14] After determining that the defendant caused the contamination on CN's property, the Court developed a two-stage test to determine the level of compensation:

1. Did CN act reasonably in remediating the property?

2. Were the remediation costs reasonable?

In addition to awarding virtually all of CN Railway's remediation costs, the Court also awarded CN Railway full legal costs rather than the usual award of approximately 30 percent of the actual legal costs. However, this ruling on legal costs remains under appeal as of the date of publication of this text.

23.3 Governmental Regulation

Governments have been forced by the electorate to focus more attention on environmental matters. The two principal goals of government are to offload remediation costs for environmental cleanup to private parties and to protect the environment from future contamination.

To accomplish these goals, governments have created a complex web of environmental regulation. At best, these regulations are confusing and difficult to apply. At worst, they represent an impossibly difficult set of ever-changing benchmarks.

(a) Legislative Jurisdiction and Legislation

Both federal and provincial levels of government have the constitutional authority to create environmental regulation. Federal, provincial, and local governments have environmental statutes, regulations, bylaws, policies, and guidelines. However, where there is a direct operational conflict between federal legislation and provincial or local legislation, federal legislation governs.

Federal environmental legislation has tended to focus on fisheries, oceans, and international and interprovincial issues, although it also regulates certain toxic substances, such as PCBs. The principal federal statutes in these areas are the *Fisheries Act*[15] and the *Canadian Environmental Protection Act (CEPA)*.[16] In addition, the federal government has enacted the *Transportation of Dangerous Goods Act*[17] and the *Canadian Environmental Assessment Act*.[18] International treaties negotiated by the federal government, such as the North American Free Trade Agreement (NAFTA) and the Kyoto Accord, also impact environmental law.

Meanwhile, provincial governments have a diverse set of environmental legislation and regulation on subjects such as air quality, boiler requirements, parks, hazardous and dangerous goods,

---

[13] For example, see *John Campbell Law Corp. v. The Owners, Strata Plan 1350,* 2001 BCSC 1342.

[14] *Canadian National Railway Co. v. A.B.C. Recycling Ltd.,* [2005] B.C.S.C. 1559.

[15] R.S.C. 1985, c. F-14.

[16] S.C. 1999, c. 33.

[17] S.C. 1992, c. 24.

[18] S.C. 1992, c. 37. See also Chapter 2, p. 11.

transportation and disposal, land drainage, environmental emergencies, environmental permits, public health, pesticides, and land use. Many provinces, including Ontario,[19] Alberta,[20] Nova Scotia,[21] and British Columbia,[22] have passed comprehensive environmental legislation.

Municipal and local governments are created by their respective provincial governments. They therefore derive their ability to pass bylaws on environmental matters through their provincial government. Local governments have always had some indirect control over environmental matters by controlling land development through zoning; but they have become more involved in the environment in recent years by passing their own environmental bylaws.

One difficulty for professionals working with environmental legislation is locating all of the relevant statutes, regulations, bylaws, policies, and guidelines, understanding them, and then keeping up to date with constant changes.

(b) Common Legislative Concepts

Many elements of environmental regulation are common to all levels of government, such as the following:

- regulation of potentially harmful conduct, including emissions and discharges
- creation of administrative systems to prevent pollution and to administer cleanup after it has occurred
- requirements to provide information, such as goods that are being transported and contamination that is identified
- requirements to undergo environmental assessments
- establishment of offences.

(c) Environmental Offences

Because of the large number of environmental statutes from both federal and provincial levels of government, the range of environmental offences in Canada is very broad. The most important offences relate to discharge of environmental contaminants. But there are also offences for failure to obtain required permits, to keep accurate books and records, to report environmental contamination, and to assist environmental officers in the conduct of an investigation.

Generally speaking, environmental offences target corporations and focus on actions and omissions of the corporation and its employees. Changes in 2004 to the *Criminal Code* established rules for attributing criminal liability to all organizations, including corporations, societies, partnerships, trade unions, and associations for acts of their directors, officers, partners, employees, members, agents, and contractors. In addition, section 217.1 of the *Criminal Code* places a duty on all persons directing work to ensure the safety of workers and the public. Prior to these amendments, most environmental offences were only punishable by fines. But since the *Criminal Code* amendments, some offences are now punishable with jail sentences.

In addition to liability of corporate directors and officers, many environmental offences are directed at individuals as well. However, individual environmental offences based on the *Criminal Code* and environmental statutes must include some element of personal fault. Individuals cannot be found guilty solely because the corporation they work for is guilty.

---

[19] *Environmental Protection Act*, R.S.O. 1990. c. E.19.

[20] *Environmental Protection and Enhancement Act*, R.S.A. 1992, c. E-12.

[21] *Environment Act*, S.N.S. 1994/95, c. 1.

[22] *Environmental Management Act*, S.B.C. 2003, c. 53.

With the exception of the *Criminal Code* provisions, environmental offences are generally **strict liability offences**, meaning that once the prohibited act has been proven to have occurred, the only defence is due diligence. **Due diligence** means that the defendant has taken all reasonable steps to satisfy statutory obligations. But what constitutes due diligence depends on the situation. A professional who follows these practical methods of reducing risk of environmental offences should be more likely to be able to prove due diligence:

- maintaining a program of environmental auditing
- ensuring that policies and systems are in place to prevent statutory breaches
- ensuring that policies and systems are in place to fulfill all reporting requirements
- ensuring that directors, partners, employees, members, agents, and contractors are all familiar with their responsibilities and the relevant policies and systems
- ensuring that all of the above steps are actively monitored and reviewed regularly.

Note that simply putting a policy in place and then assuming that it is functioning properly is not likely to constitute due diligence.

(d) Environmental Cleanup

The general principle in contamination cases is that the polluter pays for the cleanup. However, the environmental offence net can catch more than just the polluter and may pull in the following additional parties:

- the owner of the offending substance, process, or land
- the party who has charge, custody, or control of the offending substance, process, or land
- the party holding the licence required by the statute
- a party who caused or contributed to the damage
- successors to the responsible parties, such as receivers, receiver-managers, assignees, trustees, principals, and agents
- officers and directors of the company that currently owns the contaminated property
- parties involved in the cleanup of the property.

In fact, several provinces have enacted legislation imposing liability for environmental cleanup on all parties connected to a property. This means that a current owner could be responsible for cleanup costs on a property even if that owner had nothing to do with the contamination. This is yet another reason for insisting on a pre-purchase ESA.

Moreover, the search for funds to clean up "orphan" sites can extend back in time to prior property owners. Those former owners may also be held to today's standard of environmental cleanliness, even though their former actions had been both legal and acceptable at the time. For example, in the 1930s, service stations normally dumped oil from oil changes onto the street. Many hundreds of oil changes and a few decades later, the government may start looking for a party to pay for the street cleanup for a particular service station site. The fact that the contamination was within standards of the time will not likely save the service station company from liability. Hence, when specifying environmental standards for projects today, engineers and geoscientists should recognize that the regulatory standards are just a minimum and that as much as possible, they should attempt to foresee what the standards will be well into the future.

23.4 The Environmental Assessment Process

Some new projects and modifications to existing projects require environmental assessments prior to project approval. An assessment team studies the impact of these projects on the environment and considers the long-term impact on the environment, including risks to human health. In many cases, the team seeks public input as part of the process.

Under the federal *Canadian Environmental Assessment Act*[23] (CEAA), four types of assessment can be used for this purpose, depending on the nature of the project: a screening, a comprehensive study, a mediation, or a panel review. Any project that has federal implications, such as an inter-provincial pipeline, and any project that affects navigable waterways or oceans or otherwise meets criteria of the Canadian Environmental Assessment Agency must undergo some form of federal environmental assessment. Modest projects generally use screenings or comprehensive studies, whereas large or controversial projects generally use mediation or panel review. Meanwhile, provincial environmental statutes can impose other assessment standards. But while it is possible that a project could be forced to go through both a federal and a provincial assessment, generally the two jurisdictions work together to prevent duplication.

These environmental assessments can significantly delay a project. Moreover, the uncertain length and outcome of the process can cause further problems. Most parties view assessments as an annoying extra cost. But one benefit of having gone through a detailed environmental assessment, in addition to the benefit to the environment, is that regulatory authorities are less likely to be able to complain about the project in years to come.

However, sometimes the outcome of the assessment is that the project cannot proceed, or that it can proceed only on conditions imposed to minimize potential environmental harm. Parties may also have to perform extra work or include extra stages to the project to conform to the assessment's recommendations. Note that this work may or may not occur on the site of the project or even be connected to the project.

CHAPTER QUESTIONS

1. Describe the three types of environmental site assessments. At which phase is intrusive testing used?

2. Describe the difference between an environmental site assessment and an environmental audit. Describe the benefits of conducting an environmental audit.

3. LMN Co. used Assurance Realty Service to purchase a piece of land. Assurance had previously been involved in the sale of the property 20 years ago when it was still a gas station. The seller had noticed odd smells emanating from the basement of the building built on the property 15 years ago but had said nothing about it to his realtor. LMN did not do an ESA before purchasing the property. LMN later finds that the property has significant environmental contamination. Describe the potential legal actions.

---

[23] S.C. 1992, c. 37.

4. What levels of government may enact environmental laws?
    a) Only the federal government
    b) Both the federal and the provincial governments
    c) All levels (federal, provincial, local)

5. Can a prior landowner be found to be liable for environmental cleanup costs?

6. If a company causes contamination, who may be liable?
    a) The company itself
    b) The directors
    c) The previous landowner
    d) All of the above
    e) A and B

7. What is nuisance?
    a) The cost of performing an environmental audit
    b) The interference with a neighbour's use or enjoyment of land
    c) The bother of having to exercise due diligence

8. Where a consulting contract contains a confidentiality clause, the environmental assessor has these obligations:
    a) He or she cannot speak to anyone about his or her findings, without the client's consent.
    b) He or she may be required to report findings if the environmental statute requires such disclosure.
    c) He or she has an ethical obligation if the he or she is a professional engineer.
    d) B and C

ANSWERS

1. A Phase I ESA is a fact-gathering process. A Phase II ESA is a physical sampling of the property. A Phase III ESA is a detailed delineation of the chemicals on the site. Intrusive testing begins at Phase II.

2. An ESA is a program to determine the contamination of a site, whereas an environmental audit is a broader review of the systems that a corporation has in place to control environmental factors.

3. Assurance may be sued for its prior knowledge of the historical use of the site as a gas station. The seller may be sued for not providing full disclosure.

4. C

5. Yes

6. E

7. B

8. D

CHAPTER 24

ABORIGINAL LAW

Overview

Aboriginal law in Canada is both a unique Canadian creation and a mix of traditional legal principles. This chapter gives a broad overview of Aboriginal law and highlights some areas of particular relevance to architects, engineers, and geoscientists.

In Canada, Aboriginal people occupied the land and used the natural resources long before the arrival of European settlers. From an Aboriginal perspective, the terms occupation and use are incorrect, since Aboriginal people view themselves as being part of the land. However, from a Canadian legal perspective, occupation and use are the appropriate terms for these concepts.

Aboriginal claims to certain land and natural resources exist in much of Canada. For many generations, federal and provincial governments have been trying to resolve Aboriginal issues; but where they have not been successful and Aboriginal claims remain, businesses must be aware of the existence of claims and prepare for them in their contracts and their plans.

Moreover, professionals working with Aboriginal people and Aboriginal companies must be aware of Aboriginal rights, including income tax exemptions and treaty rights. It is also important to be careful when entering into contracts with Aboriginal Band Councils to ensure that the contracts are enforceable.

24.1 The Duty to Consult

In the Haida and Taku River cases,[1] the Supreme Court of Canada held that all governments have a duty to consult with and accommodate the interests of Aboriginal people who will be affected by any proposed activities. This obligation arises whether or not the Aboriginal group in question has proven its claim to the land or resources in question. Environmental impact assessments[2] generally include a consultation process with all people affected by a project; but the Supreme Court ruling suggests that obligations to Aboriginal people go beyond that form of consultation.

The degree and nature of the consultation and accommodation of Aboriginal interests depends on the circumstances. However, project managers must consider the risk of delays or changes to a project arising from this obligation and account for them in the contracts and plans.

Note that this duty to consult and accommodate only applies to government; it does not apply to private parties. However, consulting impacted Aboriginal groups is a prudent step in preventing future legal action.

24.2 Nature of Reserve Property

Because Aboriginal reserve property is generally unpatented Crown land, builders' liens and other real property charges, such as mortgages, cannot be registered against them. Leases can be registered, but only with federal government consent; and even then, the lease is registered in a separate federal government registry in Ottawa. Therefore, for parties doing business on reserve land, traditional forms of security may be unavailable, and other forms of security should be considered.

24.3 Aboriginal Participation in Projects

Aboriginal people and corporations have advantages and disadvantages with respect to participating in projects. On the one hand, Aboriginal people and corporations are not subject to income tax. If they can maintain their tax-exempt status, they obviously increase their potential for profits. Aboriginal contractors should seek specific legal advice about maintaining their status in group projects. But in general, joint ventures and partnerships maintain the tax status of Aboriginal participants, since the profits in a joint venture or partnership flow directly to the individual.

On the other hand, because of their inability to mortgage their reserve lands, and because of the potential difficulty in collecting against property on reserves, Aboriginal contractors have a difficult time getting bonded. Thus, if parties want to provide opportunities to Aboriginal contractors, they should consider deleting bonding requirements or providing alternatives to bonding, such as an increased contractual holdback.

24.4 Contracts with Band Councils

Band Councils are not incorporated entities, and their legal status has been the subject of some debate. Much like a corporation, a Band Council resolution authorizes a contract. Thus, organizations considering a project should obtain this authorization before entering into the contract and commencing work. Contracts with Band Councils are generally enforceable, but legal advice should be sought if there is any concern.

---

[1] *Haida Nation v. British Columbia (Ministry of Forests)*, 2004 SCC 73; and *Taku River Tlingit First Nation v. British Columbia (Project Assessment Director)*, 2004 SCC 74.

[2] See Chapter 23, p. 257.

CHAPTER QUESTIONS

1. If a project is planned such that it will have a potential impact on an Aboriginal group with a land claim, under what circumstances should the Aboriginal group be consulted?

2. Can a lien be filed against reserve lands? a mortgage? a lease?

3. Name one advantage and one disadvantage of Aboriginal contractors that are located on reserves.

ANSWERS

1. Governments are required to consult with Aboriginal peoples affected by a governmental project. However, it is prudent to consult with impacted Aboriginal peoples even for private projects.

2. Real property charges cannot generally be registered against reserve land.

3. Aboriginal contractors located on reserve land are generally tax-exempt, but they experience difficulty getting bonded because reserve lands are not mortgageable.

CHAPTER 25

SECURITIES LAW

Overview

Some of the rules governing the operation of public corporations are found in securities legislation. In this context, **securities** are publicly traded shares, bonds, and other investment devices. Both the federal and the provincial governments have authority over the regulation of securities; but the majority of securities legislation is provincial, under each province's securities commission. This means that provinces may have different rules regarding securities transactions and that multiple rules can sometimes govern a single transaction. Securities legislation generally covers the registration and control of people involved in the trading of securities and the disclosure of material information by people selling securities to the public. The securities commissions are responsible for protecting and promoting fair and efficient trading of securities. They do so by prohibiting *insider trading*[1] and by strictly governing the disclosure of information that is material to a potential investor's decision. Geoscientists and engineers are often involved in the disclosure of material information regarding the discoveries of natural resources, the creation of patentable inventions, and the viability of projects. This chapter focuses on the disclosure of such material information.

25.1 Information Disclosure Requirements

Securities legislation requires corporations to disclose adequate and timely information to investors and potential investors so that they can make informed decisions about the purchase and sale of securities. The most common way this is done is for corporations to publish a formal prospectus about the investment opportunity. A **prospectus** is a written document containing specific financial and technical information about a security, and it must be made available to prospective purchasers before sales of securities can take place. A prospectus

---

[1] See Chapter 5, p. 40.

must contain full, true, and plain disclosure of all material facts about the securities being issued and must be approved by the relevant securities commissions.

But disclosure requirements do not end with the prospectus. After release of the prospectus and as an ongoing process, corporations must continue to make disclosure of all material changes. They must also conform to specific rules relating to filing audited and unaudited financial statements, writing press releases, and reporting insider trading.

25.2 Technical Disclosure Guidelines

The Canadian Securities Administrators (CSA) is a forum of the individual provincial securities commissions, through which the commissions attempt to coordinate and harmonize securities regulation in Canada. The CSA has created a number of guidelines called National Instruments (NI) that are intended to create consistency across provincial boundaries. These include the Standards of Disclosure for Mineral Projects (NI 43-101), and the Standards of Disclosure for Oil and Gas Activities (NI 51-101), along with companion policies that further detail the requirements. Each province can choose not to adopt the CSA guidelines; but these guidelines still represent a common minimum standard of practice. Professionals should always ensure that they are using the latest version of the guidelines, since these documents change from time to time.

Because of a number of well-publicized problems with disclosure related to mining (the BRE-X case is the most notorious), provincial securities commissions have developed and implemented NI 43-101 (the Standards of Disclosure for Mineral Projects) and its companion policy, 43-101CP. Several additional standards apply to mining, including the Canadian Institute of Mining (CIM) Best Practices Guidelines, the CIM Standards on Mineral Resources and Reserves: Definitions and Guidelines, the CIM Guidelines for the Reporting of Diamond Exploration Results, and the Geologic Survey of Canada (GSC) Paper 88-21 on Coal Resource/Reserve Reporting.

The goal of NI 43-101 and NI 51-101 is to establish standards for the disclosure of scientific and technical information. The guidelines provide definitions and technical report requirements and specify that required scientific and technical information be based on information from a qualified person. The standards also stipulate qualifications for this qualified person and define how the qualified person must act independently. For example, for a mineral project, the qualified person must be a geoscientist or engineer with at least five years' experience relevant to the subject matter of the project and the technical report, and must also be a member in good standing with a professional association.

Responsibility for choosing the qualified person and for proper use of information provided by the qualified person rests with the officers and directors of the company. The qualified person should review the prospectus and any other disclosure to ensure that the information has been properly interpreted.

Moreover, the qualified person must follow the CSA guidelines. The guidelines require the qualified person to personally inspect the property, to sign and seal the technical report, and to sign a consent document stipulating that the technical disclosure does not contain any misrepresentations.

25.3 Common Law and Statutory Liability

Geoscientists and engineers face potential liability as a result of any misrepresentation or negligence regarding securities. If the technical report, consent letter, or prospectus reviewed by the geoscientist or engineer contains misrepresentations, the company and its investors will likely have a right of action. If the technical report was prepared negligently, including not being prepared in accordance with recognized guidelines such as NI 43-101 and NI 51-101, the company and its investors will likely have a right of action. Much like other guidelines and practice standards, the national instruments themselves do not create a cause of action; but because they set a minimum practice standard,[2] they become relevant in court. Disclaimer clauses[3] included in the reports are unlikely to have a significant chance of success in this context.

In addition, securities legislation creates another a cause of action if the contents of technical reports are used to support a prospectus or to attempt a company takeover. It is not possible to disclaim statutory liability. There is also potential direct liability for officers and directors as discussed in Chapter 5.

CHAPTER QUESTIONS

1. Do material disclosure requirements end with the filing of a prospectus? Why or why not?

2. What is the impact of the CSA's National Instruments on the liability of architects, geoscientists, and engineers?

ANSWERS

1. They do not. Material disclosure is an ongoing obligation. Continuing disclosure must be made of all material changes.

2. If a professional prepared a technical report negligently, including not writing it in accordance with recognized guidelines, such as NI 43-101 and NI 51-101, the company and its investors will likely have a right of action against that professional.

---

[2] See Chapter 13, pp. 151–153.
[3] See Chapter 15, pp. 171–172.

CHAPTER 26

PRIVACY LAW

Overview

The common law does not provide any significant protection to individuals for breaches of their right to privacy. Hence, federal and provincial governments have enacted privacy legislation in Canada to regulate the collection, use, and disclosure of personal information by organizations (corporations, associations, partnerships, persons, or trade unions).[1] However, privacy legislation does not apply to information obtained about corporations.

26.1 Federal Legislation

The federal law governing privacy is the Personal Information Protection and Electronic Documents Act (PIPEDA).[2] **PIPEDA** regulates how private sector organizations collect, use, and disclose personal information. It states that all organizations collecting personal information must do the following:

- protect personal information
- identify the purpose for which personal information is collected
- obtain the consent of the individual prior to the collection of information
- collect only required information
- limit the use of the information to the purpose for which it was collected
- ensure that the information is accurate
- ensure that adequate safeguards are in place for the protection of the information
- ensure that persons can easily access their information and the organization's privacy policy upon request.

Note that PIPEDA does not apply to collection of personal information for domestic, journalistic, artistic, or literary purposes. Nor does it apply to investigators of self-governing professions.

---

[1] See PIPEDA S.C. 2004, c. 5, s. 2.

[2] S.C. 2004, c. 5.

PIPEDA also contains a process for complaints about breaches of the statute. The PIPEDA privacy commissioner has authority to decide complaints and impose fines. The commissioner can also conduct audits of organizations for compliance with PIPEDA. PIPEDA also provides a right of action for damages by an individual against an organization for breaches of the organization's PIPEDA obligations.

However, PIPEDA also provides situations in which disclosure of personal information might sometimes be permitted. These situations include legal requirements to disclose, national security issues, personal safety situations, and the public availability of such information.

26.2 Provincial Legislation

Except for Newfoundland and Labrador, all provincial and territorial governments in Canada have privacy legislation governing the collection, use, and disclosure of personal information. The legislation varies from province to province, but typically deal with collection of, access to, and correction of personal information, and all designate a commissioner or ombudsperson who is authorized to handle complaints.

Privacy legislation typically falls into one of two categories:

1. Those that make breach of privacy a tort (*e.g.*, a private right to sue), for violation of a person's privacy. These statutes tend to be very brief, and are often titled "Privacy Act"; and

2. Those that make breach of privacy a quasi-criminal offence, and in some cases allow for serious penalties to be imposed. These statutes are similar to PIPEDA and are often titled "Personal Information Protection Act" (PIPA).

For example, Saskatchewan's *Privacy Act* contains only 11 sections, most of which are clarifications and exceptions, with the main statements of the Act outlined in the second section:

2. It is a tort, actionable without proof of damage, for a person wilfully and without a claim of right, to violate the privacy of another.[3]

Some provinces, such as British Columbia, have enacted both a Privacy Act and a PIPA.

In British Columbia, in a recent decision that highlights the potentially far-reaching effects of the legislation, the Office of the Information and Privacy Commissioner held that demanding information from customers, where that information is not necessary for the company's operations, is a breach of the PIPA statute.[4] So, too, is collection and retention of that information.

When considering provincial legislation, courts often resolve disputes on the basis of whether the complainant had a reasonable expectation of privacy. For example, in one case involving the use of a hidden camera by an employer, the Court found that the employee had no reasonable expectation of privacy in a lunchroom used by other employees.[5]

CHAPTER QUESTIONS

1. Can a consulting firm share information about an individual client with other clients of the firm?

2. Can information be disclosed where the personal safety of the individual is in serious jeopardy?

ANSWERS

1. The firm cannot share information without consent from the client.

2. Yes.

---

[3] R.S.S 1978, c. P-24, s. 2.

[4] OIPC Order P09-01, Cruz Ventures Ltd. d.b.a. Wild Coyote Club, July 21, 2009.

[5] *Richardson v. Davis Wire Industries* (1997), 33 B.C.L.R. (3d) 224 (S.C.).

CHAPTER 27

INTERNET LAW[1]

Overview

With the sudden arrival and subsequent growth of the Internet over the past three decades, the law was caught by surprise. No laws had yet been developed to deal with online issues. To face this challenge, the law has since adapted concepts from other areas of the law to apply to Internet use.

One of the principal challenges to creating these laws is that the Internet transcends national and international borders. Hence, the question of which laws and courts have jurisdiction for individual situations becomes difficult to resolve. Certainly, some countries, such as China, have attempted to regulate the Internet. But many people recognize that control of the Internet is impossible.

Current legal principles relating to torts,[2] contracts,[3] copyrights,[4] trademarks,[5] privacy,[6] and securities regulation[7] have been applied to the Internet. These concepts have also been adapted to deal with Internet-specific issues, such as websites. Internet users have had to significantly adapt traditional communications strategies to incorporate the Internet into their daily business. Ideally, these new strategies include policies related to email and other forms of electronic communication.

But with the Internet's new opportunities come new legal and business risks. Federal and provincial laws have been created to deal with issues such as electronic commerce, electronic evidence, consumer protection, and trademark infringement. As with other areas of the law, proper contracts and insurance can help to reduce risks.

---

[1] Much of this chapter was adopted with permission from *The Internet Law Handbook* by Bradley Freedman and Robert Deane of Borden Ladner Gervais LLP.

[2] See Chapter 12.

[3] See Chapter 6.

[4] See Chapter 4, pp. 29–30.

[5] See Chapter 4, pp. 30–31.

[6] See Chapter 26.

[7] See Chapter 25.

27.1 Jurisdiction

The question of which laws apply and which courts have jurisdiction, often referred to as *conflict of laws*, can be difficult to answer. For instance, if a contract to produce software is signed by one party in British Columbia and the other party in Washington State, with the work to be done in India for use in Sweden, which laws apply to the contract? For this reason, contracts must clearly define which jurisdiction's laws apply and which jurisdiction's courts will be used to resolve any disputes. Unfortunately, this alone may not be sufficient for dealing with all potential questions, since local laws regarding employment, taxation, safety, privacy, and others may also apply to the work being done.

This problem is compounded by the borderless nature of the Internet. If conflicts relate to wrongful activities involving residents in their jurisdiction, courts may simply take jurisdiction, especially those in the United States. Thus, if business occurs over the Internet, the laws and courts of multiple countries and provinces may take jurisdiction over it. This means that corporations face the risk of unknown laws and systems being applied to their business.

27.2 Torts

The torts of defamation and misrepresentation[8] can be used in relation to the Internet. **Defamation** is a communication that tends to injure another party's reputation and includes oral (slander) and written (libel) communications. The definition of defamation varies from jurisdiction to jurisdiction, but it is generally accepted that a communication that is merely unflattering, annoying, irksome, or embarrassing, or that hurts only the plaintiff's feelings, is not actionable.[9] The communication must injure the party in a significant way.

Defamatory remarks posted on websites or communicated through email can result in actions for libel or slander. Communications on websites tend to be treated by courts as if they were communicated over radio or television.

Similarly, negligent advice provided by email or Internet can result in an actionable claim in misrepresentation. Making false statements on the Internet is as actionable as it is in any other medium.

27.3 Copyright

The Internet poses significant challenges for copyright,[10] chiefly because of the ease by which high-quality copies can be made and distributed. In the United States, courts have imposed liability for copyright infringement on Internet service providers (ISPs); but generally this liability occurs only where the ISP actively engages or participates in the copyright infringements. The *Digital Millennium Copyright Act*[11] (DMCA) in the United States protects ISPs from liability for activities beyond their knowledge and control, but no similar legislation currently exists in Canada.

The only real recourse against Internet copyright infringement is the courts. Thus, the copyright holder needs to register the copyright to prove its publication. As well, the copyright holder should ensure that licensing agreements are in place and proper copyright markings are used throughout documents. Electronic fingerprinting and other technological tools can also sometimes be helpful in court.

---

[8] See Chapter 12.

[9] Sack, Robert D. *Sack on Defamation: Libel, Slander and Related Problems* (New York: Practising Law Institute, 1999), 2–9.

[10] See Chapter 4, pp. 29–30.

[11] Pub. L. No. 105-304, 112 Stat. 2860 (Oct. 28, 1998).

27.4 Trademarks

As with copyright, the Internet poses challenges and risks respecting use and protection of trademarks. Website domain names and content, together with other Internet activities, can breach trademarks. As discussed in Chapter 4, owners must actively protect their trademarks; otherwise, they lose them. While searching on the Internet may be relatively easy, the task of policing the Internet for trademark infringement can be costly and time-consuming.

Like corporate names, the registration and use of domain names is subject to applicable trademark laws. **Domain names** are the alphanumeric addresses of sites on the Internet.[12] Domain names are administered by approved domain name registrars and are generally issued on a first-come-first-served basis upon payment of a registration fee. The most notorious domain name disputes have involved the opportunistic practice of pre-emptively registering domain names that incorporate well-known trademarks or trade-names and then offering to sell them to the highest bidder (usually the party to whom the incorporated trademark properly belongs). Courts have generally held that this practice, commonly known as "cybersquatting," constitutes unlawful trademark use because the sale of a trademark in a domain name to a different organization is likely to confuse or mislead Internet users. Courts have also held that cybersquatting constitutes misrepresentation regarding the approval, authorization, or endorsement of the trademark owner.[13]

27.5 Privacy and Security

The privacy legislation discussed in Chapter 26, including PIPEDA, applies to Internet activities. The collection, use, and disclosure of personal information over the Internet is a significant privacy and confidentiality issue. These laws are especially important for e-commerce, since personal information is collected electronically and can be transmitted quickly. E-commerce users must agree to policies on the corporate use of personal information by clicking a button; but most do so without having read the detailed text. Corporate policies should be clear and fulfill the requirements of relevant privacy legislation.

27.6 Securities Regulation

The Internet is a valuable tool for securities issuers, financial service providers, and investors. Nevertheless, it also presents considerable challenges for securities market participants seeking to ensure that their Internet activities comply with applicable laws, and for regulators seeking to protect investors and the integrity of capital markets. Regulators have issued policies and guidelines regarding the application of securities laws to Internet activities; but the governing principle is that securities laws apply as much to the Internet as they do to any other medium.

27.7 Electronic Contracts[14]

Electronic contracts present specific challenges to traditional contract law principles, including rules regarding contract formation, offer and acceptance, formalities such as signatures, and proof. Canadian lawmakers have addressed those challenges by enacting electronic commerce laws and

---

[12] See previous note.

[13] See previous note.

[14] See previous note.

electronic evidence laws. These laws generally follow the *Uniform Electronic Commerce Act* and determine how electronic contracts can satisfy legal requirements, including legal requirements for the formation of valid contracts. These statutes also provide a framework for private individuals who need to make rules for their electronic contracts, as well as for lawmakers and courts. While electronic commerce statutes do not solve all electronic contract issues, they do provide certainty regarding the rules governing the formation, performance, and enforcement of electronic contracts in Canada.

Canadian courts have also confirmed that contracts formed over the Internet may be valid and binding. This includes the "I Agree" statements on websites that have been properly created. In this situation, the computer software making the offer is the electronic agent for the offeror and is considered as effective as a human agent.

The time of delivery of this electronic communication is the instant when the message enters a computer system beyond the sender's control; similarly, the time of receipt is the instant when it enters a computer system that can be accessed by the receiver. However, parties are free to set their own rules for the timing of electronic communications by contract.

Standard form electronic contracts may also be implemented as "notice-and-acceptance-by-conduct" agreements, in which users are informed that by engaging in certain conduct, such as using a website or downloading software, they are deemed to have accepted the agreement. But these agreements must be clear and unambiguous to be enforceable.

As with conventional, unsigned standard form contracts, unusual or onerous terms in electronic contracts are not enforceable unless they have been drawn to the user's attention. Online contracting parties should also verify the identity, capacity, and authority of the contracting parties, and the authenticity of the electronic communication. Technological and practical measures may be used to reduce those risks, such as certified and secure electronic signatures, cryptography, passwords, and personal identification numbers. Some companies supplement these technological measures with traditional verification techniques, such as confirmation by telephone, facsimile delivery of signed and witnessed paper contracts, or credit card verification. Whenever a company cannot entirely eliminate risks related to identity, capacity, and authority, they may address the residual risks through contractual risk allocation and insurance.

27.8 Websites

As with any business project, a proper contract for development, implementation, and maintenance of a website is recommended for reducing risk for both parties. Website developers generally begin by assessing the purposes for the website, followed by designing and planning the website, developing the website, receiving feedback from the clients, and finally deploying the finished product. But the developer's role does not end at this point, because someone has to maintain the currency of the website to reduce problems caused by outdated information. Moreover, ongoing technical support is also very important given that loss of business resulting from a nonfunctioning website can be significant.

27.9 Communications System Risk Management[15]

Modern communications systems present significant business and legal risks. Communications system misuse can reduce productivity and profitability, impose significant additional costs, and result in costly litigation and substantial legal liabilities. Thus, every organization should address these concerns by adopting and implementing a written *communications system policy* (CSP) governing the use of Internet, intranets and extranets, email, and voice mail (vmail). A CSP should educate employees, contractors, and other systems users about risks and liabilities and should define acceptable ways to use communications technology.

Email has high risks and liabilities. A CSP should carefully define proper use of email to reduce the following risks.[16]

1. Email is typically casual, conversational, and spontaneous, and tends to be created with less care than more formal communications. For this reason, emails often contain ill-considered and potentially damaging statements.

2. Email can be effortlessly reproduced, distributed, and redistributed to innumerable recipients inside and outside an organization at virtually no cost. Thus, damaging communication can spread quickly.

3. A click of the wrong button can result in immediate and irretrievable distribution of email to numerous unintended recipients. Thus, potentially damaging or embarrassing material can end up in the wrong person's inbox. Email can also be forwarded to unintended recipients.

4. Email is relatively easy to forge (known as "spoofing") or to alter before forwarding it to others. Thus, damaging communication bearing a professional's name can be created and distributed without that professional's awareness.

5. Email can be lost or delayed due to causes beyond the control of the sender or recipient.

6. Email provides more detailed information than ordinary paper communications. It details who created the email, when and to whom it was sent, and, in some cases, when it was received and read. Thus, transmittal facts are readily apparent.

7. Email is almost always recoverable and is usually more difficult to eliminate than are paper communications. Data forensics experts can find deleted emails on hard drives.

8. Email is susceptible to unauthorized access at each computer where an electronic copy of the email resides, as well as in each filing cabinet where paper copies of the email are stored.

If users can be made aware of these risks and realities, they will use email more carefully.

Similarly, vmail presents many of the same benefits and risks as email. Vmail messages can be saved, recorded, or transcribed, and used as evidence. However, they can be more easily erased than email.

A clear and comprehensive CSP can enhance communications system benefits and reduce communications system risks. A CSP should be prepared with proper legal advice to ensure that it is consistent with applicable laws and contractual obligations, including collective bargaining agreements. It should also be prepared with proper technical advice. Proper implementation of a CSP is critical, and a senior management person should be made responsible for the CSP implementation.

---

[15] See previous note.

[16] See previous note.

CHAPTER QUESTIONS

1. Describe three problems associated with email.

2. Is a person who makes a defamatory statement on a website liable for defamation?

ANSWERS

1. See the list on p. 277.

2. Yes

APPENDIX

The following sample contracts are provided in this appendix:

CCDC 2 – 2008, Stipulated Price Contract   280

CCDC 220 – 2002, Bid Bond   313

CCDC 221 – 2002, Performance Bond   314

CCDC 222 – 2002, Labour & Material Payment Bond   315

ACEC 31 – 2009, Engineering Agreement between Client and Engineer   317
Note: Schedules to ACEC Document 31 are not included in this Appendix.

Sample copies of the Canadian Construction Documents Committee (CCDC) contracts have been provided with the kind permission of CCDC. CCDC maintains copyright over such documents and fees must be paid to CCDC if these documents are used in whole or in part. Please contact the CCDC directly at **www.ccdc.org** for working copies of these documents.

The ACEC 31 document is printed with the kind permission of the Association of Consulting Engineering Companies (ACEC). The ACEC maintains copyright over such document and fees must be paid to ACEC if this document is used in whole or in part.

CCDC 2

stipulated price contract

2008

[Name of the Project]

Apply a CCDC 2 copyright seal here. The application of the seal demonstrates the intention of the party proposing the use of this document that it be an accurate and unamended form of CCDC 2 – 2008 except to the extent that any alterations, additions or modifications are set forth in supplementary conditions.

CANADIAN CONSTRUCTION DOCUMENTS COMMITTEE
CANADIAN CONSTRUCTION DOCUMENTS COMMITTEE
CANADIAN CONSTRUCTION DOCUMENTS COMMITTEE

TABLE OF CONTENTS

AGREEMENT BETWEEN OWNER AND CONTRACTOR
A-1 The Work
A-2 Agreements and Amendments
A-3 Contract Documents
A-4 Contract Price
A-5 Payment
A-6 Receipt of and Addresses for Notices in Writing
A-7 Language of the Contract
A-8 Succession

DEFINITIONS
1. Change Directive
2. Change Order
3. Construction Equipment
4. Consultant
5. Contract
6. Contract Documents
7. Contract Price
8. Contract Time
9. Contractor
10. Drawings
11. Notice in Writing
12. Owner
13. Place of the Work
14. Product
15. Project
16. Provide
17. Shop Drawings
18. Specifications
19. Subcontractor
20. Substantial Performance of the Work
21. Supplemental Instruction
22. Supplier
23. Temporary Work
24. Value Added Taxes
25. Work
26. Working Day

GENERAL CONDITIONS OF THE STIPULATED PRICE CONTRACT

PART 1 GENERAL PROVISIONS
GC 1.1 Contract Documents
GC 1.2 Law of the Contract
GC 1.3 Rights and Remedies
GC 1.4 Assignment

PART 2 ADMINISTRATION OF THE CONTRACT
GC 2.1 Authority of the Consultant
GC 2.2 Role of the Consultant
GC 2.3 Review and Inspection of the Work
GC 2.4 Defective Work

PART 3 EXECUTION OF THE WORK
GC 3.1 Control of the Work
GC 3.2 Construction by Owner or Other Contractors
GC 3.3 Temporary Work
GC 3.4 Document Review
GC 3.5 Construction Schedule
GC 3.6 Supervision
GC 3.7 Subcontractors and Suppliers
GC 3.8 Labour and Products
GC 3.9 Documents at the Site
GC 3.10 Shop Drawings
GC 3.11 Use of the Work
GC 3.12 Cutting and Remedial Work
GC 3.13 Cleanup

PART 4 ALLOWANCES
GC 4.1 Cash Allowances
GC 4.2 Contingency Allowance

PART 5 PAYMENT
GC 5.1 Financing Information Required of the Owner
GC 5.2 Applications for Progress Payment
GC 5.3 Progress Payment
GC 5.4 Substantial Performance of the Work
GC 5.5 Payment of Holdback upon Substantial Performance of the Work
GC 5.6 Progressive Release of Holdback
GC 5.7 Final Payment
GC 5.8 Withholding of Payment
GC 5.9 Non-conforming Work

PART 6 CHANGES IN THE WORK
GC 6.1 Owner's Right to Make Changes
GC 6.2 Change Order
GC 6.3 Change Directive
GC 6.4 Concealed or Unknown Conditions
GC 6.5 Delays
GC 6.6 Claims for a Change in Contract Price

PART 7 DEFAULT NOTICE
GC 7.1 Owner's Right to Perform the Work, Terminate the Contractor's Right to Continue with the Work or Terminate the Contract
GC 7.2 Contractor's Right to Suspend the Work or Terminate the Contract

PART 8 DISPUTE RESOLUTION
GC 8.1 Authority of the Consultant
GC 8.2 Negotiation, Mediation and Arbitration
GC 8.3 Retention of Rights

PART 9 PROTECTION OF PERSONS AND PROPERTY
GC 9.1 Protection of Work and Property
GC 9.2 Toxic and Hazardous Substances
GC 9.3 Artifacts and Fossils
GC 9.4 Construction Safety
GC 9.5 Mould

PART 10 GOVERNING REGULATIONS
GC 10.1 Taxes and Duties
GC 10.2 Laws, Notices, Permits, and Fees
GC 10.3 Patent Fees
GC 10.4 Workers' Compensation

PART 11 INSURANCE AND CONTRACT SECURITY
GC 11.1 Insurance
GC 11.2 Contract Security

PART 12 INDEMNIFICATION, WAIVER OF CLAIMS AND WARRANTY
GC 12.1 Indemnification
GC 12.2 Waiver of Claims
GC 12.3 Warranty

The Canadian Construction Documents Committee (CCDC) is a national joint committee responsible for the development, production and review of standard Canadian construction contracts, forms and guides. Formed in 1974 the CCDC is made up of volunteer representatives from:

> Public Sector Owners
> Private Sector Owners
> Canadian Bar Association (Ex-Officio)
> * The Association of Canadian Engineering Companies
> * The Canadian Construction Association
> * Construction Specifications Canada
> * The Royal Architectural Institute of Canada

*Committee policy and procedures are directed and approved by the four constituent national organizations.

CCDC 2 is the product of a consensus-building process aimed at balancing the interests of all parties on the construction project. It reflects recommended industry practices. CCDC 2 can have important consequences. The CCDC and its constituent member organizations do not accept any responsibility or liability for loss or damage which may be suffered as a result of the use or interpretation of CCDC 2.

AGREEMENT BETWEEN OWNER AND CONTRACTOR
For use when a stipulated price is the basis of payment.

This Agreement made on the _____ day of _____ in the year _____ .

by and between the parties

_____

hereinafter called the *"Owner"*

and

_____

hereinafter called the *"Contractor"*

The *Owner* and the *Contractor* agree as follows:

ARTICLE A-1   THE WORK

The *Contractor* shall:

1.1   perform the *Work* required by the *Contract Documents* for

_____

insert above the name of the Work

located at

_____

insert above the Place of the Work

for which the Agreement has been signed by the parties, and for which

insert above the name of the Consultant

is acting as and is hereinafter called the *"Consultant"* and

1.2   do and fulfill everything indicated by the *Contract Documents*, and

1.3   commence the *Work* by the _____ day of _____ in the year _____ and, subject to adjustment in *Contract Time* as provided for in the *Contract Documents*, attain *Substantial Performance of the Work*, by the _____ day of _____ in the year _____ .

ARTICLE A-2   AGREEMENTS AND AMENDMENTS

2.1   The *Contract* supersedes all prior negotiations, representations or agreements, either written or oral, relating in any manner to the *Work*, including the bidding documents that are not expressly listed in Article A-3 of the Agreement - CONTRACT DOCUMENTS.

2.2   The *Contract* may be amended only as provided in the *Contract Documents*.

CCDC 2 – 2008 File 005213

1

ARTICLE A-3   CONTRACT DOCUMENTS

3.1   The following are the *Contract Documents* referred to in Article A-1 of the Agreement - THE WORK:

- Agreement between *Owner* and *Contractor*
- Definitions
- The General Conditions of the Stipulated Price Contract
*

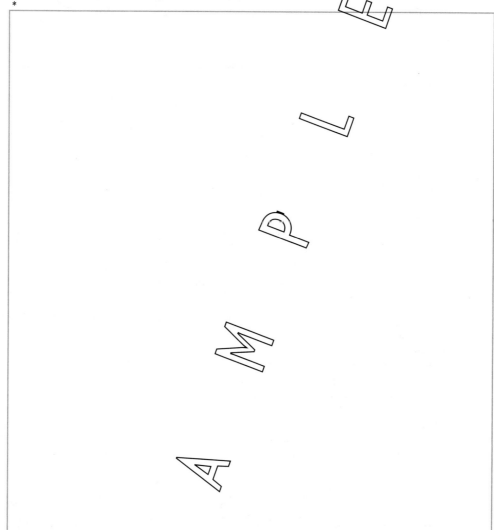

*   *(Insert here, attaching additional pages if required, a list identifying all other Contract Documents e.g. supplementary conditions; information documents; specifications, giving a list of contents with section numbers and titles, number of pages and date; material finishing schedules; drawings, giving drawing number, title, date, revision date or mark; addenda, giving title, number, date)*

CCDC 2 – 2008 File 005213

2

ARTICLE A-4   CONTRACT PRICE

4.1   The *Contract Price*, which excludes *Value Added Taxes*, is:

_____ /100 dollars $ _____

4.2   *Value Added Taxes* (of _____ %) payable by the *Owner* to the *Contractor* are:

_____ /100 dollars $ _____

4.3   Total amount payable by the *Owner* to the *Contractor* for the construction of the *Work* is:

_____ /100 dollars $ _____

4.4   These amounts shall be subject to adjustments as provided in the *Contract Documents*.

4.5   All amounts are in Canadian funds.

ARTICLE A-5   PAYMENT

5.1   Subject to the provisions of the *Contract Documents*, and in accordance with legislation and statutory regulations respecting holdback percentages and, where such legislation or regulations do not exist or apply, subject to a holdback of _____ percent (_____ %), the *Owner* shall:

   .1   make progress payments to the *Contractor* on account of the *Contract Price* when due in the amount certified by the *Consultant* together with such *Value Added Taxes* as may be applicable to such payments, and

   .2   upon *Substantial Performance of the Work*, pay to the *Contractor* the unpaid balance of the holdback amount when due together with such *Value Added Taxes* as may be applicable to such payment, and

   .3   upon the issuance of the final certificate for payment, pay to the *Contractor* the unpaid balance of the *Contract Price* when due together with such *Value Added Taxes* as may be applicable to such payment.

5.2   In the event of loss or damage occurring where payment becomes due under the property and boiler insurance policies, payments shall be made to the *Contractor* in accordance with the provisions of GC 11.1 – INSURANCE.

5.3   Interest

   .1   Should either party fail to make payments as they become due under the terms of the *Contract* or in an award by arbitration or court, interest at the following rates on such unpaid amounts shall also become due and payable until payment:

      (1) 2% per annum above the prime rate for the first 60 days.

      (2) 4% per annum above the prime rate after the first 60 days.

      Such interest shall be compounded on a monthly basis. The prime rate shall be the rate of interest quoted by

_____

(Insert name of chartered lending institution whose prime rate is to be used)

      for prime business loans as it may change from time to time.

   .2   Interest shall apply at the rate and in the manner prescribed by paragraph 5.3.1 of this Article on the settlement amount of any claim in dispute that is resolved either pursuant to Part 8 of the General Conditions – DISPUTE RESOLUTION or otherwise, from the date the amount would have been due and payable under the *Contract*, had it not been in dispute, until the date it is paid.

CCDC 2 – 2008 File 005213   3

ARTICLE A-6   RECEIPT OF AND ADDRESSES FOR NOTICES IN WRITING

6.1   *Notices in Writing* will be addressed to the recipient at the address set out below. The delivery of a *Notice in Writing* will be by hand, by courier, by prepaid first class mail, or by facsimile or other form of electronic communication during the transmission of which no indication of failure of receipt is communicated to the sender. A *Notice in Writing* delivered by one party in accordance with this *Contract* will be deemed to have been received by the other party on the date of delivery if delivered by hand or courier, or if sent by mail it shall be deemed to have been received five calendar days after the date on which it was mailed, provided that if either such day is not a *Working Day*, then the *Notice in Writing* shall be deemed to have been received on the *Working Day* next following such day. A *Notice in Writing* sent by facsimile or other form of electronic communication shall be deemed to have been received on the date of its transmission provided that if such day is not a *Working Day* or if it is received after the end of normal business hours on the date of its transmission at the place of receipt, then it shall be deemed to have been received at the opening of business at the place of receipt on the first *Working Day* next following the transmission thereof. An address for a party may be changed by *Notice in Writing* to the other party setting out the new address in accordance with this Article.

Owner

_____
name of Owner*

_____
address

_____
facsimile number   *email address*

Contractor

_____
name of Contractor*

_____
address

_____
facsimile number   *email address*

Consultant

_____
name of Consultant*

_____
address

_____
facsimile number   *email address*

* *If it is intended that the notice must be received by a specific individual, that individual's name shall be indicated.*

ARTICLE A-7   LANGUAGE OF THE CONTRACT

7.1   When the *Contract Documents* are prepared in both the English and French languages, it is agreed that in the event of any apparent discrepancy between the English and French versions, the   English   /   French   # language shall prevail.
        # Complete this statement by striking out inapplicable term.

7.2   This Agreement is drawn in English at the request of the parties hereto. La présente convention est rédigée en anglais à la demande des parties.

CCDC 2 – 2008 File 005213   4

ARTICLE A-8 SUCCESSION

8.1 The *Contract* shall enure to the benefit of and be binding upon the parties hereto, their respective heirs, legal representatives, successors, and assigns.

In witness whereof the parties hereto have executed this Agreement by the hands of their duly authorized representatives.

SIGNED AND DELIVERED
in the presence of:

WITNESS

OWNER

_____
name of owner

_____   _____
signature   *signature*

_____   _____
name of person signing   *name and title of person signing*

_____   _____
signature   *signature*

_____   _____
name of person signing   *name and title of person signing*

WITNESS

CONTRACTOR

_____
name of Contractor

_____   _____
signature   *signature*

_____   _____
name of person signing   *name and title of person signing*

_____   _____
signature   *signature*

_____   _____
name of person signing   *name and title of person signing*

N.B.   *Where legal jurisdiction, local practice or Owner or Contractor requirement calls for:*
      (a) proof of authority to execute this document, attach such proof of authority in the form of a certified copy of a resolution naming the representative(s) authorized to sign the Agreement for and on behalf of the corporation or partnership; or
      (b) the affixing of a corporate seal, this Agreement should be properly sealed.

CCDC 2 – 2008 File 005213   5

DEFINITIONS

The following Definitions shall apply to all *Contract Documents*.

1. **Change Directive**

    A *Change Directive* is a written instruction prepared by the *Consultant* and signed by the *Owner* directing the *Contractor* to proceed with a change in the *Work* within the general scope of the *Contract Documents* prior to the *Owner* and the *Contractor* agreeing upon adjustments in the *Contract Price* and the *Contract Time*.

2. **Change Order**

    A *Change Order* is a written amendment to the *Contract* prepared by the *Consultant* and signed by the *Owner* and the *Contractor* stating their agreement upon:
    - a change in the *Work*;
    - the method of adjustment or the amount of the adjustment in the *Contract Price*, if any; and
    - the extent of the adjustment in the *Contract Time*, if any.

3. **Construction Equipment**

    Construction Equipment means all machinery and equipment, either operated or not operated, that is required for preparing, fabricating, conveying, erecting, or otherwise performing the *Work* but is not incorporated into the *Work*.

4. **Consultant**

    The *Consultant* is the person or entity engaged by the *Owner* and identified as such in the Agreement. The *Consultant* is the Architect, the Engineer or entity licensed to practise in the province or territory of the *Place of the Work*. The term *Consultant* means the *Consultant* or the *Consultant*'s authorized representative.

5. **Contract**

    The *Contract* is the undertaking by the parties to perform their respective duties, responsibilities and obligations as prescribed in the *Contract Documents* and represents the entire agreement between the parties.

6. **Contract Documents**

    The *Contract Documents* consist of those documents listed in Article A-3 of the Agreement - CONTRACT DOCUMENTS and amendments agreed upon between the parties.

7. **Contract Price**

    The *Contract Price* is the amount stipulated in Article A-4 of the Agreement - CONTRACT PRICE.

8. **Contract Time**

    The *Contract Time* is the time stipulated in paragraph 1.3 of Article A-1 of the Agreement - THE WORK from commencement of the *Work* to *Substantial Performance of the Work.*

9. **Contractor**

    The *Contractor* is the person or entity identified as such in the Agreement. The term *Contractor* means the *Contractor* or the *Contractor*'s authorized representative as designated to the *Owner* in writing.

10. **Drawings**

    The *Drawings* are the graphic and pictorial portions of the *Contract Documents*, wherever located and whenever issued, showing the design, location and dimensions of the *Work*, generally including plans, elevations, sections, details, and diagrams.

11. **Notice in Writing**

    A *Notice in Writing*, where identified in the *Contract Documents*, is a written communication between the parties or between them and the *Consultant* that is transmitted in accordance with the provisions of Article A-6 of the Agreement – RECEIPT OF AND ADDRESSES FOR NOTICES IN WRITING.

12. **Owner**

    The *Owner* is the person or entity identified as such in the Agreement. The term *Owner* means the *Owner* or the *Owner*'s authorized agent or representative as designated to the *Contractor* in writing, but does not include the *Consultant*.

13. **Place of the Work**

    The *Place of the Work* is the designated site or location of the *Work* identified in the *Contract Documents*.

14. **Product**

    Product or Products means material, machinery, equipment, and fixtures forming the *Work*, but does not include *Construction Equipment*.

CCDC 2 - 2008 File 007100

6

15. Project

The *Project* means the total construction contemplated of which the *Work* may be the whole or a part.

16. Provide

Provide means to supply and install.

17. Shop Drawings

Shop Drawings are drawings, diagrams, illustrations, schedules, performance charts, brochures, *Product* data, and other data which the *Contractor* provides to illustrate details of portions of the *Work*.

18. Specifications

The *Specifications* are that portion of the *Contract Documents*, wherever located and whenever issued, consisting of the written requirements and standards for *Products*, systems, workmanship, quality, and the services necessary for the performance of the *Work*.

19. Subcontractor

A *Subcontractor* is a person or entity having a direct contract with the *Contractor* to perform a part or parts of the *Work* at the *Place of the Work*.

20. Substantial Performance of the Work

Substantial Performance of the Work is as defined in the lien legislation applicable to the *Place of the Work*. If such legislation is not in force or does not contain such definition, or if the *Work* is governed by the Civil Code of Quebec, *Substantial Performance of the Work* shall have been reached when the *Work* is ready for use or is being used for the purpose intended and is so certified by the *Consultant*.

21. Supplemental Instruction

A *Supplemental Instruction* is an instruction, not involving adjustment in the *Contract Price* or *Contract Time*, in the form of *Specifications*, *Drawings*, schedules, samples, models or written instructions, consistent with the intent of the *Contract Documents*. It is to be issued by the *Consultant* to supplement the *Contract Documents* as required for the performance of the *Work*.

22. Supplier

A *Supplier* is a person or entity having a direct contract with the *Contractor* to supply *Products*.

23. Temporary Work

Temporary Work means temporary supports, structures, facilities, services, and other temporary items, excluding *Construction Equipment*, required for the execution of the *Work* but not incorporated into the *Work*.

24. Value Added Taxes

Value Added Taxes means such sum as shall be levied upon the *Contract Price* by the Federal or any Provincial or Territorial Government and is computed as a percentage of the *Contract Price* and includes the Goods and Services Tax, the Quebec Sales Tax, the Harmonized Sales Tax, and any similar tax, the collection and payment of which have been imposed on the *Contractor* by the tax legislation.

25. Work

The *Work* means the total construction and related services required by the *Contract Documents*.

26. Working Day

Working Day means a day other than a Saturday, Sunday, statutory holiday, or statutory vacation day that is observed by the construction industry in the area of the *Place of the Work*.

CCDC 2 - 2008 File 007100

7

Standard Construction Document   CCDC 2 – 2008

GENERAL CONDITIONS OF THE STIPULATED PRICE CONTRACT

PART 1 GENERAL PROVISIONS

GC 1.1 CONTRACT DOCUMENTS

1.1.1   The intent of the *Contract Documents* is to include the labour, *Products* and services necessary for the performance of the *Work* by the *Contractor* in accordance with these documents. It is not intended, however, that the *Contractor* shall supply products or perform work not consistent with, not covered by, or not properly inferable from the *Contract Documents*.

1.1.2   Nothing contained in the *Contract Documents* shall create any contractual relationship between:

   .1   the *Owner* and a *Subcontractor*, a *Supplier*, or their agent, employee, or other person performing any portion of the *Work*.

   .2   the *Consultant* and the *Contractor*, a *Subcontractor*, a *Supplier*, or their agent, employee, or other person performing any portion of the *Work*.

1.1.3   The *Contract Documents* are complementary, and what is required by any one shall be as binding as if required by all.

1.1.4   Words and abbreviations which have well known technical or trade meanings are used in the *Contract Documents* in accordance with such recognized meanings.

1.1.5   References in the *Contract Documents* to the singular shall be considered to include the plural as the context requires.

1.1.6   Neither the organization of the *Specifications* nor the arrangement of *Drawings* shall control the *Contractor* in dividing the work among *Subcontractors* and *Suppliers*.

1.1.7   If there is a conflict within the *Contract Documents*:

   .1   the order of priority of documents, from highest to lowest, shall be

      –   the Agreement between the *Owner* and the *Contractor*,

      –   the Definitions,

      –   Supplementary Conditions,

      –   the General Conditions,

      –   Division 1 of the *Specifications*,

      –   technical *Specifications*,

      –   material and finishing schedules,

      –   the *Drawings*.

   .2   *Drawings* of larger scale shall govern over those of smaller scale of the same date.

   .3   dimensions shown on *Drawings* shall govern over dimensions scaled from *Drawings*.

   .4   later dated documents shall govern over earlier documents of the same type.

1.1.8   The *Owner* shall provide the *Contractor*, without charge, sufficient copies of the *Contract Documents* to perform the *Work*.

1.1.9   *Specifications*, *Drawings*, models, and copies thereof furnished by the *Consultant* are and shall remain the *Consultant*'s property, with the exception of the signed *Contract* sets, which shall belong to each party to the *Contract*. All *Specifications*, *Drawings* and models furnished by the *Consultant* are to be used only with respect to the *Work* and are not to be used on other work. These *Specifications*, *Drawings* and models are not to be copied or altered in any manner without the written authorization of the *Consultant*.

1.1.10   Models furnished by the *Contractor* at the *Owner*'s expense are the property of the *Owner*.

GC 1.2 LAW OF THE CONTRACT

1.2.1   The law of the *Place of the Work* shall govern the interpretation of the *Contract*.

GC 1.3 RIGHTS AND REMEDIES

1.3.1   Except as expressly provided in the *Contract Documents*, the duties and obligations imposed by the *Contract Documents* and the rights and remedies available thereunder shall be in addition to and not a limitation of any duties, obligations, rights, and remedies otherwise imposed or available by law.

1.3.2   No action or failure to act by the *Owner*, *Consultant* or *Contractor* shall constitute a waiver of any right or duty afforded any of them under the *Contract*, nor shall any such action or failure to act constitute an approval of or acquiescence in any breach thereunder, except as may be specifically agreed in writing.

GC 1.4 ASSIGNMENT

1.4.1   Neither party to the *Contract* shall assign the *Contract* or a portion thereof without the written consent of the other, which consent shall not be unreasonably withheld.

PART 2 ADMINISTRATION OF THE CONTRACT

GC 2.1 AUTHORITY OF THE CONSULTANT

2.1.1   The *Consultant* will have authority to act on behalf of the *Owner* only to the extent provided in the *Contract Documents*, unless otherwise modified by written agreement as provided in paragraph 2.1.2.

2.1.2   The duties, responsibilities and limitations of authority of the *Consultant* as set forth in the *Contract Documents* shall be modified or extended only with the written consent of the *Owner*, the *Contractor* and the *Consultant*.

2.1.3   If the *Consultant*'s employment is terminated, the *Owner* shall immediately appoint or reappoint a *Consultant* against whom the *Contractor* makes no reasonable objection and whose status under the *Contract Documents* shall be that of the former *Consultant*.

GC 2.2 ROLE OF THE CONSULTANT

2.2.1   The *Consultant* will provide administration of the *Contract* as described in the *Contract Documents*.

2.2.2   The *Consultant* will visit the *Place of the Work* at intervals appropriate to the progress of construction to become familiar with the progress and quality of the work and to determine if the *Work* is proceeding in general conformity with the *Contract Documents*.

2.2.3   If the *Owner* and the *Consultant* agree, the *Consultant* will provide at the *Place of the Work*, one or more project representatives to assist in carrying out the *Consultant*'s responsibilities. The duties, responsibilities and limitations of authority of such project representatives shall be as set forth in writing to the *Contractor*.

2.2.4   The *Consultant* will promptly inform the *Owner* of the date of receipt of the *Contractor*'s applications for payment as provided in paragraph 5.3.1.1 of GC 5.3 – PROGRESS PAYMENT.

2.2.5   Based on the *Consultant*'s observations and evaluation of the *Contractor*'s applications for payment, the *Consultant* will determine the amounts owing to the *Contractor* under the *Contract* and will issue certificates for payment as provided in Article A-5 of the Agreement - PAYMENT, GC 5.3 - PROGRESS PAYMENT and GC 5.7 - FINAL PAYMENT.

2.2.6   The *Consultant* will not be responsible for and will not have control, charge or supervision of construction means, methods, techniques, sequences, or procedures, or for safety precautions and programs required in connection with the *Work* in accordance with the applicable construction safety legislation, other regulations or general construction practice. The *Consultant* will not be responsible for the *Contractor*'s failure to carry out the *Work* in accordance with the *Contract Documents*. The *Consultant* will not have control over, charge of or be responsible for the acts or omissions of the *Contractor*, *Subcontractors*, *Suppliers*, or their agents, employees, or any other persons performing portions of the *Work*.

2.2.7   Except with respect to GC 5.1 - FINANCING INFORMATION REQUIRED OF THE OWNER, the *Consultant* will be, in the first instance, the interpreter of the requirements of the *Contract Documents*.

2.2.8   Matters in question relating to the performance of the *Work* or the interpretation of the *Contract Documents* shall be initially referred in writing to the *Consultant* by the party raising the question for interpretations and findings and copied to the other party.

2.2.9   Interpretations and findings of the *Consultant* shall be consistent with the intent of the *Contract Documents*. In making such interpretations and findings the *Consultant* will not show partiality to either the *Owner* or the *Contractor*.

2.2.10   The *Consultant*'s interpretations and findings will be given in writing to the parties within a reasonable time.

2.2.11   With respect to claims for a change in *Contract Price*, the *Consultant* will make findings as set out in GC 6.6 – CLAIMS FOR A CHANGE IN CONTRACT PRICE.

2.2.12   The *Consultant* will have authority to reject work which in the *Consultant*'s opinion does not conform to the requirements of the *Contract Documents*. Whenever the *Consultant* considers it necessary or advisable, the *Consultant* will have authority to require inspection or testing of work, whether or not such work is fabricated, installed or completed. However, neither the authority of the *Consultant* to act nor any decision either to exercise or not to exercise such authority shall give rise to any duty or responsibility of the *Consultant* to the *Contractor*, *Subcontractors*, *Suppliers*, or their agents, employees, or other persons performing any of the *Work*.

2.2.13 During the progress of the *Work* the *Consultant* will furnish *Supplemental Instructions* to the *Contractor* with reasonable promptness or in accordance with a schedule for such instructions agreed to by the *Consultant* and the *Contractor*.

2.2.14 The *Consultant* will review and take appropriate action upon *Shop Drawings*, samples and other *Contractor*'s submittals, in accordance with the *Contract Documents*.

2.2.15 The *Consultant* will prepare *Change Orders* and *Change Directives* as provided in GC 6.2 - CHANGE ORDER and GC 6.3 - CHANGE DIRECTIVE.

2.2.16 The *Consultant* will conduct reviews of the *Work* to determine the date of *Substantial Performance of the Work* as provided in GC 5.4 - SUBSTANTIAL PERFORMANCE OF THE WORK.

2.2.17 All certificates issued by the *Consultant* will be to the best of the *Consultant*'s knowledge, information and belief. By issuing any certificate, the *Consultant* does not guarantee the *Work* is correct or complete.

2.2.18 The *Consultant* will receive and review written warranties and related documents required by the *Contract* and provided by the *Contractor* and will forward such warranties and documents to the *Owner* for the *Owner*'s acceptance.

GC 2.3 REVIEW AND INSPECTION OF THE WORK

2.3.1 The *Owner* and the *Consultant* shall have access to the *Work* at all times. The *Contractor* shall provide sufficient, safe and proper facilities at all times for the review of the *Work* by the *Consultant* and the inspection of the *Work* by authorized agencies. If parts of the *Work* are in preparation at locations other than the *Place of the Work*, the *Owner* and the *Consultant* shall be given access to such work whenever it is in progress.

2.3.2 If work is designated for tests, inspections or approvals in the *Contract Documents*, or by the *Consultant*'s instructions, or by the laws or ordinances of the *Place of the Work*, the *Contractor* shall give the *Consultant* reasonable notification of when the work will be ready for review and inspection. The *Contractor* shall arrange for and shall give the *Consultant* reasonable notification of the date and time of inspections by other authorities.

2.3.3 The *Contractor* shall furnish promptly to the *Consultant* two copies of certificates and inspection reports relating to the *Work*.

2.3.4 If the *Contractor* covers, or permits to be covered, work that has been designated for special tests, inspections or approvals before such special tests, inspections or approvals are made, given or completed, the *Contractor* shall, if so directed, uncover such work, have the inspections or tests satisfactorily completed, and make good covering work at the *Contractor*'s expense.

2.3.5 The *Consultant* may order any portion or portions of the *Work* to be examined to confirm that such work is in accordance with the requirements of the *Contract Documents*. If the work is not in accordance with the requirements of the *Contract Documents*, the *Contractor* shall correct the work and pay the cost of examination and correction. If the work is in accordance with the requirements of the *Contract Documents*, the *Owner* shall pay the cost of examination and restoration.

2.3.6 The *Contractor* shall pay the cost of making any test or inspection, including the cost of samples required for such test or inspection, if such test or inspection is designated in the *Contract Documents* to be performed by the *Contractor* or is designated by the laws or ordinances applicable to the *Place of the Work*.

2.3.7 The *Contractor* shall pay the cost of samples required for any test or inspection to be performed by the *Consultant* or the *Owner* if such test or inspection is designated in the *Contract Documents*.

GC 2.4 DEFECTIVE WORK

2.4.1 The *Contractor* shall promptly correct defective work that has been rejected by the *Consultant* as failing to conform to the *Contract Documents* whether or not the defective work has been incorporated in the *Work* and whether or not the defect is the result of poor workmanship, use of defective products or damage through carelessness or other act or omission of the *Contractor*.

2.4.2 The *Contractor* shall make good promptly other contractors' work destroyed or damaged by such corrections at the *Contractor*'s expense.

2.4.3 If in the opinion of the *Consultant* it is not expedient to correct defective work or work not performed as provided in the *Contract Documents*, the *Owner* may deduct from the amount otherwise due to the *Contractor* the difference in value between the work as performed and that called for by the *Contract Documents*. If the *Owner* and the *Contractor* do not agree on the difference in value, they shall refer the matter to the *Consultant* for a determination.

PART 3 EXECUTION OF THE WORK

GC 3.1 CONTROL OF THE WORK

3.1.1 The *Contractor* shall have total control of the *Work* and shall effectively direct and supervise the *Work* so as to ensure conformity with the *Contract Documents*.

3.1.2 The *Contractor* shall be solely responsible for construction means, methods, techniques, sequences, and procedures and for co-ordinating the various parts of the *Work* under the *Contract*.

GC 3.2 CONSTRUCTION BY OWNER OR OTHER CONTRACTORS

3.2.1 The *Owner* reserves the right to award separate contracts in connection with other parts of the *Project* to other contractors and to perform work with own forces.

3.2.2 When separate contracts are awarded for other parts of the *Project*, or when work is performed by the *Owner*'s own forces, the *Owner* shall:
.1 provide for the co-ordination of the activities and work of other contractors and *Owner*'s own forces with the *Work* of the *Contract*;
.2 assume overall responsibility for compliance with the applicable health and construction safety legislation at the *Place of the Work*;
.3 enter into separate contracts with other contractors under conditions of contract which are compatible with the conditions of the *Contract*;
.4 ensure that insurance coverage is provided to the same requirements as are called for in GC 11.1 - INSURANCE and co-ordinate such insurance with the insurance coverage of the *Contractor* as it affects the *Work*; and
.5 take all reasonable precautions to avoid labour disputes or other disputes on the *Project* arising from the work of other contractors or the *Owner*'s own forces.

3.2.3 When separate contracts are awarded for other parts of the *Project*, or when work is performed by the *Owner*'s own forces, the *Contractor* shall:
.1 afford the *Owner* and other contractors reasonable opportunity to store their products and execute their work;
.2 cooperate with other contractors and the *Owner* in reviewing their construction schedules; and
.3 promptly report to the *Consultant* in writing any apparent deficiencies in the work of other contractors or of the *Owner*'s own forces, where such work affects the proper execution of any portion of the *Work*, prior to proceeding with that portion of the *Work*.

3.2.4 Where the *Contract Documents* identify work to be performed by other contractors or the *Owner*'s own forces, the *Contractor* shall co-ordinate and schedule the *Work* with the work of other contractors and the *Owner*'s own forces as specified in the *Contract Documents*.

3.2.5 Where a change in the *Work* is required as a result of the co-ordination and integration of the work of other contractors or *Owner*'s own forces with the *Work*, the changes shall be authorized and valued as provided in GC 6.1 – OWNER'S RIGHT TO MAKE CHANGES, GC 6.2 - CHANGE ORDER and GC 6.3 - CHANGE DIRECTIVE.

3.2.6 Disputes and other matters in question between the *Contractor* and other contractors shall be dealt with as provided in Part 8 of the General Conditions - DISPUTE RESOLUTION provided the other contractors have reciprocal obligations. The *Contractor* shall be deemed to have consented to arbitration of any dispute with any other contractor whose contract with the *Owner* contains a similar agreement to arbitrate.

GC 3.3 TEMPORARY WORK

3.3.1 The *Contractor* shall have the sole responsibility for the design, erection, operation, maintenance, and removal of *Temporary Work*.

3.3.2 The *Contractor* shall engage and pay for registered professional engineering personnel skilled in the appropriate disciplines to perform those functions referred to in paragraph 3.3.1 where required by law or by the *Contract Documents* and in all cases where such *Temporary Work* is of such a nature that professional engineering skill is required to produce safe and satisfactory results.

3.3.3 Notwithstanding the provisions of GC 3.1 - CONTROL OF THE WORK, paragraphs 3.3.1 and 3.3.2 or provisions to the contrary elsewhere in the *Contract Documents* where such *Contract Documents* include designs for *Temporary Work* or specify a method of construction in whole or in part, such designs or methods of construction shall be considered to be part of the design of the *Work* and the *Contractor* shall not be held responsible for that part of the design or the specified method of construction. The *Contractor* shall, however, be responsible for the execution of such design or specified method of construction in the same manner as for the execution of the *Work*.

GC 3.4 DOCUMENT REVIEW

3.4.1 The *Contractor* shall review the *Contract Documents* and shall report promptly to the *Consultant* any error, inconsistency or omission the *Contractor* may discover. Such review by the *Contractor* shall be to the best of the *Contractor*'s knowledge, information and belief and in making such review the *Contractor* does not assume any responsibility to the *Owner* or the *Consultant* for the accuracy of the review. The *Contractor* shall not be liable for damage or costs resulting from such errors, inconsistencies or omissions in the *Contract Documents*, which the *Contractor* did not discover. If the *Contractor* does discover any error, inconsistency or omission in the *Contract Documents*, the *Contractor* shall not proceed with the work affected until the *Contractor* has received corrected or missing information from the *Consultant*.

GC 3.5 CONSTRUCTION SCHEDULE

3.5.1 The *Contractor* shall:
   .1 prepare and submit to the *Owner* and the *Consultant* prior to the first application for payment, a construction schedule that indicates the timing of the major activities of the *Work* and provides sufficient detail of the critical events and their inter-relationship to demonstrate the *Work* will be performed in conformity with the *Contract Time*;
   .2 monitor the progress of the *Work* relative to the construction schedule and update the schedule on a monthly basis or as stipulated by the *Contract Documents*; and
   .3 advise the *Consultant* of any revisions required to the schedule as the result of extensions of the *Contract Time* as provided in Part 6 of the General Conditions - CHANGES IN THE WORK.

GC 3.6 SUPERVISION

3.6.1 The *Contractor* shall provide all necessary supervision and appoint a competent representative who shall be in attendance at the *Place of the Work* while work is being performed. The appointed representative shall not be changed except for valid reason.

3.6.2 The appointed representative shall represent the *Contractor* at the *Place of the Work*. Information and instructions provided by the *Consultant* to the *Contractor*'s appointed representative shall be deemed to have been received by the *Contractor*, except with respect to Article A-6 of the Agreement – RECEIPT OF AND ADDRESSES FOR NOTICES IN WRITING.

GC 3.7 SUBCONTRACTORS AND SUPPLIERS

3.7.1 The *Contractor* shall preserve and protect the rights of the parties under the *Contract* with respect to work to be performed under subcontract, and shall:
   .1 enter into contracts or written agreements with *Subcontractors* and *Suppliers* to require them to perform their work as provided in the *Contract Documents*;
   .2 incorporate the terms and conditions of the *Contract Documents* into all contracts or written agreements with *Subcontractors* and *Suppliers*; and
   .3 be as fully responsible to the *Owner* for acts and omissions of *Subcontractors*, *Suppliers* and of persons directly or indirectly employed by them as for acts and omissions of persons directly employed by the *Contractor*.

3.7.2 The *Contractor* shall indicate in writing, if requested by the *Owner*, those *Subcontractors* or *Suppliers* whose bids have been received by the *Contractor* which the *Contractor* would be prepared to accept for the performance of a portion of the *Work*. Should the *Owner* not object before signing the *Contract*, the *Contractor* shall employ those *Subcontractors* or *Suppliers* so identified by the *Contractor* in writing for the performance of that portion of the *Work* to which their bid applies.

3.7.3 The *Owner* may, for reasonable cause, at any time before the *Owner* has signed the *Contract*, object to the use of a proposed *Subcontractor* or *Supplier* and require the *Contractor* to employ one of the other subcontract bidders.

3.7.4 If the *Owner* requires the *Contractor* to change a proposed *Subcontractor* or *Supplier*, the *Contract Price* and *Contract Time* shall be adjusted by the differences occasioned by such required change.

3.7.5   The *Contractor* shall not be required to employ as a *Subcontractor* or *Supplier*, a person or firm to which the *Contractor* may reasonably object.

3.7.6   The *Owner*, through the *Consultant*, may provide to a *Subcontractor* or *Supplier* information as to the percentage of the *Subcontractor*'s or *Supplier*'s work which has been certified for payment.

GC 3.8 LABOUR AND PRODUCTS

3.8.1   The *Contractor* shall provide and pay for labour, *Products*, tools, *Construction Equipment*, water, heat, light, power, transportation, and other facilities and services necessary for the performance of the *Work* in accordance with the *Contract*.

3.8.2   Unless otherwise specified in the *Contract Documents*, *Products* provided shall be new. *Products* which are not specified shall be of a quality consistent with those specified and their use acceptable to the *Consultant*.

3.8.3   The *Contractor* shall maintain good order and discipline among the *Contractor*'s employees engaged on the *Work* and shall not employ on the *Work* anyone not skilled in the tasks assigned.

GC 3.9 DOCUMENTS AT THE SITE

3.9.1   The *Contractor* shall keep one copy of current *Contract Documents*, submittals, reports, and records of meetings at the *Place of the Work*, in good order and available to the *Owner* and the *Consultant*.

GC 3.10 SHOP DRAWINGS

3.10.1   The *Contractor* shall provide *Shop Drawings* as required in the *Contract Documents*.

3.10.2   The *Contractor* shall provide *Shop Drawings* to the *Consultant* to review in orderly sequence and sufficiently in advance so as to cause no delay in the *Work* or in the work of other contractors.

3.10.3   Upon request of the *Contractor* or the *Consultant*, they shall jointly prepare a schedule of the dates for provision, review and return of *Shop Drawings*.

3.10.4   The *Contractor* shall provide *Shop Drawings* in the form specified, or if not specified, as directed by the *Consultant*.

3.10.5   *Shop Drawings* provided by the *Contractor* to the *Consultant* shall indicate by stamp, date and signature of the person responsible for the review that the *Contractor* has reviewed each one of them.

3.10.6   The *Consultant*'s review is for conformity to the design concept and for general arrangement only.

3.10.7   *Shop Drawings* which require approval of any legally constituted authority having jurisdiction shall be provided to such authority by the *Contractor* for approval.

3.10.8   The *Contractor* shall review all *Shop Drawings* before providing them to the *Consultant*. The *Contractor* represents by this review that:
   .1 the *Contractor* has determined and verified all applicable field measurements, field construction conditions, *Product* requirements, catalogue numbers and similar data, or will do so, and
   .2 the *Contractor* has checked and co-ordinated each *Shop Drawing* with the requirements of the *Work* and of the *Contract Documents*.

3.10.9   At the time of providing *Shop Drawings*, the *Contractor* shall expressly advise the *Consultant* in writing of any deviations in a *Shop Drawing* from the requirements of the *Contract Documents*. The *Consultant* shall indicate the acceptance or rejection of such deviation expressly in writing.

3.10.10   The *Consultant*'s review shall not relieve the *Contractor* of responsibility for errors or omissions in the *Shop Drawings* or for meeting all requirements of the *Contract Documents*.

3.10.11   The *Contractor* shall provide revised *Shop Drawings* to correct those which the *Consultant* rejects as inconsistent with the *Contract Documents*, unless otherwise directed by the *Consultant*. The *Contractor* shall notify the *Consultant* in writing of any revisions to the *Shop Drawings* other than those requested by the *Consultant*.

3.10.12   The *Consultant* will review and return *Shop Drawings* in accordance with the schedule agreed upon, or, in the absence of such schedule, with reasonable promptness so as to cause no delay in the performance of the *Work*.

GC 3.11 USE OF THE WORK

3.11.1 The *Contractor* shall confine *Construction Equipment, Temporary Work*, storage of *Products*, waste products and debris, and operations of employees and *Subcontractors* to limits indicated by laws, ordinances, permits, or the *Contract Documents* and shall not unreasonably encumber the *Place of the Work*.

3.11.2 The *Contractor* shall not load or permit to be loaded any part of the *Work* with a weight or force that will endanger the safety of the *Work*.

GC 3.12 CUTTING AND REMEDIAL WORK

3.12.1 The *Contractor* shall perform the cutting and remedial work required to make the affected parts of the *Work* come together properly.

3.12.2 The *Contractor* shall co-ordinate the *Work* to ensure that the cutting and remedial work is kept to a minimum.

3.12.3 Should the *Owner*, the *Consultant*, other contractors or anyone employed by them be responsible for ill-timed work necessitating cutting or remedial work to be performed, the cost of such cutting or remedial work shall be valued as provided in GC 6.1 – OWNER'S RIGHT TO MAKE CHANGES, GC 6.2 - CHANGE ORDER and GC 6.3 - CHANGE DIRECTIVE.

3.12.4 Cutting and remedial work shall be performed by specialists familiar with the *Products* affected and shall be performed in a manner to neither damage nor endanger the *Work*.

GC 3.13 CLEANUP

3.13.1 The *Contractor* shall maintain the *Work* in a safe and tidy condition and free from the accumulation of waste products and debris, other than that caused by the *Owner*, other contractors or their employees.

3.13.2 Before applying for *Substantial Performance of the Work* as provided in GC 5.4 – SUBSTANTIAL PERFORMANCE OF THE WORK, the *Contractor* shall remove waste products and debris, other than that resulting from the work of the *Owner*, other contractors or their employees, and shall leave the *Place of the Work* clean and suitable for use or occupancy by the *Owner*. The *Contractor* shall remove products, tools, *Construction Equipment*, and *Temporary Work* not required for the performance of the remaining work.

3.13.3 Prior to application for the final payment, the *Contractor* shall remove any remaining products, tools, *Construction Equipment, Temporary Work*, and waste products and debris, other than those resulting from the work of the *Owner*, other contractors or their employees.

PART 4 ALLOWANCES

GC 4.1 CASH ALLOWANCES

4.1.1 The *Contract Price* includes the cash allowances, if any, stated in the *Contract Documents*. The scope of work or costs included in such cash allowances shall be as described in the *Contract Documents*.

4.1.2 The *Contract Price*, and not the cash allowances, includes the *Contractor*'s overhead and profit in connection with such cash allowances.

4.1.3 Expenditures under cash allowances shall be authorized by the *Owner* through the *Consultant*.

4.1.4 Where the actual cost of the *Work* under any cash allowance exceeds the amount of the allowance, the *Contractor* shall be compensated for the excess incurred and substantiated plus an amount for overhead and profit on the excess as set out in the *Contract Documents*. Where the actual cost of the *Work* under any cash allowance is less than the amount of the allowance, the *Owner* shall be credited for the unexpended portion of the cash allowance, but not for the *Contractor*'s overhead and profit on such amount. Multiple cash allowances shall not be combined for the purpose of calculating the foregoing.

4.1.5 The *Contract Price* shall be adjusted by *Change Order* to provide for any difference between the amount of each cash allowance and the actual cost of the work under that cash allowance.

4.1.6 The value of the work performed under a cash allowance is eligible to be included in progress payments.

4.1.7 The *Contractor* and the *Consultant* shall jointly prepare a schedule that shows when the *Consultant* and *Owner* must authorize ordering of items called for under cash allowances to avoid delaying the progress of the *Work*.

GC 4.2 CONTINGENCY ALLOWANCE

4.2.1   The *Contract Price* includes the contingency allowance, if any, stated in the *Contract Documents*.

4.2.2   The contingency allowance includes the *Contractor*'s overhead and profit in connection with such contingency allowance.

4.2.3   Expenditures under the contingency allowance shall be authorized and valued as provided in GC 6.1 – OWNER'S RIGHT TO MAKE CHANGES, GC 6.2 - CHANGE ORDER and GC 6.3 - CHANGE DIRECTIVE.

4.2.4   The *Contract Price* shall be adjusted by *Change Order* to provide for any difference between the expenditures authorized under paragraph 4.2.3 and the contingency allowance.

PART 5 PAYMENT

GC 5.1 FINANCING INFORMATION REQUIRED OF THE OWNER

5.1.1   The *Owner* shall, at the request of the *Contractor*, before signing the *Contract*, and promptly from time to time thereafter, furnish to the *Contractor* reasonable evidence that financial arrangements have been made to fulfill the *Owner*'s obligations under the *Contract*.

5.1.2   The *Owner* shall give the *Contractor Notice in Writing* of any material change in the *Owner*'s financial arrangements to fulfill the *Owner*'s obligations under the *Contract* during the performance of the *Contract*.

GC 5.2 APPLICATIONS FOR PROGRESS PAYMENT

5.2.1   Applications for payment on account as provided in Article A-5 of the Agreement - PAYMENT may be made monthly as the *Work* progresses.

5.2.2   Applications for payment shall be dated the last day of each payment period, which is the last day of the month or an alternative day of the month agreed in writing by the parties.

5.2.3   The amount claimed shall be for the value, proportionate to the amount of the *Contract*, of *Work* performed and *Products* delivered to the *Place of the Work* as of the last day of the payment period.

5.2.4   The *Contractor* shall submit to the *Consultant*, at least 15 calendar days before the first application for payment, a schedule of values for the parts of the *Work*, aggregating the total amount of the *Contract Price*, so as to facilitate evaluation of applications for payment.

5.2.5   The schedule of values shall be made out in such form and supported by such evidence as the *Consultant* may reasonably direct and when accepted by the *Consultant*, shall be used as the basis for applications for payment, unless it is found to be in error.

5.2.6   The *Contractor* shall include a statement based on the schedule of values with each application for payment.

5.2.7   Applications for payment for *Products* delivered to the *Place of the Work* but not yet incorporated into the *Work* shall be supported by such evidence as the *Consultant* may reasonably require to establish the value and delivery of the *Products*.

GC 5.3 PROGRESS PAYMENT

5.3.1   After receipt by the *Consultant* of an application for payment submitted by the *Contractor* in accordance with GC 5.2 - APPLICATIONS FOR PROGRESS PAYMENT:
    .1   the *Consultant* will promptly inform the *Owner* of the date of receipt of the *Contractor*'s application for payment,
    .2   the *Consultant* will issue to the *Owner* and copy to the *Contractor*, no later than 10 calendar days after the receipt of the application for payment, a certificate for payment in the amount applied for, or in such other amount as the *Consultant* determines to be properly due. If the *Consultant* amends the application, the *Consultant* will promptly advise the *Contractor* in writing giving reasons for the amendment,
    .3   the *Owner* shall make payment to the *Contractor* on account as provided in Article A-5 of the Agreement - PAYMENT on or before 20 calendar days after the later of:
        -    receipt by the *Consultant* of the application for payment, or
        -    the last day of the monthly payment period for which the application for payment is made.

GC 5.4 SUBSTANTIAL PERFORMANCE OF THE WORK

5.4.1 When the *Contractor* considers that the *Work* is substantially performed, or if permitted by the lien legislation applicable to the *Place of the Work* a designated portion thereof which the *Owner* agrees to accept separately is substantially performed, the *Contractor* shall, within one *Working Day*, deliver to the *Consultant* and to the *Owner* a comprehensive list of items to be completed or corrected, together with a written application for a review by the *Consultant* to establish *Substantial Performance of the Work* or substantial performance of the designated portion of the *Work*. Failure to include an item on the list does not alter the responsibility of the *Contractor* to complete the *Contract*.

5.4.2 The *Consultant* will review the *Work* to verify the validity of the application and shall promptly and in any event, no later than 20 calendar days after receipt of the *Contractor's* list and application:

   .1 advise the *Contractor* in writing that the *Work* or the designated portion of the *Work* is not substantially performed and give reasons why, or

   .2 state the date of *Substantial Performance of the Work* or a designated portion of the *Work* in a certificate and issue a copy of that certificate to each of the *Owner* and the *Contractor*.

5.4.3 Immediately following the issuance of the certificate of *Substantial Performance of the Work*, the *Contractor*, in consultation with the *Consultant*, shall establish a reasonable date for finishing the *Work*.

GC 5.5 PAYMENT OF HOLDBACK UPON SUBSTANTIAL PERFORMANCE OF THE WORK

5.5.1 After the issuance of the certificate of *Substantial Performance of the Work*, the *Contractor* shall:

   .1 submit an application for payment of the holdback amount,

   .2 submit CCDC 9A 'Statutory Declaration' to state that all accounts for labour, subcontracts, *Products*, *Construction Equipment*, and other indebtedness which may have been incurred by the *Contractor* in the *Substantial Performance of the Work* and for which the *Owner* might in any way be held responsible have been paid in full, except for amounts properly retained as a holdback or as an identified amount in dispute.

5.5.2 After the receipt of an application for payment from the *Contractor* and the statement as provided in paragraph 5.5.1, the *Consultant* will issue a certificate for payment of the holdback amount.

5.5.3 Where the holdback amount required by the applicable lien legislation has not been placed in a separate holdback account, the *Owner* shall, 10 calendar days prior to the expiry of the holdback period stipulated in the lien legislation applicable to the *Place of the Work*, place the holdback amount in a bank account in the joint names of the *Owner* and the *Contractor*.

5.5.4 In the common law jurisdictions, the holdback amount authorized by the certificate for payment of the holdback amount is due and payable on the first calendar day following the expiration of the holdback period stipulated in the lien legislation applicable to the *Place of the Work*. Where lien legislation does not exist or apply, the holdback amount shall be due and payable in accordance with other legislation, industry practice or provisions which may be agreed to between the parties. The *Owner* may retain out of the holdback amount any sums required by law to satisfy any liens against the *Work* or, if permitted by the lien legislation applicable to the *Place of the Work* other third party monetary claims against the *Contractor* which are enforceable against the *Owner*.

5.5.5 In the Province of Quebec, the holdback amount authorized by the certificate for payment of the holdback amount is due and payable 30 calendar days after the issuance of the certificate. The *Owner* may retain out of the holdback amount any sums required to satisfy any legal hypothecs that have been taken, or could be taken, against the *Work* or other third party monetary claims against the *Contractor* which are enforceable against the *Owner*.

GC 5.6 PROGRESSIVE RELEASE OF HOLDBACK

5.6.1 In the common law jurisdictions, where legislation permits and where, upon application by the *Contractor*, the *Consultant* has certified that the work of a *Subcontractor* or *Supplier* has been performed prior to *Substantial Performance of the Work*, the *Owner* shall pay the *Contractor* the holdback amount retained for such subcontract work, or the *Products* supplied by such *Supplier*, on the first calendar day following the expiration of the holdback period for such work stipulated in the lien legislation applicable to the *Place of the Work*. The *Owner* may retain out of the holdback amount any sums required by law to satisfy any liens against the *Work* or, if permitted by the lien legislation applicable to the *Place of the Work*, other third party monetary claims against the *Contractor* which are enforceable against the *Owner*.

5.6.2 In the Province of Quebec, where, upon application by the *Contractor*, the *Consultant* has certified that the work of a *Subcontractor* or *Supplier* has been performed prior to *Substantial Performance of the Work*, the *Owner* shall pay the *Contractor* the holdback amount retained for such subcontract work, or the *Products* supplied by such *Supplier*, no later than 30 calendar days after such certification by the *Consultant*. The *Owner* may retain out of the holdback amount any sums required to satisfy any legal hypothecs that have been taken, or could be taken, against the *Work* or other third party monetary claims against the *Contractor* which are enforceable against the *Owner*.

5.6.3 Notwithstanding the provisions of the preceding paragraphs, and notwithstanding the wording of such certificates, the *Contractor* shall ensure that such subcontract work or *Products* are protected pending the issuance of a final certificate for payment and be responsible for the correction of defects or work not performed regardless of whether or not such was apparent when such certificates were issued.

GC 5.7 FINAL PAYMENT

5.7.1 When the *Contractor* considers that the *Work* is completed, the *Contractor* shall submit an application for final payment.

5.7.2 The *Consultant* will, no later than 10 calendar days after the receipt of an application from the *Contractor* for final payment, review the *Work* to verify the validity of the application and advise the *Contractor* in writing that the application is valid or give reasons why it is not valid.

5.7.3 When the *Consultant* finds the *Contractor*'s application for final payment valid, the *Consultant* will promptly issue a final certificate for payment.

5.7.4 Subject to the provision of paragraph 10.4.1 of GC 10.4 - WORKERS' COMPENSATION, and any lien legislation applicable to the *Place of the Work*, the *Owner* shall, no later than 5 calendar days after the issuance of a final certificate for payment, pay the *Contractor* as provided in Article A-5 of the Agreement - PAYMENT.

GC 5.8 WITHHOLDING OF PAYMENT

5.8.1 If because of climatic or other conditions reasonably beyond the control of the *Contractor*, there are items of work that cannot be performed, payment in full for that portion of the *Work* which has been performed as certified by the *Consultant* shall not be withheld or delayed by the *Owner* on account thereof, but the *Owner* may withhold, until the remaining portion of the *Work* is finished, only such an amount that the *Consultant* determines is sufficient and reasonable to cover the cost of performing such remaining work.

GC 5.9 NON-CONFORMING WORK

5.9.1 No payment by the *Owner* under the *Contract* nor partial or entire use or occupancy of the *Work* by the *Owner* shall constitute an acceptance of any portion of the *Work* or *Products* which are not in accordance with the requirements of the *Contract Documents*.

PART 6 CHANGES IN THE WORK

GC 6.1 OWNER'S RIGHT TO MAKE CHANGES

6.1.1 The *Owner*, through the *Consultant*, without invalidating the *Contract*, may make:
.1 changes in the *Work* consisting of additions, deletions or other revisions to the *Work* by *Change Order* or *Change Directive*, and
.2 changes to the *Contract Time* for the *Work*, or any part thereof, by *Change Order*.

6.1.2 The *Contractor* shall not perform a change in the *Work* without a *Change Order* or a *Change Directive*.

GC 6.2 CHANGE ORDER

6.2.1 When a change in the *Work* is proposed or required, the *Consultant* will provide the *Contractor* with a written description of the proposed change in the *Work*. The *Contractor* shall promptly present, in a form acceptable to the *Consultant*, a method of adjustment or an amount of adjustment for the *Contract Price*, if any, and the adjustment in the *Contract Time*, if any, for the proposed change in the *Work*.

6.2.2 When the *Owner* and *Contractor* agree to the adjustments in the *Contract Price* and *Contract Time* or to the method to be used to determine the adjustments, such agreement shall be effective immediately and shall be recorded in a *Change Order*. The value of the work performed as the result of a *Change Order* shall be included in the application for progress payment.

GC 6.3 CHANGE DIRECTIVE

6.3.1 If the *Owner* requires the *Contractor* to proceed with a change in the *Work* prior to the *Owner* and the *Contractor* agreeing upon the corresponding adjustment in *Contract Price* and *Contract Time*, the *Owner*, through the *Consultant*, shall issue a *Change Directive*.

6.3.2 A *Change Directive* shall only be used to direct a change in the *Work* which is within the general scope of the *Contract Documents*.

6.3.3 A *Change Directive* shall not be used to direct a change in the *Contract Time* only.

6.3.4 Upon receipt of a *Change Directive*, the *Contractor* shall proceed promptly with the change in the *Work*.

6.3.5 For the purpose of valuing *Change Directives*, changes in the *Work* that are not substitutions or otherwise related to each other shall not be grouped together in the same *Change Directive*.

6.3.6 The adjustment in the *Contract Price* for a change carried out by way of a *Change Directive* shall be determined on the basis of the cost of the *Contractor*'s actual expenditures and savings attributable to the *Change Directive*, valued in accordance with paragraph 6.3.7 and as follows:

   .1 If the change results in a net increase in the *Contractor*'s cost, the *Contract Price* shall be increased by the amount of the net increase in the *Contractor*'s cost, plus the *Contractor*'s percentage fee on such net increase.

   .2 If the change results in a net decrease in the *Contractor*'s cost, the *Contract Price* shall be decreased by the amount of the net decrease in the *Contractor*'s cost, without adjustment for the *Contractor*'s percentage fee.

   .3 The *Contractor*'s fee shall be as specified in the *Contract Documents* or as otherwise agreed by the parties.

6.3.7 The cost of performing the work attributable to the *Change Directive* shall be limited to the actual cost of the following:

   .1 salaries, wages and benefits paid to personnel in the direct employ of the *Contractor* under a salary or wage schedule agreed upon by the *Owner* and the *Contractor*, or in the absence of such a schedule, actual salaries, wages and benefits paid under applicable bargaining agreement, and in the absence of a salary or wage schedule and bargaining agreement, actual salaries, wages and benefits paid by the *Contractor*, for personnel

      (1) stationed at the *Contractor*'s field office, in whatever capacity employed;

      (2) engaged in expediting the production or transportation of material or equipment, at shops or on the road;

      (3) engaged in the preparation or review of *Shop Drawings*, fabrication drawings, and coordination drawings; or

      (4) engaged in the processing of changes in the *Work*.

   .2 contributions, assessments or taxes incurred for such items as employment insurance, provincial or territorial health insurance, workers' compensation, and Canada or Quebec Pension Plan, insofar as such cost is based on wages, salaries or other remuneration paid to employees of the *Contractor* and included in the cost of the *Work* as provided in paragraph 6.3.7.1;

   .3 travel and subsistence expenses of the *Contractor*'s personnel described in paragraph 6.3.7.1;

   .4 all *Products* including cost of transportation thereof;

   .5 materials, supplies, *Construction Equipment*, *Temporary Work*, and hand tools not owned by the workers, including transportation and maintenance thereof, which are consumed in the performance of the *Work*; and cost less salvage value on such items used but not consumed, which remain the property of the *Contractor*;

   .6 all tools and *Construction Equipment*, exclusive of hand tools used in the performance of the *Work*, whether rented from or provided by the *Contractor* or others, including installation, minor repairs and replacements, dismantling, removal, transportation, and delivery cost thereof;

   .7 all equipment and services required for the *Contractor*'s field office;

   .8 deposits lost;

   .9 the amounts of all subcontracts;

   .10 quality assurance such as independent inspection and testing services;

   .11 charges levied by authorities having jurisdiction at the *Place of the Work*;

   .12 royalties, patent licence fees and damages for infringement of patents and cost of defending suits therefor subject always to the *Contractor*'s obligations to indemnify the *Owner* as provided in paragraph 10.3.1 of GC 10.3 - PATENT FEES;

   .13 any adjustment in premiums for all bonds and insurance which the *Contractor* is required, by the *Contract Documents*, to purchase and maintain;

   .14 any adjustment in taxes, other than *Value Added Taxes*, and duties for which the *Contractor* is liable;

   .15 charges for long distance telephone and facsimile communications, courier services, expressage, and petty cash items incurred in relation to the performance of the *Work*;

   .16 removal and disposal of waste products and debris; and

   .17 safety measures and requirements.

6.3.8   Notwithstanding any other provisions contained in the General Conditions of the *Contract*, it is the intention of the parties that the cost of any item under any cost element referred to in paragraph 6.3.7 shall cover and include any and all costs or liabilities attributable to the *Change Directive* other than those which are the result of or occasioned by any failure on the part of the *Contractor* to exercise reasonable care and diligence in the *Contractor*'s attention to the *Work*. Any cost due to failure on the part of the *Contractor* to exercise reasonable care and diligence in the *Contractor*'s attention to the *Work* shall be borne by the *Contractor*.

6.3.9   The *Contractor* shall keep full and detailed accounts and records necessary for the documentation of the cost of performing the *Work* attributable to the *Change Directive* and shall provide the *Consultant* with copies thereof when requested.

6.3.10   For the purpose of valuing *Change Directives*, the *Owner* shall be afforded reasonable access to all of the *Contractor*'s pertinent documents related to the cost of performing the *Work* attributable to the *Change Directive*.

6.3.11   Pending determination of the final amount of a *Change Directive*, the undisputed value of the *Work* performed as the result of a *Change Directive* is eligible to be included in progress payments.

6.3.12   If the *Owner* and the *Contractor* do not agree on the proposed adjustment in the *Contract Time* attributable to the change in the *Work*, or the method of determining it, the adjustment shall be referred to the *Consultant* for determination.

6.3.13   When the *Owner* and the *Contractor* reach agreement on the adjustment to the *Contract Price* and to the *Contract Time*, this agreement shall be recorded in a *Change Order*.

GC 6.4 CONCEALED OR UNKNOWN CONDITIONS

6.4.1   If the *Owner* or the *Contractor* discover conditions at the *Place of the Work* which are:
    .1   subsurface or otherwise concealed physical conditions which existed before the commencement of the *Work* which differ materially from those indicated in the *Contract Documents*; or
    .2   physical conditions, other than conditions due to weather, that are of a nature which differ materially from those ordinarily found to exist and generally recognized as inherent in construction activities of the character provided for in the *Contract Documents*,
   then the observing party shall give *Notice in Writing* to the other party of such conditions before they are disturbed and in no event later than 5 *Working Days* after first observance of the conditions.

6.4.2   The *Consultant* will promptly investigate such conditions and make a finding. If the finding is that the conditions differ materially and this would cause an increase or decrease in the *Contractor*'s cost or time to perform the *Work*, the *Consultant*, with the *Owner*'s approval, will issue appropriate instructions for a change in the *Work* as provided in GC 6.2 - CHANGE ORDER or GC 6.3 - CHANGE DIRECTIVE.

6.4.3   If the *Consultant* finds that the conditions at the *Place of the Work* are not materially different or that no change in the *Contract Price* or the *Contract Time* is justified, the *Consultant* will report the reasons for this finding to the *Owner* and the *Contractor* in writing.

6.4.4   If such concealed or unknown conditions relate to toxic and hazardous substances and materials, artifacts and fossils, or mould, the parties will be governed by the provisions of GC 9.2 - TOXIC AND HAZARDOUS SUBSTANCES, GC 9.3 - ARTIFACTS AND FOSSILS and GC 9.5 – MOULD.

GC 6.5 DELAYS

6.5.1   If the *Contractor* is delayed in the performance of the *Work* by an action or omission of the *Owner*, *Consultant* or anyone employed or engaged by them directly or indirectly, contrary to the provisions of the *Contract Documents*, then the *Contract Time* shall be extended for such reasonable time as the *Consultant* may recommend in consultation with the *Contractor*. The *Contractor* shall be reimbursed by the *Owner* for reasonable costs incurred by the *Contractor* as the result of such delay.

6.5.2   If the *Contractor* is delayed in the performance of the *Work* by a stop work order issued by a court or other public authority and providing that such order was not issued as the result of an act or fault of the *Contractor* or any person employed or engaged by the *Contractor* directly or indirectly, then the *Contract Time* shall be extended for such reasonable time as the *Consultant* may recommend in consultation with the *Contractor*. The *Contractor* shall be reimbursed by the *Owner* for reasonable costs incurred by the *Contractor* as the result of such delay.

6.5.3 If the *Contractor* is delayed in the performance of the *Work* by:
   .1 labour disputes, strikes, lock-outs (including lock-outs decreed or recommended for its members by a recognized contractors' association, of which the *Contractor* is a member or to which the *Contractor* is otherwise bound),
   .2 fire, unusual delay by common carriers or unavoidable casualties,
   .3 abnormally adverse weather conditions, or
   .4 any cause beyond the *Contractor*'s control other than one resulting from a default or breach of *Contract* by the *Contractor*,
   then the *Contract Time* shall be extended for such reasonable time as the *Consultant* may recommend in consultation with the *Contractor*. The extension of time shall not be less than the time lost as the result of the event causing the delay, unless the *Contractor* agrees to a shorter extension. The *Contractor* shall not be entitled to payment for costs incurred by such delays unless such delays result from actions by the *Owner*, *Consultant* or anyone employed or engaged by them directly or indirectly.

6.5.4 No extension shall be made for delay unless *Notice in Writing* of the cause of delay is given to the *Consultant* not later than 10 *Working Days* after the commencement of the delay. In the case of a continuing cause of delay only one *Notice in Writing* shall be necessary.

6.5.5 If no schedule is made under paragraph 2.2.13 of GC 2.2 - ROLE OF THE CONSULTANT, then no request for extension shall be made because of failure of the *Consultant* to furnish instructions until 10 *Working Days* after demand for such instructions has been made.

GC 6.6 CLAIMS FOR A CHANGE IN CONTRACT PRICE

6.6.1 If the *Contractor* intends to make a claim for an increase to the *Contract Price*, or if the *Owner* intends to make a claim against the *Contractor* for a credit to the *Contract Price*, the party that intends to make the claim shall give timely *Notice in Writing* of intent to claim to the other party and to the *Consultant*.

6.6.2 Upon commencement of the event or series of events giving rise to a claim, the party intending to make the claim shall:
   .1 take all reasonable measures to mitigate any loss or expense which may be incurred as a result of such event or series of events, and
   .2 keep such records as may be necessary to support the claim.

6.6.3 The party making the claim shall submit within a reasonable time to the *Consultant* a detailed account of the amount claimed and the grounds upon which the claim is based.

6.6.4 Where the event or series of events giving rise to the claim has a continuing effect, the detailed account submitted under paragraph 6.6.3 shall be considered to be an interim account and the party making the claim shall, at such intervals as the *Consultant* may reasonably require, submit further interim accounts giving the accumulated amount of the claim and any further grounds upon which it is based. The party making the claim shall submit a final account after the end of the effects resulting from the event or series of events.

6.6.5 The *Consultant's* findings, with respect to a claim made by either party, will be given by *Notice in Writing* to both parties within 30 *Working Days* after receipt of the claim by the *Consultant*, or within such other time period as may be agreed by the parties.

6.6.6 If such finding is not acceptable to either party, the claim shall be settled in accordance with Part 8 of the General Conditions - DISPUTE RESOLUTION.

PART 7 DEFAULT NOTICE

GC 7.1 OWNER'S RIGHT TO PERFORM THE WORK, TERMINATE THE CONTRACTOR'S RIGHT TO CONTINUE WITH THE WORK OR TERMINATE THE CONTRACT

7.1.1 If the *Contractor* is adjudged bankrupt, or makes a general assignment for the benefit of creditors because of the *Contractor*'s insolvency, or if a receiver is appointed because of the *Contractor*'s insolvency, the *Owner* may, without prejudice to any other right or remedy the *Owner* may have, terminate the *Contractor*'s right to continue with the *Work*, by giving the *Contractor* or receiver or trustee in bankruptcy *Notice in Writing* to that effect.

7.1.2 If the *Contractor* neglects to prosecute the *Work* properly or otherwise fails to comply with the requirements of the *Contract* to a substantial degree and if the *Consultant* has given a written statement to the *Owner* and *Contractor* that sufficient cause exists to justify such action, the *Owner* may, without prejudice to any other right or remedy the *Owner* may have, give the *Contractor Notice in Writing* that the *Contractor* is in default of the *Contractor*'s contractual obligations and instruct the *Contractor* to correct the default in the 5 *Working Days* immediately following the receipt of such *Notice in Writing*.

7.1.3 If the default cannot be corrected in the 5 *Working Days* specified or in such other time period as may be subsequently agreed in writing by the parties, the *Contractor* shall be in compliance with the *Owner*'s instructions if the *Contractor*:

.1 commences the correction of the default within the specified time, and

.2 provides the *Owner* with an acceptable schedule for such correction, and

.3 corrects the default in accordance with the *Contract* terms and with such schedule.

7.1.4 If the *Contractor* fails to correct the default in the time specified or in such other time period as may be subsequently agreed in writing by the parties, without prejudice to any other right or remedy the *Owner* may have, the *Owner* may:

.1 correct such default and deduct the cost thereof from any payment then or thereafter due the *Contractor* provided the *Consultant* has certified such cost to the *Owner* and the *Contractor*, or

.2 terminate the *Contractor*'s right to continue with the *Work* in whole or in part or terminate the *Contract*.

7.1.5 If the *Owner* terminates the *Contractor*'s right to continue with the *Work* as provided in paragraphs 7.1.1 and 7.1.4, the *Owner* shall be entitled to:

.1 take possession of the *Work* and *Products* at the *Place of the Work*; subject to the rights of third parties, utilize the *Construction Equipment* at the *Place of the Work*; finish the *Work* by whatever method the *Owner* may consider expedient, but without undue delay or expense, and

.2 withhold further payment to the *Contractor* until a final certificate for payment is issued, and

.3 charge the *Contractor* the amount by which the full cost of finishing the *Work* as certified by the *Consultant*, including compensation to the *Consultant* for the *Consultant*'s additional services and a reasonable allowance as determined by the *Consultant* to cover the cost of corrections to work performed by the *Contractor* that may be required under GC 12.3 - WARRANTY, exceeds the unpaid balance of the *Contract Price*; however, if such cost of finishing the *Work* is less than the unpaid balance of the *Contract Price*, the *Owner* shall pay the *Contractor* the difference, and

.4 on expiry of the warranty period, charge the *Contractor* the amount by which the cost of corrections to the *Contractor*'s work under GC 12.3 - WARRANTY exceeds the allowance provided for such corrections, or if the cost of such corrections is less than the allowance, pay the *Contractor* the difference.

7.1.6 The *Contractor*'s obligation under the *Contract* as to quality, correction and warranty of the work performed by the *Contractor* up to the time of termination shall continue after such termination of the *Contract*.

GC 7.2 CONTRACTOR'S RIGHT TO SUSPEND THE WORK OR TERMINATE THE CONTRACT

7.2.1 If the *Owner* is adjudged bankrupt, or makes a general assignment for the benefit of creditors because of the *Owner*'s insolvency, or if a receiver is appointed because of the *Owner*'s insolvency, the *Contractor* may, without prejudice to any other right or remedy the *Contractor* may have, terminate the *Contract* by giving the *Owner* or receiver or trustee in bankruptcy *Notice in Writing* to that effect.

7.2.2 If the *Work* is suspended or otherwise delayed for a period of 20 *Working Days* or more under an order of a court or other public authority and providing that such order was not issued as the result of an act or fault of the *Contractor* or of anyone directly or indirectly employed or engaged by the *Contractor*, the *Contractor* may, without prejudice to any other right or remedy the *Contractor* may have, terminate the *Contract* by giving the *Owner Notice in Writing* to that effect.

7.2.3 The *Contractor* may give *Notice in Writing* to the *Owner*, with a copy to the *Consultant*, that the *Owner* is in default of the *Owner*'s contractual obligations if:

.1 the *Owner* fails to furnish, when so requested by the *Contractor*, reasonable evidence that financial arrangements have been made to fulfill the *Owner*'s obligations under the *Contract*, or

.2 the *Consultant* fails to issue a certificate as provided in GC 5.3 - PROGRESS PAYMENT, or

.3 the *Owner* fails to pay the *Contractor* when due the amounts certified by the *Consultant* or awarded by arbitration or court, or

.4 the *Owner* violates the requirements of the *Contract* to a substantial degree and the *Consultant*, except for GC 5.1 - FINANCING INFORMATION REQUIRED OF THE OWNER, confirms by written statement to the *Contractor* that sufficient cause exists.

7.2.4 The *Contractor*'s *Notice in Writing* to the *Owner* provided under paragraph 7.2.3 shall advise that if the default is not corrected within 5 *Working Days* following the receipt of the *Notice in Writing*, the *Contractor* may, without prejudice to any other right or remedy the *Contractor* may have, suspend the *Work* or terminate the *Contract*.

7.2.5 If the *Contractor* terminates the *Contract* under the conditions set out above, the *Contractor* shall be entitled to be paid for all work performed including reasonable profit, for loss sustained upon *Products* and *Construction Equipment*, and such other damages as the *Contractor* may have sustained as a result of the termination of the *Contract*.

PART 8 DISPUTE RESOLUTION

GC 8.1 AUTHORITY OF THE CONSULTANT

8.1.1 Differences between the parties to the *Contract* as to the interpretation, application or administration of the *Contract* or any failure to agree where agreement between the parties is called for, herein collectively called disputes, which are not resolved in the first instance by findings of the *Consultant* as provided in GC 2.2 - ROLE OF THE CONSULTANT, shall be settled in accordance with the requirements of Part 8 of the General Conditions - DISPUTE RESOLUTION.

8.1.2 If a dispute arises under the *Contract* in respect of a matter in which the *Consultant* has no authority under the *Contract* to make a finding, the procedures set out in paragraph 8.1.3 and paragraphs 8.2.3 to 8.2.8 of GC 8.2 - NEGOTIATION, MEDIATION AND ARBITRATION, and in GC 8.3 - RETENTION OF RIGHTS apply to that dispute with the necessary changes to detail as may be required.

8.1.3 If a dispute is not resolved promptly, the *Consultant* will give such instructions as in the *Consultant's* opinion are necessary for the proper performance of the *Work* and to prevent delays pending settlement of the dispute. The parties shall act immediately according to such instructions, it being understood that by so doing neither party will jeopardize any claim the party may have. If it is subsequently determined that such instructions were in error or at variance with the *Contract Documents*, the *Owner* shall pay the *Contractor* costs incurred by the *Contractor* in carrying out such instructions which the *Contractor* was required to do beyond what the *Contract Documents* correctly understood and interpreted would have required, including costs resulting from interruption of the *Work*.

GC 8.2 NEGOTIATION, MEDIATION AND ARBITRATION

8.2.1 In accordance with the Rules for Mediation of Construction Disputes as provided in CCDC 40 in effect at the time of bid closing, the parties shall appoint a Project Mediator
  .1   within 20 *Working Days* after the *Contract* was awarded, or
  .2   if the parties neglected to make an appointment within the 20 *Working Days*, within 10 *Working Days* after either party by *Notice in Writing* requests that the Project Mediator be appointed.

8.2.2 A party shall be conclusively deemed to have accepted a finding of the *Consultant* under GC 2.2 - ROLE OF THE CONSULTANT and to have expressly waived and released the other party from any claims in respect of the particular matter dealt with in that finding unless, within 15 *Working Days* after receipt of that finding, the party sends a *Notice in Writing* of dispute to the other party and to the *Consultant*, which contains the particulars of the matter in dispute and the relevant provisions of the *Contract Documents*. The responding party shall send a *Notice in Writing* of reply to the dispute within 10 *Working Days* after receipt of such *Notice in Writing* setting out particulars of this response and any relevant provisions of the *Contract Documents*.

8.2.3 The parties shall make all reasonable efforts to resolve their dispute by amicable negotiations and agree to provide, without prejudice, frank, candid and timely disclosure of relevant facts, information and documents to facilitate these negotiations.

8.2.4 After a period of 10 *Working Days* following receipt of a responding party's *Notice in Writing* of reply under paragraph 8.2.2, the parties shall request the Project Mediator to assist the parties to reach agreement on any unresolved dispute. The mediated negotiations shall be conducted in accordance with the Rules for Mediation of Construction Disputes as provided in CCDC 40 in effect at the time of bid closing.

8.2.5 If the dispute has not been resolved within 10 *Working Days* after the Project Mediator was requested under paragraph 8.2.4 or within such further period agreed by the parties, the Project Mediator shall terminate the mediated negotiations by giving *Notice in Writing* to the *Owner*, the *Contractor* and the *Consultant*.

8.2.6 By giving a *Notice in Writing* to the other party and the *Consultant*, not later than 10 *Working Days* after the date of termination of the mediated negotiations under paragraph 8.2.5, either party may refer the dispute to be finally resolved by arbitration under the Rules for Arbitration of Construction Disputes as provided in CCDC 40 in effect at the time of bid closing. The arbitration shall be conducted in the jurisdiction of the *Place of the Work*.

8.2.7 On expiration of the 10 *Working Days*, the arbitration agreement under paragraph 8.2.6 is not binding on the parties and, if a *Notice in Writing* is not given under paragraph 8.2.6 within the required time, the parties may refer the unresolved dispute to the courts or to any other form of dispute resolution, including arbitration, which they have agreed to use.

8.2.8 If neither party, by *Notice in Writing*, given within 10 *Working Days* of the date of *Notice in Writing* requesting arbitration in paragraph 8.2.6, requires that a dispute be arbitrated immediately, all disputes referred to arbitration as provided in paragraph 8.2.6 shall be

.1 held in abeyance until

 (1) *Substantial Performance of the Work,*

 (2) the *Contract* has been terminated, or

 (3) the *Contractor* has abandoned the *Work,*

 whichever is earlier; and

.2 consolidated into a single arbitration under the rules governing the arbitration under paragraph 8.2.6.

GC 8.3 RETENTION OF RIGHTS

8.3.1 It is agreed that no act by either party shall be construed as a renunciation or waiver of any rights or recourses, provided the party has given the *Notice in Writing* required under Part 8 of the General Conditions - DISPUTE RESOLUTION and has carried out the instructions as provided in paragraph 8.1.3 of GC 8.1 – AUTHORITY OF THE CONSULTANT.

8.3.2 Nothing in Part 8 of the General Conditions - DISPUTE RESOLUTION shall be construed in any way to limit a party from asserting any statutory right to a lien under applicable lien legislation of the jurisdiction of the *Place of the Work* and the assertion of such right by initiating judicial proceedings is not to be construed as a waiver of any right that party may have under paragraph 8.2.6 of GC 8.2 – NEGOTIATION, MEDIATION AND ARBITRATION to proceed by way of arbitration to adjudicate the merits of the claim upon which such a lien is based.

PART 9 PROTECTION OF PERSONS AND PROPERTY

GC 9.1 PROTECTION OF WORK AND PROPERTY

9.1.1 The *Contractor* shall protect the *Work* and the *Owner's* property and property adjacent to the *Place of the Work* from damage which may arise as the result of the *Contractor's* operations under the *Contract*, and shall be responsible for such damage, except damage which occurs as the result of:

.1 errors in the *Contract Documents*;

.2 acts or omissions by the *Owner*, the *Consultant*, other contractors, their agents and employees.

9.1.2 Before commencing any work, the *Contractor* shall determine the location of all underground utilities and structures indicated in the *Contract Documents* or that are reasonably apparent in an inspection of the *Place of the Work*.

9.1.3 Should the *Contractor* in the performance of the *Contract* damage the *Work*, the *Owner's* property or property adjacent to the *Place of the Work*, the *Contractor* shall be responsible for making good such damage at the *Contractor's* expense.

9.1.4 Should damage occur to the *Work* or *Owner's* property for which the *Contractor* is not responsible, as provided in paragraph 9.1.1, the *Contractor* shall make good such damage to the *Work* and, if the *Owner* so directs, to the *Owner's* property. The *Contract Price* and *Contract Time* shall be adjusted as provided in GC 6.1 – OWNER'S RIGHT TO MAKE CHANGES, GC 6.2 - CHANGE ORDER and GC 6.3 - CHANGE DIRECTIVE.

GC 9.2 TOXIC AND HAZARDOUS SUBSTANCES

9.2.1 For the purposes of applicable legislation related to toxic and hazardous substances, the *Owner* shall be deemed to have control and management of the *Place of the Work* with respect to existing conditions.

9.2.2 Prior to the *Contractor* commencing the *Work*, the *Owner* shall,

.1 take all reasonable steps to determine whether any toxic or hazardous substances are present at the *Place of the Work*, and

.2 provide the *Consultant* and the *Contractor* with a written list of any such substances that are known to exist and their locations.

9.2.3 The *Owner* shall take all reasonable steps to ensure that no person's exposure to any toxic or hazardous substances exceeds the time weighted levels prescribed by applicable legislation at the *Place of the Work* and that no property is damaged or destroyed as a result of exposure to, or the presence of, toxic or hazardous substances which were at the *Place of the Work* prior to the *Contractor* commencing the *Work*.

9.2.4 Unless the *Contract* expressly provides otherwise, the *Owner* shall be responsible for taking all necessary steps, in accordance with applicable legislation in force at the *Place of the Work*, to dispose of, store or otherwise render harmless toxic or hazardous substances which were present at the *Place of the Work* prior to the *Contractor* commencing the *Work*.

9.2.5 If the *Contractor*
   .1   encounters toxic or hazardous substances at the *Place of the Work*, or
   .2   has reasonable grounds to believe that toxic or hazardous substances are present at the *Place of the Work*,
which were not brought to the *Place of the Work* by the *Contractor* or anyone for whom the *Contractor* is responsible and which were not disclosed by the *Owner* or which were disclosed but have not been dealt with as required under paragraph 9.2.4, the *Contractor* shall
   .3   take all reasonable steps, including stopping the *Work*, to ensure that no person's exposure to any toxic or hazardous substances exceeds any applicable time weighted levels prescribed by applicable legislation at the *Place of the Work*, and
   .4   immediately report the circumstances to the *Consultant* and the *Owner* in writing.

9.2.6 If the *Owner* and *Contractor* do not agree on the existence, significance of, or whether the toxic or hazardous substances were brought onto the *Place of the Work* by the *Contractor* or anyone for whom the *Contractor* is responsible, the *Owner* shall retain and pay for an independent qualified expert to investigate and determine such matters. The expert's report shall be delivered to the *Owner* and the *Contractor*.

9.2.7 If the *Owner* and *Contractor* agree or if the expert referred to in paragraph 9.2.6 determines that the toxic or hazardous substances were not brought onto the place of the *Work* by the *Contractor* or anyone for whom the *Contractor* is responsible, the *Owner* shall promptly at the *Owner's* own expense:
   .1   take all steps as required under paragraph 9.2.4;
   .2   reimburse the *Contractor* for the costs of all steps taken pursuant to paragraph 9.2.5;
   .3   extend the *Contract* time for such reasonable time as the *Consultant* may recommend in consultation with the *Contractor* and the expert referred to in 9.2.6 and reimburse the *Contractor* for reasonable costs incurred as a result of the delay; and
   .4   indemnify the *Contractor* as required by GC 12.1 - INDEMNIFICATION.

9.2.8 If the *Owner* and *Contractor* agree or if the expert referred to in paragraph 9.2.6 determines that the toxic or hazardous substances were brought onto the place of the *Work* by the *Contractor* or anyone for whom the *Contractor* is responsible, the *Contractor* shall promptly at the *Contractor's* own expense:
   .1   take all necessary steps, in accordance with applicable legislation in force at the *Place of the Work*, to safely remove and dispose the toxic or hazardous substances;
   .2   make good any damage to the *Work*, the *Owner's* property or property adjacent to the place of the *Work* as provided in paragraph 9.1.3 of GC 9.1 – PROTECTION OF WORK AND PROPERTY;
   .3   reimburse the *Owner* for reasonable costs incurred under paragraph 9.2.6; and
   .4   indemnify the Owner as required by GC 12.1 - INDEMNIFICATION.

9.2.9 If either party does not accept the expert's findings under paragraph 9.2.6, the disagreement shall be settled in accordance with Part 8 of the General Conditions - Dispute Resolution. If such disagreement is not resolved promptly, the parties shall act immediately in accordance with the expert's determination and take the steps required by paragraph 9.2.7 or 9.2.8 it being understood that by so doing, neither party will jeopardize any claim that party may have to be reimbursed as provided by GC 9.2 – TOXIC AND HAZARDOUS SUBSTANCES.

GC 9.3 ARTIFACTS AND FOSSILS

9.3.1 Fossils, coins, articles of value or antiquity, structures and other remains or things of scientific or historic interest discovered at the *Place or Work* shall, as between the *Owner* and the *Contractor*, be deemed to be the absolute property of the *Owner*.

9.3.2 The *Contractor* shall take all reasonable precautions to prevent removal or damage to discoveries as identified in paragraph 9.3.1, and shall advise the *Consultant* upon discovery of such items.

9.3.3 The *Consultant* will investigate the impact on the *Work* of the discoveries identified in paragraph 9.3.1. If conditions are found that would cause an increase or decrease in the *Contractor's* cost or time to perform the *Work*, the *Consultant*, with the *Owner's* approval, will issue appropriate instructions for a change in the *Work* as provided in GC 6.2 - CHANGE ORDER or GC 6.3 CHANGE DIRECTIVE.

GC 9.4 CONSTRUCTION SAFETY

9.4.1 Subject to paragraph 3.2.2.2 of GC 3.2 - CONSTRUCTION BY OWNER OR OTHER CONTRACTORS, the *Contractor* shall be solely responsible for construction safety at the *Place of the Work* and for compliance with the rules, regulations and practices required by the applicable construction health and safety legislation and shall be responsible for initiating, maintaining and supervising all safety precautions and programs in connection with the performance of the *Work*.

GC 9.5 MOULD

9.5.1 If the *Contractor* or *Owner* observes or reasonably suspects the presence of mould at the *Place of the Work*, the remediation of which is not expressly part of the *Work*,

   .1 the observing party shall promptly report the circumstances to the other party in writing, and

   .2 the *Contractor* shall promptly take all reasonable steps, including stopping the *Work* if necessary, to ensure that no person suffers injury, sickness or death and that no property is damaged as a result of exposure to or the presence of the mould, and

   .3 if the *Owner* and *Contractor* do not agree on the existence, significance or cause of the mould or as to what steps need be taken to deal with it, the *Owner* shall retain and pay for an independent qualified expert to investigate and determine such matters. The expert's report shall be delivered to the *Owner* and *Contractor*.

9.5.2 If the *Owner* and *Contractor* agree, or if the expert referred to in paragraph 9.5.1.3 determines that the presence of mould was caused by the *Contractor*'s operations under the *Contract*, the *Contractor* shall promptly, at the *Contractor*'s own expense:

   .1 take all reasonable and necessary steps to safely remediate or dispose of the mould, and

   .2 make good any damage to the *Work*, the *Owner*'s property or property adjacent to the *Place of the Work* as provided in paragraph 9.1.3 of GC 9.1 - PROTECTION OF WORK AND PROPERTY, and

   .3 reimburse the *Owner* for reasonable costs incurred under paragraph 9.5.1.3, and

   .4 indemnify the *Owner* as required by GC 12.1 - INDEMNIFICATION.

9.5.3 If the *Owner* and *Contractor* agree, or if the expert referred to in paragraph 9.5.1.3 determines that the presence of mould was not caused by the *Contractor*'s operations under the *Contract*, the *Owner* shall promptly, at the *Owner*'s own expense:

   .1 take all reasonable and necessary steps to safely remediate or dispose of the mould, and

   .2 reimburse the *Contractor* for the cost of taking the steps under paragraph 9.5.1.2 and making good any damage to the *Work* as provided in paragraph 9.1.4 of GC 9.1 - PROTECTION OF WORK AND PROPERTY, and

   .3 extend the *Contract Time* for such reasonable time as the *Consultant* may recommend in consultation with the *Contractor* and the expert referred to in paragraph 9.5.1.3 and reimburse the *Contractor* for reasonable costs incurred as a result of the delay, and

   .4 indemnify the *Contractor* as required by GC 12.1 - INDEMNIFICATION.

9.5.4 If either party does not accept the expert's finding under paragraph 9.5.1.3, the disagreement shall be settled in accordance with Part 8 of the General Conditions - DISPUTE RESOLUTION. If such disagreement is not resolved promptly, the parties shall act immediately in accordance with the expert's determination and take the steps required by paragraphs 9.5.2 or 9.5.3, it being understood that by so doing neither party will jeopardize any claim the party may have to be reimbursed as provided by GC 9.5 - MOULD.

PART 10 GOVERNING REGULATIONS

GC 10.1 TAXES AND DUTIES

10.1.1 The *Contract Price* shall include all taxes and customs duties in effect at the time of the bid closing except for *Value Added Taxes* payable by the *Owner* to the *Contractor* as stipulated in Article A-4 of the Agreement - CONTRACT PRICE.

10.1.2 Any increase or decrease in costs to the *Contractor* due to changes in such included taxes and duties after the time of the bid closing shall increase or decrease the *Contract Price* accordingly.

GC 10.2 LAWS, NOTICES, PERMITS, AND FEES

10.2.1 The laws of the *Place of the Work* shall govern the *Work*.

10.2.2 The *Owner* shall obtain and pay for development approvals, building permit, permanent easements, rights of servitude, and all other necessary approvals and permits, except for the permits and fees referred to in paragraph 10.2.3 or for which the *Contract Documents* specify as the responsibility of the *Contractor*.

10.2.3 The *Contractor* shall be responsible for the procurement of permits, licences, inspections, and certificates, which are necessary for the performance of the *Work* and customarily obtained by contractors in the jurisdiction of the *Place of the Work* after the issuance of the building permit. The *Contract Price* includes the cost of these permits, licences, inspections, and certificates, and their procurement.

10.2.4 The *Contractor* shall give the required notices and comply with the laws, ordinances, rules, regulations, or codes which are or become in force during the performance of the *Work* and which relate to the *Work*, to the preservation of the public health, and to construction safety.

10.2.5 The *Contractor* shall not be responsible for verifying that the *Contract Documents* are in compliance with the applicable laws, ordinances, rules, regulations, or codes relating to the *Work*. If the *Contract Documents* are at variance therewith, or if, subsequent to the time of bid closing, changes are made to the applicable laws, ordinances, rules, regulations, or codes which require modification to the *Contract Documents*, the *Contractor* shall advise the *Consultant* in writing requesting direction immediately upon such variance or change becoming known. The *Consultant* will make the changes required to the *Contract Documents* as provided in GC 6.1 - OWNER'S RIGHT TO MAKE CHANGES, GC 6.2 - CHANGE ORDER and GC 6.3 - CHANGE DIRECTIVE.

10.2.6 If the *Contractor* fails to advise the *Consultant* in writing; and fails to obtain direction as required in paragraph 10.2.5; and performs work knowing it to be contrary to any laws, ordinances, rules, regulations, or codes; the *Contractor* shall be responsible for and shall correct the violations thereof; and shall bear the costs, expenses and damages attributable to the failure to comply with the provisions of such laws, ordinances, rules, regulations, or codes.

10.2.7 If, subsequent to the time of bid closing, changes are made to applicable laws, ordinances, rules, regulations, or codes of authorities having jurisdiction which affect the cost of the *Work*, either party may submit a claim in accordance with the requirements of GC 6.6 – CLAIMS FOR A CHANGE IN CONTRACT PRICE.

GC 10.3 PATENT FEES

10.3.1 The *Contractor* shall pay the royalties and patent licence fees required for the performance of the *Contract*. The *Contractor* shall hold the *Owner* harmless from and against claims, demands, losses, costs, damages, actions, suits, or proceedings arising out of the *Contractor*'s performance of the *Contract* which are attributable to an infringement or an alleged infringement of a patent of invention by the *Contractor* or anyone for whose acts the *Contractor* may be liable.

10.3.2 The *Owner* shall hold the *Contractor* harmless against claims, demands, losses, costs, damages, actions, suits, or proceedings arising out of the *Contractor*'s performance of the *Contract* which are attributable to an infringement or an alleged infringement of a patent of invention in executing anything for the purpose of the *Contract*, the model, plan or design of which was supplied to the *Contractor* as part of the *Contract Documents*.

GC 10.4 WORKERS' COMPENSATION

10.4.1 Prior to commencing the *Work*, again with the *Contractor*'s application for payment of the holdback amount following *Substantial Performance of the Work* and again with the *Contractor*'s application for final payment, the *Contractor* shall provide evidence of compliance with workers' compensation legislation at the *Place of the Work*, including payments due thereunder.

10.4.2 At any time during the term of the *Contract*, when requested by the *Owner*, the *Contractor* shall provide such evidence of compliance by the *Contractor* and *Subcontractors*.

PART 11 INSURANCE AND CONTRACT SECURITY

GC 11.1 INSURANCE

11.1.1 Without restricting the generality of GC 12.1 - INDEMNIFICATION, the *Contractor* shall provide, maintain and pay for the following insurance coverages, the minimum requirements of which are specified in CCDC 41 – CCDC Insurance Requirements in effect at the time of bid closing except as hereinafter provided:

  .1 General liability insurance in the name of the *Contractor* and include, or in the case of a single, blanket policy, be endorsed to name, the *Owner* and the *Consultant* as insureds but only with respect to liability, other than legal liability arising out of their sole negligence, arising out of the operations of the *Contractor* with regard to the *Work*. General liability insurance shall be maintained from the date of commencement of the *Work* until one year from the date of *Substantial Performance of the Work*. Liability coverage shall be provided for completed operations hazards from the date of *Substantial Performance of the Work*, as set out in the certificate of *Substantial Performance of the Work*, on an ongoing basis for a period of 6 years following *Substantial Performance of the Work*.

  .2 Automobile Liability Insurance from the date of commencement of the *Work* until one year after the date of *Substantial Performance of the Work*.

  .3 Aircraft or Watercraft Liability Insurance when owned or non-owned aircraft or watercraft are used directly or indirectly in the performance of the *Work*

  .4 "Broad form" property insurance in the joint names of the *Contractor*, the *Owner* and the *Consultant*. The policy shall include as insureds all *Subcontractors*. The "Broad form" property insurance shall be provided from the date of commencement of the *Work* until the earliest of:

    (1) 10 calendar days after the date of *Substantial Performance of the Work*;

      (2) on the commencement of use or occupancy of any part or section of the *Work* unless such use or occupancy is for construction purposes, habitational, office, banking, convenience store under 465 square metres in area, or parking purposes, or for the installation, testing and commissioning of equipment forming part of the *Work*;

      (3) when left unattended for more than 30 consecutive calendar days or when construction activity has ceased for more than 30 consecutive calendar days.

.5 Boiler and machinery insurance in the joint names of the *Contractor*, the *Owner* and the *Consultant*. The policy shall include as insureds all *Subcontractors*. The coverage shall be maintained continuously from commencement of use or operation of the boiler and machinery objects insured by the policy and until 10 calendar days after the date of *Substantial Performance of the Work*.

.6 The "Broad form" property and boiler and machinery policies shall provide that, in the case of a loss or damage, payment shall be made to the *Owner* and the *Contractor* as their respective interests may appear. In the event of loss or damage:

      (1) the *Contractor* shall act on behalf of the *Owner* for the purpose of adjusting the amount of such loss or damage payment with the insurers. When the extent of the loss or damage is determined, the *Contractor* shall proceed to restore the *Work*. Loss or damage shall not affect the rights and obligations of either party under the *Contract* except that the *Contractor* shall be entitled to such reasonable extension of *Contract Time* relative to the extent of the loss or damage as the *Consultant* may recommend in consultation with the *Contractor*;

      (2) the *Contractor* shall be entitled to receive from the *Owner*, in addition to the amount due under the *Contract*, the amount which the *Owner*'s interest in restoration of the *Work* has been appraised, such amount to be paid as the restoration of the *Work* proceeds in accordance with the progress payment provisions. In addition the *Contractor* shall be entitled to receive from the payments made by the insurer the amount of the *Contractor*'s interest in the restoration of the *Work*; and

      (3) to the *Work* arising from the work of the *Owner*, the *Owner*'s own forces or another contractor, the *Owner* shall, in accordance with the *Owner*'s obligations under the provisions relating to construction by *Owner* or other contractors, pay the *Contractor* the cost of restoring the *Work* as the restoration of the *Work* proceeds and as in accordance with the progress payment provisions.

.7 Contractors' Equipment Insurance from the date of commencement of the *Work* until one year after the date of *Substantial Performance of the Work*.

11.1.2 Prior to commencement of the *Work* and upon the placement, renewal, amendment, or extension of all or any part of the insurance, the *Contractor* shall promptly provide the *Owner* with confirmation of coverage and, if required, a certified true copy of the policies certified by an authorized representative of the insurer together with copies of any amending endorsements applicable to the *Work*.

11.1.3 The parties shall pay their share of the deductible amounts in direct proportion to their responsibility in regards to any loss for which the above policies are required to pay, except where such amounts may be excluded by the terms of the *Contract*.

11.1.4 If the *Contractor* fails to provide or maintain insurance as required by the *Contract Documents*, then the *Owner* shall have the right to provide and maintain such insurance and give evidence to the *Contractor* and the *Consultant*. The *Contractor* shall pay the cost thereof to the *Owner* on demand or the *Owner* may deduct the cost from the amount which is due or may become due to the *Contractor*.

11.1.5 All required insurance policies shall be with insurers licensed to underwrite insurance in the jurisdiction of the *Place of the Work*.

11.1.6 If a revised version of CCDC 41 – INSURANCE REQUIREMENTS is published, which specifies reduced insurance requirements, the parties shall address such reduction, prior to the *Contractor*'s insurance policy becoming due for renewal, and record any agreement in a *Change Order*.

11.1.7 If a revised version of CCDC 41 – INSURANCE REQUIREMENTS is published, which specifies increased insurance requirements, the *Owner* may request the increased coverage from the Contractor by way of a *Change Order*.

11.1.8 A *Change Directive* shall not be used to direct a change in the insurance requirements in response to the revision of CCDC 41 – INSURANCE REQUIREMENTS.

GC 11.2 CONTRACT SECURITY

11.2.1 The *Contractor* shall, prior to commencement of the *Work* or within the specified time, provide to the *Owner* any *Contract* security specified in the *Contract Documents*.

11.2.2 If the *Contract Documents* require surety bonds to be provided, such bonds shall be issued by a duly licensed surety company authorized to transact the business of suretyship in the province or territory of the *Place of the Work* and shall be maintained in good standing until the fulfillment of the *Contract*. The form of such bonds shall be in accordance with the latest edition of the CCDC approved bond forms.

PART 12 INDEMNIFICATION, WAIVER OF CLAIMS AND WARRANTY

GC 12.1 INDEMNIFICATION

12.1.1 Without restricting the parties' obligation to indemnify as described in paragraphs 12.1.4 and 12.1.5, the *Owner* and the *Contractor* shall each indemnify and hold harmless the other from and against all claims, demands, losses, costs, damages, actions, suits, or proceedings whether in respect to losses suffered by them or in respect to claims by third parties that arise out of, or are attributable in any respect to their involvement as parties to this *Contract*, provided such claims are:

.1 caused by:

    (1) the negligent acts or omissions of the party from whom indemnification is sought or anyone for whose acts or omissions that party is liable, or

    (2) a failure of the party to the *Contract* from whom indemnification is sought to fulfill its terms or conditions; and

.2 made by *Notice in Writing* within a period of 6 years from the date of *Substantial Performance of the Work* as set out in the certificate of *Substantial Performance of the Work* issued pursuant to paragraph 5.4.2.2 of GC 5.4 – SUBSTANTIAL PERFORMANCE OF THE WORK or within such shorter period as may be prescribed by any limitation statute of the province or territory of the *Place of the Work*.

The parties expressly waive the right to indemnity for claims other than those provided for in this *Contract*.

12.1.2 The obligation of either party to indemnify as set forth in paragraph 12.1.1 shall be limited as follows:

.1 In respect to losses suffered by the *Owner* and the *Contractor* for which insurance is to be provided by either party pursuant to GC 11.1 – INSURANCE, the general liability insurance limit for one occurrence as referred to in CCDC 41 in effect at the time of bid closing.

.2 In respect to losses suffered by the *Owner* and the *Contractor* for which insurance is not required to be provided by either party in accordance with GC 11.1 – INSURANCE, the greater of the *Contract Price* as recorded in Article A-4 – CONTRACT PRICE or $2,000,000, but in no event shall the sum be greater than $20,000,000.

.3 In respect to claims by third parties for direct loss resulting from bodily injury, sickness, disease or death, or to injury to or destruction of tangible property, the obligation to indemnify is without limit. In respect to all other claims for indemnity as a result of claims advanced by third parties, the limits of indemnity set forth in paragraphs 12.1.2.1 and 12.1.2.2 shall apply.

12.1.3 The obligation of either party to indemnify the other as set forth in paragraphs 12.1.1 and 12.1.2 shall be inclusive of interest and all legal costs.

12.1.4 The *Owner* and the *Contractor* shall indemnify and hold harmless the other from and against all claims, demands, losses, costs, damages, actions, suits, or proceedings arising out of their obligations described in GC 9.2 – TOXIC AND HAZARDOUS SUBSTANCES.

12.1.5 The *Owner* shall indemnify and hold harmless the *Contractor* from and against all claims, demands, losses, costs, damages, actions, suits, or proceedings:

.1 as described in paragraph 10.3.2 of GC 10.3 – PATENT FEES, and

.2 arising out of the *Contractor*'s performance of the *Contract* which are attributable to a lack of or defect in title or an alleged lack of or defect in title to the *Place of the Work*.

12.1.6 In respect to any claim for indemnity or to be held harmless by the *Owner* or the *Contractor*:

.1 *Notice in Writing* of such claim shall be given within a reasonable time after the facts upon which such claim is based became known;

.2 should any party be required as a result of its obligation to indemnify another to pay or satisfy a final order, judgment or award made against the party entitled by this *Contract* to be indemnified, then the indemnifying party upon assuming all liability for any costs that might result shall have the right to appeal in the name of the party against whom such final order or judgment has been made until such rights of appeal have been exhausted.

GC 12.2 WAIVER OF CLAIMS

12.2.1 Subject to any lien legislation applicable to the *Place of the Work*, as of the fifth calendar day before the expiry of the lien period provided by the lien legislation applicable at the *Place of the Work*, the *Contractor* waives and releases the *Owner* from all claims which the *Contractor* has or reasonably ought to have knowledge of that could be advanced by the *Contractor* against the *Owner* arising from the *Contractor*'s involvement in the *Work*, including, without limitation, those arising from negligence or breach of contract in respect to which the cause of action is based upon acts or omissions which occurred prior to or on the date of *Substantial Performance of the Work*, except as follows:

  .1 claims arising prior to or on the date of *Substantial Performance of the Work* for which *Notice in Writing* of claim has been received by the *Owner* from the *Contractor* no later than the sixth calendar day before the expiry of the lien period provided by the lien legislation applicable at the *Place of the Work*,

  .2 indemnification for claims advanced against the *Contractor* by third parties for which a right of indemnification may be asserted by the *Contractor* against the *Owner* pursuant to the provisions of this *Contract*;

  .3 claims for which a right of indemnity could be asserted by the *Contractor* pursuant to the provisions of paragraphs 12.1.4 or 12.1.5 of GC 12.1 – INDEMNIFICATION; and

  .4 claims resulting from acts or omissions which occur after the date of *Substantial Performance of the Work*.

12.2.2 The *Contractor* waives and releases the *Owner* from all claims referenced in paragraph 12.2.1.4 except for those referred in paragraphs 12.2.1.2 and 12.2.1.3 and claims for which *Notice in Writing* of claim has been received by the *Owner* from the *Contractor* within 395 calendar days following the date of *Substantial Performance of the Work*.

12.2.3 Subject to any lien legislation applicable to the *Place of the Work*, as of the fifth calendar day before the expiry of the lien period provided by the lien legislation applicable at the *Place of the Work*, the *Owner* waives and releases the *Contractor* from all claims which the *Owner* has or reasonably ought to have knowledge of that could be advanced by the *Owner* against the *Contractor* arising from the *Owner*'s involvement in the *Work*, including, without limitation, those arising from negligence or breach of contract in respect to which the cause of action is based upon acts or omissions which occurred prior to or on the date of *Substantial Performance of the Work*, except as follows:

  .1 claims arising prior to or on the date of *Substantial Performance of the Work* for which *Notice in Writing* of claim has been received by the *Contractor* from the *Owner* no later than the sixth calendar day before the expiry of the lien period provided by the lien legislation applicable at the *Place of the Work*;

  .2 indemnification for claims advanced against the *Owner* by third parties for which a right of indemnification may be asserted by the *Owner* against the *Contractor* pursuant to the provisions of this *Contract*;

  .3 claims for which a right of indemnity could be asserted by the *Owner* against the *Contractor* pursuant to the provisions of paragraph 12.1.4 of GC 12.1 - INDEMNIFICATION;

  .4 damages arising from the *Contractor*'s actions which result in substantial defects or deficiencies in the *Work*. "Substantial defects or deficiencies" mean those defects or deficiencies in the *Work* which affect the *Work* to such an extent or in such a manner that a significant part or the whole of the *Work* is unfit for the purpose intended by the *Contract Documents*;

  .5 claims arising pursuant to GC 12.3 - WARRANTY; and

  .6 claims arising from acts or omissions which occur after the date of *Substantial Performance of the Work*.

12.2.4 The *Owner* waives and releases the *Contractor* from all claims referred to in paragraph 12.2.3.4 except claims for which *Notice in Writing* of claim has been received by the *Contractor* from the *Owner* within a period of six years from the date of *Substantial Performance of the Work* should any limitation statute of the Province or Territory of the *Place of the Work* permit such agreement. If the applicable limitation statute does not permit such agreement, within such shorter period as may be prescribed by:

  .1 any limitation statute of the Province or Territory of the *Place of the Work*; or

  .2 if the *Place of the Work* is the Province of Quebec, then Article 2118 of the Civil Code of Quebec.

12.2.5 The *Owner* waives and releases the *Contractor* from all claims referenced in paragraph 12.2.3.6 except for those referred in paragraph 12.2.3.2, 12.2.3.3 and those arising under GC 12.3 – WARRANTY and claims for which *Notice in Writing* has been received by the *Contractor* from the *Owner* within 395 calendar days following the date of *Substantial Performance of the Work*.

12.2.6 "*Notice in Writing* of claim" as provided for in GC 12.2 – WAIVER OF CLAIMS to preserve a claim or right of action which would otherwise, by the provisions of GC 12.2 – WAIVER OF CLAIMS, be deemed to be waived, must include the following:

  .1 a clear and unequivocal statement of the intention to claim;

  .2 a statement as to the nature of the claim and the grounds upon which the claim is based; and

  .3 a statement of the estimated quantum of the claim.

12.2.7 The party giving "*Notice in Writing* of claim" as provided for in GC 12.2 – WAIVER OF CLAIMS shall submit within a reasonable time a detailed account of the amount claimed.

12.2.8 Where the event or series of events giving rise to a claim made under paragraphs 12.2.1 or 12.2.3 has a continuing effect, the detailed account submitted under paragraph 12.2.7 shall be considered to be an interim account and the party making the claim shall submit further interim accounts, at reasonable intervals, giving the accumulated amount of the claim and any further grounds upon which it is based. The party making the claim shall submit a final account after the end of the effects resulting from the event or series of events.

12.2.9 If a *Notice in Writing* of claim pursuant to paragraph 12.2.1.1 is received on the seventh or sixth calendar day before the expiry of the lien period provided by the lien legislation applicable at the *Place of the Work*, the period within which *Notice in Writing* of claim shall be received pursuant to paragraph 12.2.3.1 shall be extended to two calendar days before the expiry of the lien period provided by the lien legislation applicable at the *Place of the Work*.

12.2.10 If a *Notice in Writing* of claim pursuant to paragraph 12.2.3.1 is received on the seventh or sixth calendar day before the expiry of the lien period provided by the lien legislation applicable at the *Place of the Work*, the period within which *Notice in Writing* of claim shall be received pursuant to paragraph12.2.1.1 shall be extended to two calendar days before the expiry of the lien period provided by the lien legislation applicable at the *Place of the Work*.

GC 12.3 WARRANTY

12.3.1 Except for extended warranties as described in paragraph 12.3.6, the warranty period under the *Contract* is one year from the date of *Substantial Performance of the Work*.

12.3.2 The *Contractor* shall be responsible for the proper performance of the *Work* to the extent that the design and *Contract Documents* permit such performance.

12.3.3 The *Owner*, through the *Consultant*, shall promptly give the *Contractor Notice in Writing* of observed defects and deficiencies which occur during the one year warranty period.

12.3.4 Subject to paragraph 12.3.2, the *Contractor* shall correct promptly, at the *Contractor*'s expense, defects or deficiencies in the *Work* which appear prior to and during the one year warranty period.

12.3.5 The *Contractor* shall correct or pay for damage resulting from corrections made under the requirements of paragraph 12.3.4.

12.3.6 Any extended warranties required beyond the one year warranty period as described in paragraph 12.3.1, shall be as specified in the *Contract Documents*. Extended warranties shall be issued by the warrantor to the benefit of the *Owner*. The *Contractor*'s responsibility with respect to extended warranties shall be limited to obtaining any such extended warranties from the warrantor. The obligations under such extended warranties are solely the responsibilities of the warrantor.

CCÐC

75 Albert Street
Suite 400
Ottawa, Ont. K1P 5E7

Tel: (613) 236-9455
Fax: (613) 236-9526
info@ccdc.org

CANADIAN CONSTRUCTION ÐOCUMENTS COMMITTEE
CANADIAN CONSTRUCTION ÐOCUMENTS COMMITTEE
CANADIAN CONSTRUCTION ÐOCUMENTS COMMITTEE

CCDC 41
CCDC INSURANCE REQUIREMENTS

PUBLICATION DATE: JANUARY 21, 2008

1. General liability insurance shall be with limits of not less than $5,000,000 per occurrence, an aggregate limit of not less than $5,000,000 within any policy year with respect to completed operations, and a deductible not exceeding $5,000. The insurance coverage shall not be less than the insurance provided by IBC Form 2100 (including an extension for a standard provincial and territorial form of non-owned automobile liability policy) and IBC Form 2320. To achieve the desired limit, umbrella or excess liability insurance may be used. Subject to satisfactory proof of financial capability by the *Contractor*, the *Owner* may agree to increase the deductible amounts.

2. Automobile liability insurance in respect of vehicles that are required by law to be insured under a contract by a Motor Vehicle Liability Policy, shall have limits of not less than $5,000,000 inclusive per occurrence for bodily injury, death and damage to property, covering all vehicles owned or leased by the *Contractor*. Where the policy has been issued pursuant to a government-operated automobile insurance system, the *Contractor* shall provide the *Owner* with confirmation of automobile insurance coverage for all automobiles registered in the name of the *Contractor*.

3. Aircraft and watercraft liability insurance with respect to owned or non-owned aircraft and watercraft (if used directly or indirectly in the performance of the *Work*), including use of additional premises, shall have limits of not less than $5,000,000 inclusive per occurrence for bodily injury, death and damage to property including loss of use thereof and limits of not less than $5,000,000 for aircraft passenger hazard. Such insurance shall be in a form acceptable to the *Owner*.

4. "Broad form" property insurance shall have limits of not less than the sum of 1.1 times *Contract Price* and the full value, as stated in the *Contract*, of *Products* and design services that are specified to be provided by the *Owner* for incorporation into the *Work*, with a deductible not exceeding $5,000. The insurance coverage shall not be less than the insurance provided by IBC Forms 4042 and 4047 (excluding flood and earthquake) or their equivalent replacement. Subject to satisfactory proof of financial capability by the *Contractor*, the *Owner* may agree to increase the deductible amounts.

Association
of Canadian
Engineering
Companies

5. Boiler and machinery insurance shall have limits of not less than the replacement value of the permanent or temporary boilers and pressure vessels, and other insurable objects forming part of the *Work*. The insurance coverage shall not be less than the insurance provided by a comprehensive boiler and machinery policy.

Canadian
Construction
Association

6. "Broad form" contractors' equipment insurance coverage covering *Construction Equipment* used by the *Contractor* for the performance of the *Work*, shall be in a form acceptable to the *Owner* and shall not allow subrogation claims by the insurer against the *Owner*. Subject to satisfactory proof of financial capability by the *Contractor* for self-insurance, the *Owner* may agree to waive the equipment insurance requirement.

Construction
Specifications
Canada

7. Standard Exclusions

   7.1 In addition to the broad form property exclusions identified in IBC forms 4042(1995), and 4047(2000), the *Contractor* is not required to provide the following insurance coverage:
   - Asbestos
   - Cyber Risk
   - Mould
   - Terrorism

The Royal
Architectural
Institute of Canada

www.ccdc.org

BID BOND

Standard Construction Document

CCDC 220 - 2002

No. _____

Bond Amount $_____

_____ as Principal, hereinafter called the Principal, and

_____ a corporation created and existing under the laws

of _____ and duly authorized to transact the business of Suretyship in _____ as Surety, hereinafter

called the Surety. are held and firmly bound unto _____ as

Obligee. hereinafter called the Obligee. in the amount of _____

_____ Dollars ($ _____) lawful money of Canada, for the payment

of which sum the Principal and the Surety bind themselves, their heirs, executors, administrators, successors and assigns. jointly and severally.

WHEREAS, the Principal has submitted a written bid to the Obligee, dated _____ day of _____, in the year _____

for _____

_____

The condition of this obligation is such that if the Principal shall have the bid accepted within the time period prescribed in the Obligee's bid documents, or, if no time period is specified in the Obligee's bid documents, within _____ ( ) days from the closing date as specified in the Obligee's bid documents, and the Principal enters into a formal contract and gives the specified security, then this obligation shall be void; otherwise, provided the Obligee takes all reasonable steps to mitigate the amount of such excess costs, the Principal and the Surety will

y to the Obligee the difference in money between the amount of the bid of the Principal and the amount for which the Obligee legally contracts

ith another party to perform the work if the latter amount be in excess of the former.

The Principal and Surety shall not be liable for a greater sum than the Bond Amount.

It is a condition of this bond that any suit or action must be commenced within seven (7) months of the date of this Bond.

No right of action shall accrue hereunder to or for the use of any person or corporation other than the Obligee named herein, or the heirs, executors, administrators or successors of the Obligee.

IN WITNESS WHEREOF, the Principal and the Surety have Signed and Sealed this Bond dated _____ day of _____, in the year _____

SIGNED and SEALED

Principal

in the presence of

ATTORNEY IN FACT

_____

Signature

_____

Name of person signing

Surety

_____

Signature

_____

Name of person signing

PERFORMANCE BOND

Standard Construction Document

CCDC 221 - 2002

o _____

Bond Amount $_____

_____ as Principal. hereinafter called the Principal, and

_____ a corporation created and existing under the laws

of _____ and duly authorized to transact the business of Suretyship in _____ as Surety, hereinafter

called the Surety, are held and firmly bound unto _____ as

Obligee. hereinafter called the Obligee, in the amount of _____

_____ Dollars ($ _____) lawful money of Canada, for the payment

of which sum the Principal and the Surety bind themselves, their heirs, executors, administrators, successors and assigns, jointly and severally.

WHEREAS. the Principal has entered into a written contract with the Obligee, dated _____ day of _____, in the year _____

for _____

hereinafter referred to as the Contract

The condition of this obligation is such that if the Principal shall promptly and faithfully perform the Contract then this obligation shall be null and void; otherwise it shall remain in full force and effect.

Whenever the Principal shall be, and declared by the Obligee to be, in default under the Contract, the Obligee having performed the Obligee's obligations thereunder, the Surety shall promptly:
1) remedy the default, or;
2) complete the Contract in accordance with its terms and conditions or;
3) obtain a bid or bids for submission to the Obligee for completing the Contract in accordance with its terms and conditions and upon determination by the Obligee and the Surety of the lowest responsible bidder, arrange for a contract between such bidder and the Obligee and make available as work progresses (even though there should be a default, or a succession of defaults, under the contract or contracts of completion, arranged under this paragraph) sufficient funds to pay to complete the Principal's obligations in accordance with the terms and conditions of the Contract and to pay those expenses incurred by the Obligee as a result of the Principal's default relating directly to the performance of the work under the Contract. less the balance of the Contract price; but not exceeding the Bond Amount. The balance of the Contract price is the total amount payable by the Obligee to the Principal under the Contract, less the amount properly paid by the Obligee to the Principal, or;
4) pay the Obligee the lesser of (1) the Bond Amount or (2) the Obligee's proposed cost of completion, less the balance of Contract price.

It is a condition of this bond that any suit or action must be commenced before the expiration of two (2) years from the earlier of (1) the date of Substantial Performance of the Contract as defined in the lien legislation where the work under the Contract is taking place, or, if no such definition exists. the date when the work is ready for use or is being used for the purpose intended, or (2) the date on which the Principal is declared in default by the Obligee

The Surety shall not be liable for a greater sum than the Bond Amount.

No right of action shall accrue on this Bond, to or for the use of, any person or corporation other than the Obligee named herein, or the heirs, executors, administrators or successors of the Obligee.

IN WITNESS WHEREOF, the Principal and the Surety have Signed and Sealed this Bond dated _____ day of _____,

in the year _____ .

SIGNED and SEALED

Principal

in the presence of

ATTORNEY IN FACT

_____
Signature

_____
Name of person signing

Surety

_____
Signature

_____
Name of person signing

LABOUR & MATERIAL PAYMENT BOND
(Trustee Form)

Standard Construction Document

CCDC 222 - 2002

No _____   Bond Amount $_____

_____ as Principal. hereinafter called the Principal, and

_____ a corporation created and existing under the laws

of _____ and duly authorized to transact the business of Suretyship in _____ as Surety, hereinafter

called the Surety, are held and firmly bound unto _____ as

Obligee. hereinafter called the Obligee, in the amount of _____

_____ dollars ($ _____) lawful money of Canada, for the payment

of which sum the Principal and the Surety bind themselves, their heirs, executors, administrators, successors and assigns, jointly and severally.

WHEREAS, the Principal has entered into a written contract with the Obligee, dated _____ day of _____, in the year _____

for _____

_____

in accordance with the Contract Documents submitted, and which are by reference made part hereof and are hereinafter referred to as the Contract

The Condition of this obligation is such that. if the Principal shall make payment to all Claimants for all labour and material used or reasonably required for use in the performance of the Contract, then this obligation shall be null and void; otherwise it shall remain in full force and effect, subject, however, to the following conditions:

. A Claimant for the purpose of this Bond is defined as one having a direct contract with the Principal for labour, material, or both, used or reasonably required for use in the performance of the Contract, labour and material being construed to include that part of water. gas, power, light, heat, oil, gasoline, telephone service or rental equipment directly applicable to the Contract provided that a person, firm or corporation who rents equipment to the Principal to be used in the performance of the Contract under a contract which provides that all or any part of the rent is to be applied towards the purchase price thereof, shall only be a Claimant to the extent of the prevailing industrial rental value of such equipment for the period during which the equipment was used in the performance of the Contract. The prevailing industrial rental value of equipment shall be determined, insofar as it is practical to do so, by the prevailing rates in the equipment marketplace in which the work is taking place

2. The Principal and the Surety, hereby jointly and severally agree with the Obligee, as Trustee, that every Claimant who has not been paid as provided for under the terms of its contract with the Principal, before the expiration of a period of ninety (90) days after the date on which the last of such Claimant's work or labour was done or performed or materials were furnished by such Claimant, may as a beneficiary of the trust herein provided for, sue on this Bond. prosecute the suit to final judgment for such sum or sums as may be justly due to such Claimant under the terms of its contract with the Principal and have execution thereon. Provided that the Obligee is not obliged to do or take any act, action or proceeding against the Surety on behalf of the Claimants, or any of them, to enforce the provisions of this Bond. If any act, action or proceeding is taken either in the name of the Obligee or by joining the Obligee as a party to such proceeding, then such act, action or proceeding, shall be taken on the understanding and basis that the Claimants, or any of them, who take such act, action or proceeding shall indemnify and save harmless the Obligee against all costs. charges and expenses or liabilities incurred thereon and any loss or damage resulting to the Obligee by reason thereof. Provided still further that, subject to the foregoing terms and conditions, the Claimants, or any of them may use the name of the Obligee to sue on and enforce the provisions of this Bond

3 It is a condition precedent to the liability of the Surety under this Bond that such Claimant shall have given written notice as hereinafter set forth to each of the Principal, the Surety and the Obligee, stating with substantial accuracy the amount claimed, and that such Claimant shall have brought suit or action in accordance with this Bond, as set out in sub-clauses 3 (b) and 3 (c) below, Accordingly, no suit or action shall be commenced hereunder by any Claimant:

a) unless such notice shall be served by mailing the same by registered mail to the Principal, the Surety and the Obligee, at any place where an office is regularly maintained for the transaction of business by such persons or served in any manner in which legal process may be served in the Province or Territory in which the subject matter of the Contract is located. Such notice shall be given.

CCDC 222 – 2002

i) in respect of any claim for the amount or any portion thereof, required to be held back from the Claimant by the Principal, under either the terms of the Claimant's contract with the Principal, or under the lien Legislation applicable to the Claimant's contract with the Principal, whichever is the greater, within one hundred and twenty (120) days after such Claimant should have been paid in full under the Claimant's contract with the Principal;

ii) in respect of any claim other than for the holdback, or portion thereof, referred to above, within one hundred and twenty (120) days after the date upon which such Claimant did, or performed, the last of the work or labour or furnished the last of the materials for which such claim is made under the Claimant's contract with the Principal;

b) after the expiration of one (1) year following the date on which the Principal ceased work on the Contract, including work performed under the guarantees provided in the Contract;

c) other than in a Court of competent jurisdiction in the Province or Territory in which the work described in the Contract is to be installed or delivered as the case may be and not elsewhere, and the parties hereto agree to submit to the jurisdiction of such Court.

4. The Surety agrees not to take advantage of Article 2365 of the Civil Code of the Province of Quebec in the event that. by an act or an omission of a Claimant, the Surety can no longer be subrogated in the rights, hypothec and privileges of said Claimant.

5. Any material change in the contract between the Principal and the Obligee shall not prejudice the rights or interest of any Claimant under this Bond, who is not instrumental in bringing about or has not caused such change.

6. The amount of this Bond shall be reduced by, and to the extent of any payment or payments made in good faith, and in accordance with the provisions hereof, inclusive of the payment by the Surety of claims made under the applicable lien legislation or legislation relating to legal hypothecs, whether or not such claim is presented under and against this Bond.

7. The Surety shall not be liable for a greater sum than the Bond Amount.

IN WITNESS WHEREOF, the Principal and the Surety have Signed and Sealed this Bond dated _____ day of _____,

in the year _____ .

SIGNED and SEALED

in the presence of

Principal

ATTORNEY IN FACT

_____
Signature

_____
Name of person signing

Surety

_____
Signature

_____
Name of person signing

Copyright 2002
Canadian Construction Documents Committee

(CCDC 222 – 2002 has been approved by the Surety Association of Canada)

ASSOCIATION OF CONSULTING ENGINEERING COMPANIES - CANADA

DOCUMENT NO. 31 - 2009

ENGINEERING AGREEMENT BETWEEN CLIENT AND ENGINEER

1981
Revised 1991
Addendum 1996
Revised 2009

ACEC Document 31 – 2009

TABLE OF CONTENTS

AGREEMENT 1
A-1   The Services...1
A-2   Agreement and Amendments......................2
A-3   ENGINEERING Agreement Documents............2
A-4   Fees and Reimbursable Expenses 2
A-5   Payment..3
A-6   Notices...3
A-7   Language of the Contract...................................4
A-8   Succession..4

DEFINITIONS 6
1.   Construction Contract 6
2.   Construction Administration Services...............6
3.   Construction Contract Documents 6
4.   Construction Contract Time.............................6
5.   Construction Cost.......................................6
6.   Consultant or Consultant of the Client...............7
7.   Contractor 7
8.   Coordinate or Coordination.............................7
9.   Engineering Agreement or Agreement...............7
10.   Engineering Documents..................................7
11.   Fees..7
12.   Hazardous Substances...................................7
13.   Notice..7
14.   Place of the Work.......................................8
15.   Project..8
16.   Project Budget...8
17.   Reimbursable Expenses.................................8
18.   Services..8
19.   Shop Drawings...8
20.   Sub-Consultant or Sub-Consultant of the
    Engineer..8
21.   Substantial Performance of the Work 8
22.   Suspension Expenses 8

23.   Termination Expenses 8
24.   Value Added Taxes 9
25.   Work...9
26.   Working Day 9

GENERAL CONDITIONS 10
Part 1 Agreement Documents.................................. 10
Part 2 Law of the Contract 10
Part 3 Rights and Remedies..................................... 10
Part 4 Assignment 11
Part 5 Engineer's Responsibilities............................ 11
Part 6 Client's Responsibilities 13
Part 7 Construction Administration........................... 14
Part 8 Certifications by the Engineer.......................... 16
Part 9 Construction Cost and Contract Time Estimates.... 17
Part 10 Termination and Suspension.......................... 17
Part 11 Ownership and Use of Documents, Patents and
    Trademarks.. 18
Part 12 Building Codes and By-Laws 19
Part 13 Project Ownership, Identification and
    Confidentiality.. 19
Part 14 Insurance and Liability................................. 20
Part 15 Dispute Resolution...................................... 21
Part 16 Payment 22
Part 17 Severability 23

**SCHEDULE A - ENGINEER'S SCOPE OF
SERVICES A-1**

**SCHEDULE B – FEES AND REIMBURSABLE
EXPENSES B-1**

ACEC Document 31 – 2009   1 of 23

ENGINEERING AGREEMENT BETWEEN CLIENT AND ENGINEER

dated as of the _____ day of _____ , 20____ .

by and between:

_____

hereinafter called the *"Client"*   *(Insert legal name and address)*

and:

_____

hereinafter called the *"Engineer"*.   *(Insert legal name and address)*

AGREEMENT

The *Client* and *Engineer* agree as follows:

A-1   THE SERVICES

1.1   The *Engineer* will provide *Services* in connection with the following *Project*:

_____

                                             (Insert a short description of the Project)

The location of the *Project* (the *"Place of the Work"*) is as follows:

_____

                (Insert the address, location or legal description of the site of the Work)

1.2   The *Engineer* will provide *Services* for the *Project* in accordance with Schedule A –
       ENGINEER'S SCOPE OF SERVICES.

1.3   Any change to the *Services* listed in Schedule A – ENGINEER'S SCOPE OF SERVICES will be
       made by written order signed by both parties identifying the change plus adjustments, if any, to
       the *Engineer's Fees* and *Reimbursable Expenses* and time for completion of the *Services*.

ACEC Document 31 – 2009

A-2   AGREEMENT AND AMENDMENTS

2.1   This *Engineering Agreement* constitutes the entire agreement between the *Client* and the *Engineer* relating to the *Project*, and supersedes all prior agreements between them, whether written or oral, respecting the *Services*. No other terms, conditions or warranties, whether express or implied, form a part of this *Engineering Agreement*.

2.2   This *Engineering Agreement* may be amended only by a written document signed by both the *Client* and the *Engineer*.

A-3   ENGINEERING AGREEMENT DOCUMENTS

The following sections and documents form part of and are incorporated into the *Engineering Agreement*:

In this *Engineering Agreement*:

- Agreement
- Definitions
- General Conditions
- Schedule A - ENGINEER'S SCOPE OF SERVICES
- Schedule B - FEES AND REIMBURSABLE EXPENSES

Other documents:

*

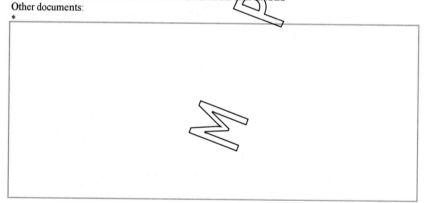

(Insert here, attaching additional pages if required, a list of all other sections and documents, including any supplementary conditions, other schedules and lists that are to be incorporated into the Engineering Agreement.)

A-4   FEES AND REIMBURSABLE EXPENSES

4.1   The *Fees* for the *Services* of the *Engineer* are set forth in Schedule B – FEES AND REIMBURSABLE EXPENSES.

4.2   *Reimbursable Expenses* are the costs and charges identified in Schedule B – FEES AND REIMBURSABLE EXPENSES that are incurred by the *Engineer* in performing the *Services*.

ACEC Document 31 – 2009

A-5   PAYMENT

5.1   The *Client* will pay to the *Engineer* the *Fees* and *Reimbursable Expenses* set out in this *Engineer*ing Agreement.

5.2   The *Engineer* will issue monthly invoices for *Fees* and *Reimbursable Expenses*, together with applicable *Value Added Taxes*.

5.3   The *Engineer's* invoices are due when presented. Invoices unpaid by the *Client* 30 days after presentation will bear interest of _____ % per annum calculated monthly.

A-6   NOTICES

6.1   A *Notice* will be addressed to the recipient at the address set out below. The delivery of a *Notice* will be by personal delivery, receipted courier delivery or by facsimile. A *Notice* delivered by one party in accordance with this *Engineering Agreement* will be deemed to have been received by the other party on the first W*orking Day* after actual delivery. An address for a party may be changed by *Notice* to the other party setting out the new address in accordance with this Article.

6.2   Although the parties may use electronic communications for the purposes of general communication, e-mail will not be used for delivery of a *Notice*.

6.3   The addresses for the parties are as follows:

*Client**

_____

*(name of Client)**

_____

(address)

_____

(facsimile number)

*Engineer**

_____

*(name of Engineer)**

_____

(address)

_____

(facsimile number)

*(If it is intended that a specific individual or officer must receive the Notice, indicate that individual's name and/or office.)

ACEC Document 31 – 2009

A-7   LANGUAGE OF THE CONTRACT

7.1   (*For use in the Province of Quebec.*) The parties confirm their wish that this *Engineering Agreement* as well as any other related documents including future amendments, *Notices* and correspondence be drawn in English.   Parts of the *Engineering Agreement* may be included as available in English or in French or both, according to the language or languages in which they originally were drawn.

Les parties confirment leur volonté que cette convention de même que tous les documents s'y rattachant, y compris tous amendements, avis et correspondance futures, soient rédigés en anglais. Des portions de la Convention d'ingénierie sont incluses telles que disponibles, soit en français ou en anglais ou les deux, selon la langue ou les langues dans lesquelles la portion pertinente de la Convention d'ingénierie aura été rédigée à l'origine.

A-8   SUCCESSION

8.1   This *Engineering Agreement* will inure to the benefit of and be binding upon the parties, and upon their executors, administrators, successors and permitted assigns.

(*Signatures next follow*)

IN WITNESS WHEREOF the parties hereto have executed this *Engineering Agreement* as of the day and year first above written.

CLIENT

_____
name of Client

_____
signature

_____
name and title of person signing

_____
signature

_____
name and title of person signing

ENGINEER

_____
name of Engineer

_____
signature

_____
name and title of person signing

_____
signature

_____
name and title of person signing

WITNESS
(only required where the Client is an individual)

_____
signature

_____
name and title of person signing

_____
signature

_____
name and title of person signing

WITNESS
(only required where the Engineer is an individual)

_____
signature

_____
name and title of person signing

_____
signature

_____
name and title of person signing

Where legal jurisdiction, local practice, or Client or Engineer requirements calls for:

   (a)   *proof of authority to execute this document, attach such proof of authority in the form of a certified copy of a resolution naming the representative(s) authorized to sign the Engineering Agreement for and on behalf of the corporation or partnership; or*

   (b)   *the affixing of a corporate seal, this Engineering Agreement should be properly sealed.*

DEFINITIONS

1. **Construction Contract**
    Construction Contract means the contract between the *Client* and the *Contractor* for the performance of the *Work* by the *Contractor*.

2. **Construction Administration Services**
    Construction Administration Services means those services, if any, which relate to the administration of the *Construction Contract* and which are identified as such in Schedule A – ENGINEER'S SCOPE OF SERVICES and which form part of the *Services*.

3. **Construction Contract Documents**
    Construction Contract Documents means all documents relating to the *Work* issued by or through the *Engineer* that are incorporated into the *Construction Contract* and all variations and modifications issued by or approved by the *Engineer*.

4. **Construction Contract Time**
    Construction Contract Time means the period from the *Notice* to proceed with the *Work* issued to the *Contractor* to the completion date of the *Work* in accordance with the *Construction Contract*.

5. **Construction Cost**
    Construction Cost means the total cost to the *Client* of the *Work*, and includes:

    (a) all materials, equipment, labour, *Value Added Taxes*, *Contractor's* overhead and profit provided in accordance with the *Construction Contract Documents*;

    (b) the cost of all installations for the *Project* carried out by parties other than the *Contractor*;

    (c) the cost of all *Work* carried out under the *Construction Contract*;

    (d) refunds or sales tax exemptions on any materials or equipment, or both;

    (e) the cost of *Work* carried out by direct labour or direct purchase of materials or equipment by the *Client* at prevailing prices;

    (f) the value of new or old materials provided by the *Client*;

    (g) the value of all deletions made by the *Client* from the *Work* after the *Engineer* has completed a design for the deleted items as a part of the *Work*; and

    (h) the value of any monetary damages or set offs retained by the *Client* from the *Contractor* with respect to the *Work*;

    but does not include:

    (i) *Fees* and *Reimbursable Expenses* of the *Engineer*;

    (j) the fees and reimbursable expenses of *Consultant of the Client*;

ACEC Document 31 – 2009   7 of 23

    (k)   the salary of the *Client's* representative or other salary and administrative costs of the *Client*;

    (l)   the cost of land and any related rights or easements; or

    (m)   the costs of items, such as equipment, furniture or fixtures, that do not form a part of the *Construction Contract.*

6.   Consultant or Consultant of the Client
Consultant or *Consultant of the Client* means a registered or licensed professional engineer, architect, or other specialist engaged directly by the *Client* other than the *Engineer* or *Sub-Consultants of the Engineer.*

7.   Contractor
Contractor means a person or entity contracting with the *Client* to perform some or all of the *Work.*

8.   Coordinate or Coordination
Coordinate or *Coordination*, when referring to the *Services* of the *Engineer*, means the management and supervision of communications between the *Engineer* and a *Sub-Consultant* or a *Consultant of the Client.*

9.   Engineering Agreement or Agreement
Engineering Agreement or *Agreement* means this agreement between the *Client* and the *Engineer*, including all of the documents identified in Article A-3 ENGINEERING AGREEMENT DOCUMENTS and any amendments thereto.

10.   Engineering Documents
Engineering Documents means drawings, plans, models, designs, specifications, reports, photographs, computer software if proprietary to the *Engineer*, surveys, calculations and other data, including computer print outs, contained in the *Construction Contract Documents* or which are otherwise used in connection with the *Project*, and which were prepared by or on behalf of the *Engineer* and are instruments of service for the execution of the *Work.*

11.   Fees
Fees means those fees that are identified in Schedule B – FEES AND REIMBURSABLE EXPENSES and which are payable by the *Client* to the *Engineer.*

12.   Hazardous Substances
Hazardous Substances means any toxic or hazardous solid, liquid, gaseous, thermal, or electromagnetic irritant or contaminant, and includes, without limitation, pollutants, moulds, and hazardous and special materials and wastes whether or not defined as such in any federal, provincial, territorial, or municipal laws, statutes, or regulations.

13.   Notice
Notice means a written communication between the parties that is delivered in accordance with the provisions of Article A-6 – RECEIPT OF AND ADDRESSES FOR NOTICES. Use of the verb "**to notify**" means to send a *Notice* in the above manner.

14. **Place of the Work**
     Place of the Work means the designated site or location of the *Work* identified in this *Engineering Agreement*.

15. **Project**
     Project means the total endeavour contemplated in this *Engineering Agreement* of which the *Services* and the *Work* may be the whole or a part.

16. **Project Budget**
     Project Budget means the estimated cost of the *Work*, including the *Services* and other professional services, but excluding expenses relating to site acquisition, promotion and marketing.

17. **Reimbursable Expenses**
     Reimbursable Expenses means those expenses that are identified in Schedule B – FEES AND REIMBURSABLE EXPENSES and which are payable by the *Client* to the *Engineer*.

18. **Services**
     Services means those services that are identified in Schedule A – ENGINEER'S SCOPE OF SERVICES.

19. **Shop Drawings**
     Shop Drawings means drawings, diagrams, illustrations, schedules, performance charts, technical brochures, and other data that are to be provided by the *Contractor* or by others to illustrate details of a portion of the *Work*.

20. **Sub-Consultant or Sub-Consultant of the Engineer**
     Sub-Consultant or *Sub-Consultant of the Engineer* means any registered or licensed professional engineer, architect, or other specialist engaged by the *Engineer* to perform a discreet scope of services in connection with the *Project*, but does not include employees of the *Engineer* or consultants working under a personal services agreement with the *Engineer*.

21. **Substantial Performance of the Work**
     Substantial Performance of the Work means, where defined in the lien legislation applicable to the *Place of the Work*, the meaning given to that term in the lien legislation. If such legislation is not in force or does not contain such definition or if the *Work* is governed by the Civil Code of Quebec, *Substantial Performance of the Work* will have been reached when the *Work* is ready for use or is being used for the purpose intended and is so certified by the *Engineer* or by the certifier, if any, appointed under the *Construction Contract*, as the case may be.

22. **Suspension Expenses**
     Suspension Expenses means expenses incurred by the *Engineer*, including demobilization and remobilization expenses, which are directly attributable to suspension of the *Services* by the *Client*.

23. **Termination Expenses**
     Termination Expenses means expenses incurred by the *Engineer* which are directly attributable to termination of the *Services* and include the *Engineer's* expenses reasonably and necessarily incurred in winding down the *Services*.

24. **Value Added Taxes**

Value Added Taxes means such sum as levied upon the *Fee, Reimbursable Expenses* and the *Work* by a Federal, Provincial or Territorial Government and is computed as a percentage of the same and includes the Goods and Services Tax, the Quebec Sales Tax, the Harmonized Sales Tax, and any similar tax, the payment or collection of which is imposed by legislation.

25. **Work**

Work means the total construction and related services required by the *Construction Contract*.

26. **Working Day**

Working Day means a day other than a Saturday, Sunday, statutory holiday or statutory vacation day that is observed by the construction industry in the area of the *Place of the Work*. Reference to a day, other than a *Working Day*, indicates a calendar day.

GENERAL CONDITIONS

PART 1   AGREEMENT DOCUMENTS

GC 1.1   If there is a conflict within the *Engineering Agreement*, the order of priority of the documents which make up the *Engineering Agreement*, from highest to lowest, will be:

(a)   Agreement;

(b)   Definitions;

(c)   Any supplementary conditions to the General Conditions;

(d)   General Conditions;

(e)   Schedule A – ENGINEER'S SCOPE OF SERVICES;

(f)   Schedule B – FEES AND REIMBURSABLE EXPENSES;

(g)   Other schedules to the *Engineering Agreement*.

GC 1.2   The documents which make up the *Engineering Agreement* are complementary, and what is required by any one will be as binding as if required by all.

GC 1.3   Words and abbreviations with well known technical or trade meanings are used in the *Engineering Agreement Documents* in accordance with such recognized meanings.

GC 1.4   References in the *Engineering Agreement Documents* to the singular will be considered to include the plural as the context requires.

GC 1.5   References in the *Engineering Agreement Documents* to regulations and codes are considered to be references to the latest published version as of the signature date of the *Engineering Agreement*, unless otherwise indicated.

PART 2   LAW OF THE CONTRACT

GC 2.1   The law of the *Place of the Work* will govern the interpretation of the *Engineering Agreement*.

GC 2.2   The *Client* acknowledges receipt of sufficient information from the *Engineer*, including information concerning the *Fees and Services* of the *Engineer*, so as to allow the *Client* to assess the nature, extent and cost of the *Services* of the *Engineer* and the obligations which the *Client* assumes under this *Engineering Agreement*.

PART 3   RIGHTS AND REMEDIES

GC 3.1   Except as expressly provided in the *Engineering Agreement Documents*, the duties and obligations imposed by the *Engineering Agreement Documents* and the rights and remedies available thereunder will be in addition to and not a limitation of any duties, obligations, rights, and remedies otherwise imposed or available by law.

GC 3.2   No action or failure to act by the *Client* or *Engineer* will constitute a waiver of a right or duty afforded or imposed under this *Engineering Agreement*, except as may be specifically specified in writing.

PART 4   ASSIGNMENT

GC 4.1   Neither party may assign this *Engineering Agreement* in whole or part without the written consent of the other, which consent will not be unreasonably withheld.

PART 5   ENGINEER'S RESPONSIBILITIES

GC 5.1   The *Engineer* is bound by the legislation governing the *Engineer's* profession. Nothing in this *Engineering Agreement* requires the *Engineer* to derogate from obligations prescribed by law that are binding upon the *Engineer*.

GC 5.2   The *Engineer* will provide the *Services* in accordance with this *Engineering Agreement* and with the degree of care, skill, and diligence normally provided by engineers in the performance of comparable services in respect of projects of a similar nature to that contemplated by this *Engineering Agreement*.

GC 5.3   The *Engineer* will maintain records of *Reimbursable Expenses* and time records for *Services* performed for which the *Fee* is computed on an hourly basis. These records will be maintained to acceptable accounting standards and made available to the *Client* at mutually convenient times during the term of this Engineering Agreement and for a period not exceeding one year following completion of the *Services*.

GC 5.4   The *Engineer* will:

(a)   not be responsible for the performance by the *Contractor*, subcontractors, suppliers or any other contractors of the *Work* or for the failure of any of them to carry out the *Work* in accordance with the *Construction Contract*;

(b)   not be responsible for, nor control, direct or supervise, the construction methods, means, techniques, sequences or procedures of the *Contractor*, subcontractors, suppliers, or any other contractors;

(c)   not be responsible for acts or omissions of the *Consultant of the Client*, or the *Contractor*, subcontractors, suppliers, or any other contractor;

(d)   not be responsible for safety precautions and programs required in connection with the *Work* or for general site safety at the *Place of the Work* under applicable health and construction safety legislation at the *Place of the Work*;

(e)   not be responsible for the advice of any independent expert engaged either by the *Client* or the *Contractor*, whether or not recommended by the *Engineer*; and

(f)   not be responsible to make exhaustive or continuous on-site reviews.

GC 5.5   The *Engineer* may engage *Sub-Consultants* to enable the *Engineer* to provide the *Services*. Should the *Client* reasonably object to a *Sub-Consultant* engaged by the *Engineer*, the *Client* may request the *Engineer* to replace the *Sub-Consultant*. In this event, the *Client* will pay all costs resulting from termination and replacement of that *Sub-Consultant* and the parties will adjust the *Fees* and time for completion of the *Services* to take into account the termination and replacement.

GC 5.6   The *Engineer* will *coordinate* the activities of its *Sub-Consultants*.

GC 5.7   The *Engineer* has discretion, where the *Client* provides equipment or materials for the *Project*, to request the *Client* to arrange that items to be used or installed in the *Work* first be tested or verified before being used for the purposes intended by the *Client* or be validated by an appropriate certificate of compliance.

Upon receipt of the requested test or verification reports or certificate of compliance, the *Engineer* will *notify* the *Client* of the *Engineer*'s acceptance or refusal of equipment or materials concerned, with or without such reservations as the *Engineer* considers to be appropriate. If the *Client* insists upon using an item to which the *Engineer* has objected or expressed reservations in writing or if the *Client* declines to arrange to test, verify or certify an item as requested by the *Engineer*, the *Client* will be considered to have waived any recourse against the *Engineer* resulting from the use of such item or from a defect or inadequacy in such item.

GC 5.8   The *Engineer* is entitled to rely upon the accuracy and completeness of information and data furnished by the *Client*, including information and data originating from a *Consultant of the Client*, whether such *Consultant* is engaged at the request of the *Engineer*, the *Client* or otherwise.

GC 5.9   The *Engineer* is entitled to rely upon the accuracy and completeness of records, information, data and specifications furnished by:

(a)   government authorities and public utilities; and

(b)   by manufacturers and suppliers of equipment, material or supplies.

Should such records, information, data, and specifications prove to be erroneous or inaccurate, the *Engineer* is entitled to make the necessary changes to the *Engineering Documents* at the expense of the *Client*.

GC 5.10   The *Engineer* is not responsible for manufacturing defects in equipment, material or supplies specified or recommended by ~~the Engineer~~

GC 5.11   The *Engineer* will not accept a commission or other compensation from a manufacturer, supplier or contractor involved in the *Project*. The *Engineer* will have no financial interest in the materials or equipment specified or recommended by the *Engineer* as part of the *Services*. However, ownership of less than 1% of the securities issued by a company whose securities are traded on a recognized securities exchange will not be deemed to constitute a financial interest.

GC 5.12   Where the *Engineer* does not provide *Construction Administration Services* under this *Engineering Agreement* but the *Client* nevertheless requests the *Engineer* to attend at the

ACEC Document 31 – 2009   13 of 23

Place of the Work for any reason, the *Engineer* will not incur any liability to the *Client* for having attended at the *Place of the Work* unless the *Client* makes a specific request to the *Engineer* in writing stating why the *Client* has requested the *Engineer's* attendance and the *Engineer* has agreed to attend for that sole purpose. In such event, the only responsibility of the *Engineer* will be to respond to the *Client's* specific request provided such request falls within the mandate and competence of the *Engineer.*

PART 6   CLIENT'S RESPONSIBILITIES

GC 6.1   The *Client* will promptly fulfill all of the *Client's* responsibilities so as not to impede the *Engineer's* orderly performance of the *Services.*

GC 6.2   The *Client* will fully advise the *Engineer* in writing of the *Client's* requirements in connection with the *Project*, including the *Project Budget* and time constraints of the *Client.*

GC 6.3   The *Client*, when so *notified* by the *Engineer*, will make available to the *Engineer* all information or data pertinent to the *Project* which is required by the *Engineer* to perform the *Services.*

GC 6.4   The *Client*, when so *notified* by the *Engineer*, will directly engage the services of a specialist to provide information or to perform ancillary services that are necessary to enable the *Engineer* to carry out the *Services*. Ancillary services may include, but are not limited to, topographic surveys and mapping of the *Place of the Work*, site services reports, technical investigations, geotechnical reports, quantity surveys and testing services. The parties will jointly agree on the selection of any such specialist.

GC 6.5   Should the *Client* not provide the information required by the *Engineer* to perform the *Services* as mentioned in GC 6.3 or not accept the request of the *Engineer* to engage a specialist as mentioned in GC 6.4, the *Engineer* will be entitled at the *Engineer's* option and upon a further *Notice* to the *Client* either to terminate this *Engineering Agreement* or to be relieved of any responsibility for the consequences of the *Client's* decision not to provide the information or to engage a specialist as requested by the *Engineer.*

GC 6.6   The *Client* will ensure that *Consultants of the Client* have adequate professional liability insurance, commensurate with the services they will provide for the *Project* and the *Work.*

GC 6.7   Should the *Engineer* be required to act as the agent of the *Client* in order to perform some of the *Services*, the *Client* will authorize the *Engineer* in writing to act as the *Client's* agent for such purposes as may be necessary. Where the *Engineer* acts as the *Client's* agent pursuant to a written authorization, the *Client* is responsible for the authorized actions of the *Engineer* as agent of the *Client*. The *Client* will indemnify the *Engineer* for damages and expenses incurred by the *Engineer*, including reasonable legal fees, when acting as agent of the *Client.*

GC 6.8   The *Client* will promptly consider requests by the *Engineer* for directions or decisions and diligently inform the *Engineer* of the *Client's* direction or decision within a reasonable time so as not to delay the *Services.*

GC 6.9   The *Client* will pay the *Engineer* as provided in this *Engineering Agreement.*

GC 6.10   The *Client*, at the request of the *Engineer*, will furnish reasonable evidence to the *Engineer* that financial arrangements have been made to fulfill the *Client*'s payment obligations under this *Engineering Agreement* before signing the *Engineering Agreement*, and promptly from time to time thereafter.

GC 6.11   The *Client* will *notify* the *Engineer* of any material change in the *Client*'s financial arrangements that affect the *Client*'s ability to fulfill the *Client*'s payment obligations under this *Engineering Agreement*.

GC 6.12   The *Client* will provide those legal, accounting, insurance, bonding and other counselling services which are necessary for the preparation of tenders or requests for proposals and the like or for the performance of other *Services* of the *Engineer*. If the *Client* is unable to provide such counselling services and requests the *Engineer* to do so, the *Client* will reimburse the *Engineer* for expenses incurred in securing any such counselling services.

GC 6.13   The *Client* is responsible for obtaining legal advice regarding tenders, requests for a proposal or information, bids, contract awards and the like, regarding the *Project*. The *Client* is responsible for decisions relating to the issuance, validity or award of tenders, proposals or bids and for the resulting consequences, even where the *Services* require the *Engineer* to review or assist in the preparation of tenders, proposals or bids and the like or to make recommendations regarding them or regarding the qualification or selection of bidders.

GC 6.14   The *Client* will arrange where necessary for the *Engineer*'s access to the *Place of the Work* or other required locations to enable the *Engineer* to perform the *Services*.

GC 6.15   The *Client* will designate in writing an individual to act as the *Client's* representative who will have authority to transmit instructions to and receive information from the *Engineer*.

GC 6.16   The *Client* will promptly *notify* the *Engineer* whenever the *Client* or the *Client's* representative becomes aware of any defects or deficiencies in the *Services*, the *Engineering Documents* or in the *Construction Contract Documents*.

GC 6.17   The *Client* will obtain required approvals, licences, and permits from municipal, governmental or other authorities having jurisdiction over the *Project* so as not to delay the *Engineer* in the performance of the *Services*.

GC 6.18   The *Client* will not enter into contracts with *Consultants of the Client* or *Contractors* that are incompatible or inconsistent with the *Services* to be provided under this *Engineering Agreement*.

PART 7   CONSTRUCTION ADMINISTRATION

GC 7.1   This PART 7 CONSTRUCTION ADMINISTRATION applies only when and to the extent that the *Engineer* provides *Construction Administration Services* under Schedule A – ENGINEER'S SCOPE OF SERVICES.

GC 7.2   *Construction Administration Services* provided by the *Engineer* are for the benefit of the *Client*.

GC 7.3   The *Engineer* will have authority to act on behalf of the Client but only to the extent provided in the *Construction Administration Services*.

GC 7.4   The *Client* may modify or extend the duties, responsibilities, and authority of the *Engineer* as set forth in the *Construction Administration Services* with the written consent of the *Engineer*.

GC 7.5   *Notices*, instructions, requests, claims, or other communications between the *Client* and the *Contractor* and between the *Client* and any *Consultants of the Client* will be made by or through the *Engineer*, unless the *Client notifies* the *Engineer* otherwise.

GC 7.6   The *Engineer*, in the first instance, will be the interpreter of the requirements of the *Engineering Documents* and will make findings on all claims made by either the *Client* or the *Contractor* under the *Construction Contract*, and on all matters relating to the interpretation of the *Engineering Documents*, unless otherwise provided in the *Construction Contract*.

GC 7.7   The *Engineer*, if specified in the *Construction Administration Services* and in the contracts among the *Client* and its *Consultants*, will *coordinate* the activities of the *Consultants of the Client*.

GC 7.8   The *Engineer* will visit the *Place of the Work* at such intervals as the *Engineer*, in the *Engineer's* judgment, considers to be appropriate relative to the progress of construction in order to enable the *Engineer* to assess whether the *Contractor* is carrying out the *Work* in general conformity with the *Engineering Documents*. Only *Work* which the *Engineer* has reviewed during the construction will be considered to have been assessed. Should the *Engineer* comment on parts of the *Work* which the *Engineer* has not reviewed, the comments of the *Engineer* must be construed as being assumptions only and must not be relied upon unless the *Client notifies* the *Engineer* to review, and the *Engineer* reviews, the parts of the *Work* in question.

GC 7.9   The *Engineer* is not responsible for performance of the *Construction Contract*. The *Contractor* is solely responsible for the execution, quality, schedule and cost of the *Work*.

GC 7.10   The *Engineer* is not responsible to the *Client*, the *Contractor* or any *Consultant of the Client* for the means, methods, techniques, sequences, procedures and use of equipment for the *Project*, whether or not reviewed by the *Engineer*, which are employed by the *Contractor* or by a *Consultant of the Client* in executing, designing or administering the *Work*; or for the services of a *Consultant of the Client*; or for commissioning and start-up of any facility or equipment; or for health and safety precautions and programs incidental to the *Project* or to the commissioning and start-up of any facility or equipment.

GC 7.11   No acceptance by the *Engineer of the Work* or of the services of the *Consultants of the Client*, whether express or implied, will relieve the *Contractor* or the *Consultants of the Client* from their responsibility to the *Client* for the proper performance of the *Work* or their services.

GC 7.12   Unless otherwise specifically stated within the *Engineering Documents* or included in the *Construction Administration Services*, the *Contractor's Shop Drawings* will be reviewed by the *Engineer* only for the limited purpose of checking for general conformance with information given and the design concept expressed in the *Construction Contract Documents*. The *Engineer's* review of *Shop Drawings* is not for the purpose of determining the feasibility or constructability of the *Work* detailed within the *Shop Drawings* or the accuracy or completeness of:

(a)   details such as dimension and quantities;

(b)   instructions for installation or performance of equipment or systems;

(c)   *Contractor's* construction means, methods, techniques, sequences or procedures; or

(d)   safety precautions for those engaged in the *Work* or others at the *Place of the Work*.

GC 7.13   Where required by the *Services*, at the end of the *Project* the *Engineer* will compile and deliver to the *Client* a reproducible set of record documents showing significant changes made to the *Work*, based upon, without additional verification on the part of the *Engineer*, updated record drawings, as-built and other data provided by the *Contractor*, *Consultants of the Client*, or other parties.

PART 8   CERTIFICATIONS BY THE ENGINEER

GC 8.1   This PART 8 CERTIFICATIONS BY THE ENGINEER applies only when and to the extent that the *Engineer* is required to issue certifications under Schedule A – ENGINEER'S SCOPE OF SERVICES.

GC 8.2   The *Engineer* will issue those certifications which the *Engineer* is required to give as part of the *Services* with the degree of care, skill, and diligence normally provided by engineers issuing comparable certifications in respect of projects of a similar nature to that contemplated by this *Engineering Agreement*, based upon data reasonably available to the *Engineer*.

GC 8.3   If included in the *Construction Administration Services*, the *Engineer*'s issuance of a certificate for payment constitutes a representation by the *Engineer* to the *Client*, based on the *Construction Administration Services* performed by the *Engineer* and on review of the *Contractor's* schedule of values and applications for payment, that, to the best of the *Engineer's* information and belief:

(a)   the *Work* has progressed to the value indicated;

(b)   *Work* observed by the *Engineer* while performing *Construction Administration Services* conforms generally with the *Construction Contract Documents*; and

(c)   the *Contractor* is entitled to payment in the amount certified.

GC 8.4   The *Engineer's* issuance of a certificate for payment is subject to:

(a)   review and evaluation of the *Work*, to the extent specified in the *Services*, as it progresses for general conformity with the *Construction Contract Documents*;

(b)   the results of any subsequent tests required by the *Construction Contract Documents*;

(c)   correction of deviations from the *Construction Contract Documents* detected prior to completion or after completion, as the case may be; and

(d)   any specific qualifications stated in the certificate for payment.

GC 8.5   The *Engineer*'s issuance of a certificate for payment is not a representation that the *Engineer* has inquired into the *Contractor*'s:

(a)   use or allocation of monies paid on account of the contract price specified in the *Construction Contract*; or

(b)   compliance with obligations imposed on the *Contractor* by law, including requirements of workplace health and safety legislation at the *Place of the Work*.

PART 9   CONSTRUCTION COST AND CONTRACT TIME ESTIMATES

GC 9.1   This PART 9 - CONSTRUCTION COST AND CONTRACT TIME ESTIMATES applies only in the event the *Services* require the *Engineer* to provide the *Client* with an estimate of the probable *Construction Cost* or *Construction Contract Time*, whether to assist the *Client* with a call for tenders for the *Work* or otherwise.

GC 9.2   The parties acknowledge that an estimate of probable *Construction Cost* and an estimate of *Construction Contract Time* provided by the *Engineer* are subject to change and are contingent upon factors, including market forces, over which the *Engineer* has no control. The *Engineer* does not guarantee the accuracy of such estimates nor does the *Engineer* represent that bids, negotiated prices or the time for performance will not vary from such estimates.   More definitive estimates regarding costs and time for performance may be assessed only when bids and negotiated prices are received for the *Work*.

PART 10   TERMINATION AND SUSPENSION

GC 10.1   This *Engineering Agreement* is terminated on the earliest of:

(a)   the date when the *Engineer* has performed all of the *Services*; or

(b)   the date of termination if termination occurs in accordance with this GC 10 TERMINATION AND SUSPENSION.

GC 10.2   If the *Engineer* is a natural person practicing alone (and not part of a company or a partnership) and should the *Engineer* die or become seriously incapacitated before having supplied all of the *Services*, either the *Client* or the estate or legal representative of the *Engineer* may terminate this *Engineering Agreement* upon *Notice* to the other, with effect from the date of decease or, in the case of serious incapacity, from the date of the *Notice* of termination.

GC 10.3   If the *Engineer* is in material default in the performance of any of the *Engineer's* obligations under this *Engineering Agreement*, the *Client* will *notify* the *Engineer* that the default must be corrected.   If the *Engineer* does not correct the default within 30 days after receipt of such *Notice* or if the *Engineer* does not take reasonable steps to correct the default if the default is not susceptible of immediate correction, the *Client* may terminate this *Engineering Agreement* upon further *Notice* to the *Engineer*, without prejudice to any other rights or recourses of the *Client*. Such termination will not release the *Client* from its obligation to pay all *Fees* and *Reimbursable Expenses* incurred by the *Engineer* up to the date of termination in the manner provided in this *Engineering Agreement.*

GC 10.4   If the *Client* is in material default in the performance of any of the *Client's* obligations set forth in this *Engineering Agreement,* including but not limited to the non-payment of *Fees* and *Reimbursable Expenses* of the *Engineer* in the manner specified in this *Engineering Agreement,* the *Engineer* will *notify* the *Client* that the default must be corrected. If the *Client* does not correct the default within 30 days after receipt of such *Notice,* the *Engineer* may terminate this *Engineering Agreement* upon further *Notice* to the *Client.* In such event, the *Client* will promptly pay the *Fees* and *Reimbursable Expenses* of the *Engineer* that are incurred and unpaid as of the date of such termination, plus the *Termination Expenses,* without prejudice to any other rights or recourses of the *Engineer.*

GC 10.5   If the *Client* is unwilling or unable to proceed with the *Project,* the *Client* may suspend or terminate this *Engineering Agreement* by *Notice* of 30 days to the *Engineer.* Upon receipt of such *Notice,* the *Engineer* will perform no further *Services* other than those reasonably necessary to suspend or terminate that portion of the *Project* for which the *Engineer* is responsible. In such event, the *Client* will pay all of the *Fees* and *Reimbursable Expenses* incurred by the *Engineer* up to the date of suspension or termination, plus the *Suspension Expenses* or *Termination Expenses,* as the case may be, in the manner provided for in this *Engineering Agreement.*

GC 10.6   If the *Client* suspends performance of the *Services* at any time for more than 30 consecutive or non-consecutive days through no fault of the *Engineer,* then the *Engineer* may choose to terminate this *Engineering Agreement* upon *Notice* to the *Client.* In this event, the *Client* will promptly pay the *Fees* and *Reimbursable Expenses* of the *Engineer* that are incurred and unpaid as of the date of such termination, plus the *Termination Expenses,* without prejudice to any other rights or recourses of the *Engineer.*

PART 11   OWNERSHIP AND USE OF DOCUMENTS, PATENTS AND TRADEMARKS

GC 11.1   The *Engineering Documents* are the property of the *Engineer,* whether the *Work* is executed or not. The *Engineer* reserves the copyright therein and in the *Work* executed therefrom. The *Client* is entitled to keep a copy of the *Engineering Documents* for its records.

GC 11.2   The *Engineer* retains ownership of all patents, trademarks, copyrights, industrial or other intellectual property rights resulting from the *Services* or from concepts, products, or processes which are developed or first reduced to practice by the *Engineer* in performing the *Services.* The *Client* will not use, infringe or appropriate such proprietary rights without the prior consent and compensation of the *Engineer.*

GC 11.3   Provided the *Fees* and *Reimbursable Expenses* of the *Engineer* are paid, the *Client* will have a non-exclusive license to use any proprietary concept, product or process of the *Engineer* which relates to or results from the *Services* for the life of the *Project* and solely for purposes of its maintenance and repair.

GC 11.4   The *Engineer* warrants that the designs, drawings, and calculations developed by the *Engineer* under this *Engineering Agreement* will not infringe the patent, copyright, trade mark or other intellectual property rights of another person.

GC 11.5   The *Engineer* will retain the original of the *Engineering Documents* and of those parts of the *Construction Contract Documents* which are generated by the *Engineer*, including computer-generated designs relating thereto, but excluding any models or graphic presentations specifically commissioned and paid for by the *Client*.

GC 11.6   Should the *Client* use the *Engineering Documents* or provide them to third parties for purposes other than in connection with the *Project* without *notifying* the *Engineer* and without the *Engineer's* prior written consent, the *Engineer* will be entitled either to compensation for such improper use or to prevent such improper use, or to both. The *Client* will indemnify the *Engineer* against claims and costs (including legal costs) associated with such improper use. In no event will the *Engineer* be responsible for the consequences of any such improper use.

GC 11.7   Should the *Client* alter the *Engineering Documents* without *notifying* the *Engineer* and without the *Engineer's* prior written consent, the *Client* will indemnify the *Engineer* against claims and costs (including legal costs) associated with such improper alteration. In no event will the *Engineer* be responsible for the consequences of any such improper alteration.

GC 11.8   The *Client* may not use the *Engineering Documents* without having paid the *Fees* and *Reimbursable Expenses* of the *Engineer*. The *Engineer* is entitled to injunctive relief should the *Engineering Documents* be used without payment of the *Fees* and *Reimbursement Expenses* provided for in this *Engineering Agreement*.

GC 11.9   The *Engineering Documents* are not to be used on any other project without the prior written consent and compensation of the *Engineer*.

PART 12   BUILDING CODES AND BY-LAWS

GC 12.1   The *Engineer* will interpret building codes and by-laws as they apply to the *Project* at the time of design to the best of the *Engineer's* ability. As the *Work* progresses, building codes and by-laws may change or the interpretation by an authority having jurisdiction may differ from the interpretation of the *Engineer*. In this event, the *Client* will compensate the *Engineer* for any additional *Services* of the *Engineer* that are required in order to have the *Work* conform to such changes or interpretations.

PART 13   PROJECT OWNERSHIP, IDENTIFICATION AND CONFIDENTIALITY

GC 13.1   The *Client* represents to the *Engineer* that the *Client* is the owner of the *Place of the Work*. If the *Client* is not the owner, the *Client* will *notify* the *Engineer* of the identity of the owner before signature of this *Engineering Agreement*.

GC 13.2   The *Engineer* will be identified on *Project* signage and promotional material whenever other *Project* design professionals are mentioned. The *Engineer* may refer to the *Project* in the *Engineer's* promotional material.

GC 13.3   Information regarding the design, functionality, equipment, management, costs, or progress of the *Project* is confidential where one party has *notified* the other party of the confidential or proprietary nature of such information and where such information is not public knowledge. The parties agree not to disclose confidential information to third parties, except to the extent required for performance of the *Services* or where required by law or by mutual consent of the parties.

PART 14   INSURANCE AND LIABILITY

GC 14.1   The *Engineer* will carry professional liability insurance of $250,000 per claim and $500,000 in the aggregate within any policy year. Coverage will be maintained continuously from the commencement of the *Services* until completion or termination of the *Services* and, subject to availability at reasonable cost, for 2 years after completion or termination of the *Services*.

GC 14.2   The *Client* may choose to increase the amount or the coverage of the *Engineer's* professional liability insurance above that provided in GC 14.1 so as to obtain additional insurance that is specific to the *Project*. The *Engineer* will cooperate with the *Client* to obtain such additional insurance, at the *Client's* expense.

GC 14.3   If the *Engineer* carries professional liability insurance for amounts greater than those specified in GC 14.1, such insurance will be available under this *Engineering Agreement* only up to the amount specified in GC 14.1 plus, if applicable, the amount of additional insurance obtained under GC 14.2.

GC 14.4   Where the *Project* involves construction, the *Client* will provide or arrange for *Project* specific liability (wrap-up) insurance and property ("broad form"/builder's risk) insurance in respect of the *Work* and include the *Engineer* thereunder as an additional insured.

GC 14.5   The *Engineer's* liability for claims which the *Client* has or may have against the *Engineer* or the *Engineer's* employees, agents, representatives and *Sub-Consultants* under this *Agreement*, whether these claims arise in contract, tort, negligence or under any other theory of liability, will be limited, notwithstanding any other provision of this *Engineering Agreement*:

(a)   to claims brought within the limitation period prescribed by law in the jurisdiction in which the *Project* is located or, where permitted by law, within 2 years of completion or termination of the *Services*, whichever occurs first; and

(b)   to re-performance of defective *Services* by the *Engineer*, plus:

(i)   where claims are covered by insurance under section GC 14.1, and, if applicable, by any additional insurance under section GC 14.2 - to the amount of such insurance; or

(ii)   where claims are not covered by insurance under section GC 14.1, and, if applicable, by any additional insurance under section GC 14.2 - to the amount of $250,000.

GC 14.6   The *Engineer* will not be liable for the failure of any manufactured product or any manufactured or factory assembled system of components to perform in accordance with the manufacturer's specifications, product literature or written documentation.

GC 14.7   Where the *Engineer* is a corporation or partnership, the *Client* and *Consultants of the Client* will limit any claim they may have to the corporation or partnership, without liability on the part of any officer, director, member, employee, or agent of such corporation or partnership.

ACEC Document 31 – 2009   21 of 23

GC 14.8   The liability of each party with respect to a claim against each other is limited to direct damages only and neither party will have any liability whatsoever for consequential or indirect loss or damage (such as, but not limited to, claims for loss of profit, revenue, production, business, contracts or opportunity and increased cost of capital, financing or overhead) incurred by the other party.

GC 14.9   The *Engineer* is not responsible for the identification, reporting, analysis, evaluation, presence, handling, removal or disposal of *Hazardous Substances* at or adjacent to the *Place of the Work*, unless specified in Schedule A – ENGINEER'S SCOPE OF SERVICES, or for the exposure of persons, property or the environment to *Hazardous Substances* at or adjacent to the *Place of the Work*.

GC 14.10   Subject to the limitations of liability set out in this *Engineering Agreement*, each party will indemnify the other party, to the extent of the fault or negligence of the indemnifying party, for damages and costs (including reasonable legal fees) resulting from:

           (a)   claims of third parties; or

           (b)   a breach of contractual obligations under this *Engineering Agreement* by the indemnifying party or anyone for whom that party is responsible; or

           (c)   negligent or faulty acts or omissions of the indemnifying party or anyone for whom that party is responsible.

PART 15   DISPUTE RESOLUTION

GC 15.1   The parties will make reasonable efforts to resolve disputes arising under this *Engineering Agreement* by amicable negotiations. They agree to provide frank, candid and timely disclosure of relevant facts, information and documents to facilitate these negotiations, without prejudice to their rights and recourses.

GC 15.2   If a dispute has not been resolved by negotiations, either party may *notify* the other party that it wishes the dispute to be resolved by mediation. If the parties are unable to agree upon the choice of a mediator, either party may apply to a superior court in the jurisdiction where the *Project* is located to appoint a mediator.

GC 15.3   Should mediation not resolve the dispute, a party may refer the unresolved dispute to the courts or, upon mutual agreement, to any other form of dispute resolution, including binding arbitration.

GC 15.4   Unless the parties otherwise agree, any mediation or arbitration under this *Agreement* will be conducted in accordance with the latest edition of CCDC 40 - Rules for Mediation and Arbitration of Construction Disputes, as applied to and compatible with this *Engineering Agreement*, save that arbitration will be limited to a single arbitrator.

GC 15.5   Any endeavour to resolve disputes arising out of this *Engineering Agreement* by negotiation, mediation or other means of dispute resolution, including arbitration, will be conducted on a confidential basis.

GC 15.6   The parties agree to submit to the exclusive jurisdiction of the courts in *Place of the Work* if a dispute is to be resolved by the courts, or to mediation or arbitration at the *Place of the Work* if a dispute is to be resolved by mediation or arbitration.

PART 16   PAYMENT

GC 16.1   The *Client* will pay to the *Engineer* the amount of the *Fees* and *Reimbursable Expenses* of the *Engineer* together with applicable *Value Added Taxes*, when invoiced by the *Engineer* for *Services* which have been rendered, in accordance with Article A5 – PAYMENT and Schedule B – FEES AND REIMBURSABLE EXPENSES.

GC 16.2   In the event the *Client* disputes in good faith a portion of the *Fees* and *Reimbursable Expenses* invoiced by the *Engineer*, the *Client* will pay the uncontested portion within the prescribed time.

GC 16.3   Disputes regarding *Fees* and *Reimbursable Expenses* of the *Engineer* will be resolved in the manner specified in PART 15 - DISPUTE RESOLUTION.

GC 16.4   Where the *Engineer* provides *Construction Administration Services* which extend beyond the period contemplated at the time this *Engineering Agreement* was signed, the *Engineer* will *notify* the *Client* and, upon mutual agreement of the parties, the *Fees* of the *Engineer* will be increased in order to take into account the extended time required for providing the *Construction Administration Services*.

GC 16.5   Should the *Client* request a change to the *Project* or *Work* which requires the *Engineer* to provide additional *Services* beyond those contemplated at the time the *Engineering Agreement* is signed, before undertaking such additional *Services* the *Client* and the *Engineer* will agree in writing upon the *Engineer*'s remuneration and time for providing the additional *Services*. Failing an agreement with the *Client*, the *Client* will pay the *Engineer* for the additional *Services* at the hourly rates set out in Schedule B – FEES AND REIMBURSABLE EXPENSES and any additional *Reimbursable Expenses* incurred, and grant a reasonable extension of time to the *Engineer* for the performance of the additional *Services*.

GC 16.6   Should the *Client* request a change to the *Project* or *Work* which renders useless a part of the *Services* already provided, the *Client* nonetheless will pay the *Engineer* in accordance with this *Engineering Agreement* for *Services* already provided which the change has rendered useless.

GC 16.7   Should it prove necessary for the *Engineer* to rework or revise the plans and specifications forming part of the *Services* for reasons which the *Engineer* could not reasonably foresee when the *Engineering Agreement* was signed, or owing to the default or the insolvency of the *Client* or the *Contractor* or a subcontractor, or as a result of the *Client's* suspension of the *Services* or *Work* on the *Project*, or because of damage to the *Project* by fire or some other cause, the *Client* will pay the *Engineer* for any reworked or revised plans and specifications at the hourly rates set out in Schedule B – FEES AND REIMBURSABLE EXPENSES.

PART 17   SEVERABILITY

GC 17.1   If any provision of this *Engineering Agreement* is declared by a court of competent jurisdiction to be invalid, illegal, or unenforceable, such provision will be severed from this *Engineering Agreement* and the other provisions of this *Engineering Agreement* will remain in full force and effect.

GLOSSARY

Note: All glossary terms used in definitions have been italicized.

An **Acceleration** is an increase in labour or equipment to shorten the time for completion or mitigate the effects of an impact or a delay.   **p. 216**

An **Acceleration Claim** is a claim for the cost of an acceleration.   **p. 220**

An **Acceptance** is an unequivocal agreement to an offer.   **p. 45**

ADR. See Alternative Dispute Resolution.   **p. 159**

An **Agent** is a person who is authorized to act on behalf of another party, known as the principal.   **p. 68**

An **Agreement** is the same as a contract.   **p. 44**

Air Rights are rights that extend above the physical property to a reasonable level above the property.   **p. 26**

An **Air Space Parcel** is the creation of an artificial layer or layers of property above the physical property.   **p. 26**

An **Alliance Agreement** is typically an agreement involving all of the major parties, including the owner, significant contractors, and design professionals.   **p. 113**

An **All-Risk Policy** is the same as a builders' risk policy. **p. 183**

Alternative Dispute Resolution (ADR) is any process of resolving a dispute other than litigation, such as arbitration and mediation.   **p. 159**

An **Amendment** is a change or alteration to a contract. **p. 52**

Anticipatory Breach is when one party to a contract communicates to the other party that it intends to breach an obligation of the contract when the time for performance of the obligation has not yet occurred or occurs when a party finds itself in a position in which it is impossible for it to perform.   **p. 62**

Apparent Authority, also called ostensible authority or implied authority, is the authority given by a principal to an agent through representations (*e.g.*, statements) made by the principal to the third party.   **p. 68**

Arbitration is a private litigation or trial process in which the parties can set the rules and choose a judge, called the arbitrator.   **p. 159**

An **Arbitrator** is a private judge.   **p. 159**

A **Bankable Contract** is a contract with terms acceptable to the lender financing one of the project participants.   **p. 128**

A **Bid**, sometimes called a tender, is an offer made in compliance with a fixed set of contract terms, usually in the context of a competitive process.   **p. 46**

A **Bid Bond** is a bond in favour of a project owner, such that if the owner accepts a bid by the contractor in question, and the contractor fails to enter into the contract that is the subject matter of the bid, the surety will pay the penalty specified in the bond.   **p. 193**

A **Bidder** is a party that submits a bid in response to a call for tenders.   **p. 84**

Bidding, often called tendering, is the process of awarding a contract through a call for tenders.   **p. 92**

A **Bond** is a special form of contract, whereby one party (the surety) guarantees the performance by another party (the principal) of certain obligations.   **p. 191**

A **Bonus Clause** is a clause in a contract that entitles one party to additional payment if its performance exceeds what has been promised, such as finishing prior to the contractual completion date.   **p. 60**

A **Breach of Contract** is the failure by one party to a contract to perform his or her obligations.   **p. 57**

A **Builders' Lien** is the same as a construction lien.   **p. 203**

A **Builders' Risk Policy**, sometimes called an all-risk policy, is a property insurance policy that covers all damage to the insured property except those perils specifically excluded.   **p. 183**

A **Business Name** is a name under which the individual carries on business.   **p. 34**

A **Call for Tenders**, sometimes call an invitation or invitation to tender, is a request by a buyer of goods or services (often called the owner) to a group of potential sellers of those goods or services (often called bidders) to submit a proposal or bid for delivering a defined set of goods or services.   **p. 84**

A **Cause of Action** is a right to sue through litigation. **p. 203**

The **CCDC** is the Canadian Construction Documents Committee, which creates a number of standard form construction contracts and documents.   **p. 114**

A **Change Directive** is an order by the owner that the contractor proceed with a change to change the work to be performed under a *contract*, that is generally paid for on a *cost-plus* basis.   **p. 54**

A **Change Order** is an agreement to change the work to be performed under a *contract* .   **p. 54**

The **Charter** (the *Canadian Charter of Rights and Freedoms*) is a section of the Constitution that establishes individual rights that apply to all Canadians, and in many cases to non-citizens who are in Canada.   **pp. 1, 3**

A **Chattel** is the same as tangible personal property.   **p. 23**

Civil Law is a system of law based on the Napoleonic Code of France and is applicable in Quebec and Louisiana.   **p. 4**

A **Claims-Made Policy** is an insurance policy that covers claims that are made during the coverage period and does not depend on when the work was done.   **p. 180**

Clear Title is title to property for which there are no other interests in the property or charges against the property.   **p. 26**

A **Closed Shop** is a workplace in which membership in the union is a precondition to hiring.   **p. 230**

A **Closely Held Corporation** is the same as a private corporation.   **p. 38**

A **Commercial General Liability (CGL) Policy** is an insurance policy that covers claims by others for bodily injury and property damage.   **p. 186**

Common Law is a body of law which traces its roots to England. Common law is created by judges based upon principles of law and equity established and modified over hundreds of years.   **p. 4**

A **Company** is the same as a corporation.   **p. 34**

A **Compensable Delay** is a delay for which one party is contractually entitled to recover damages caused by the delay of another party.   **p. 218**

A **Concurrent Delay** is a combination of compensable, excusable, and contractor-caused delays, all occurring at the same time and delaying the same activities.   **p. 219**

Confidential Information generally means information imparted in circumstances in which an obligation of confidence arises. In the employment context, this means information obtained through the employment relationship.   **p. 238**

A **Conflict of Interest** is a situation in which there is a conflict between a professional's obligation to the public, the client, the employer, and him- or herself.   **p. 14**

Consequential Damages is a phrase which has different meanings in different contracts and different contexts, but is generally understood to mean indirect losses, such as loss of business.   **p. 60**

Consideration means something of value, however small, given or promised by each party to the contract.   **p. 46**

A **Construction Lien**, sometimes called a builders' lien, is a charge or claim in respect of real property that has been improved by construction.   **p. 203**

Construction Management is a process in which the owner contracts directly with all of the trades rather than hiring a general contractor.   **p. 109**

Constructive Acceleration occurs when a contractor has been forced into an acceleration, without any acknowledgement by the owner that the contractor is being asked to accelerate.   **p. 220**

Constructive Changes are changes to the work to be performed under a contract where an owner refuses to acknowledge that any change has occurred.   **p. 54**

Constructive Dismissal is an indirect form of dismissal in which an employer unilaterally changes the employment contract to the detriment of an employee, such that the employee refuses to accept the changes and terminates the contract.   **p. 242**

A **Contract**, sometimes called an agreement, is an enforceable voluntary agreement between two or more parties.   **pp. 6, 44**

A **Contractor-Caused Delay** is a delay that is caused by the contractor, or by a party for whom the contractor is responsible, and is generally neither an excusable delay nor a compensable delay.   **p. 218**

Contra Proferentem is a rule of contractual interpretation that states that wherever a contract is ambiguous and there are two alternate interpretations, the courts will use the interpretation that favours the party that did not write the contract.   **p. 65**

Copyright protects the expression of words and data in original literary, musical, dramatic, and artistic works. Copyrighted expression must be written, performed, recorded, or communicated in a form such as radio or television.   **p. 29**

A **Corporation**, also known as a company, is a separate legal person created pursuant to federal or provincial statutes.   **p. 34**

A **Cost-Plus Contract**, sometimes called a cost reimbursable contract, provides a contractor with payment for the labour, equipment, and material expenses incurred on the project plus a percentage or amount for profit and overhead, sometimes to a guaranteed maximum.   **p. 113**

A **Cost Reimbursable Contract** is the same as a cost-plus contract.   **p. 113**

Coverage Period is the length of time during which insurance coverage is in place under an insurance policy.   **p. 175**

Damages are monetary compensation awarded by a court that are payable to an injured party by the party or parties that caused the injury.   **p. 59**

Debt is an obligation to pay or render something, generally money, to someone else.   **p. 38**

A **Declaratory Order** is a court's opinion that does not require either of the parties to do anything.   **p. 59**

A **Deductible** is the portion of the costs of defending or paying for a claim under an insurance policy that the insured must pay. Also known as *self-retention*.   **p. 175**

Defamation is a communication that tends to injure another party's reputation and includes oral (slander) and written (libel) communications. **p. 274**

The **Defendants** are the parties against whom a remedy is sought through litigation. **p. 156**

A **Delay** is an event that causes extended time to complete all or part of a project. **p. 216**

Design-Bid-Build is a process in which a project owner and its consultant conceptualize and design a project in its entirety, the owner then hires a contractor to build it (generally through a tendering process), the consultant inspects the construction, and the owner finances the project and puts it into service. **p. 86**

Design-Build is a process in which a project owner conceptualizes the project, with or without a consultant; then hires a contractor to design it, build it, and inspect it, with or without independent secondary inspection; and then finances the project and puts it into service. **p. 86**

Directors are individuals who direct the business of a corporation. **p. 38**

Discipline is the process of charging a member of a profession with professional misconduct and then proceeding to a hearing to determine guilt. **p. 12**

A **Disclaimer** is an exclusion clause or a limitation clause. **p. 171**

Discovery of Documents is an ongoing process of identifying documents that are in the possession of a party that are relevant to the litigation and disclosing them to the other parties, and often continues until shortly before trial. **p. 158**

A **Dispute Resolution Provision** is a clause that defines the process to be followed in resolving disputes between the parties to a contract. **p. 127**

A **Disruption** is the same as an impact. **p. 216**

Domain Names are the alphanumeric addresses of sites on the Internet. **p. 275**

Due Diligence means that all reasonable steps have been taken to satisfy statutory obligations. **pp. 40, 262**

Duress is improper pressure, threats, or coercion used to induce a party to enter into a contract. **p. 51**

Duty of Care is an element of negligence and is the requirement in law that every person must take reasonable care for others based upon the concept of reasonable foreseeability. **p. 132**

The **Duty of Good Faith** is the insured's obligation to disclose to the insurer any and all facts that could influence the insurer's decisions about providing coverage. **p. 181**

A **Duty to Cooperate** is a duty owed by an insured to cooperate with an insurer in the defence of a claim. **p. 182**

The **Duty to Defend** is the obligation to pay for legal fees and other costs of defending a claim under an insurance policy. **p. 177**

The **Duty to Indemnify** is the duty to pay claims under an insurance policy. **p. 177**

An **Easement** is the right to use a neighbouring piece of property, generally for the purpose of crossing it. **p. 27**

Enforceability refers to whether a court will enforce a contract or a portion of a contract in the event of a dispute. **p. 44**

An **Enforceable Contract** is one that a court will uphold as being effective. **p. 44**

Enforcement is a professional organization's process of charging a non-member with either using the protected professional title in breach of the right to title or practicing the profession in breach of the exclusive scope of practice. **p. 12**

An **Environmental Audit** is a systematic process of objectively obtaining and evaluating evidence about a verifiable environmental matter, ascertaining the degree of correspondence between the assertion and established criteria, and then communicating the results to the client. **p. 258**

An **Environmental Site Assessment** (ESA) is an assessment of a property for possible contamination and to recommend a remediation protocol if contamination is discovered. **p. 257**

Equity is residual value of a property or business after deducting any mortgage or other liability. **p. 38**

Errors and Omissions Insurance is a claims-made policy that covers claims made by third parties for loss or damage caused by negligence of the insured. **p. 187**

An **ESA** is the same as an environmental site assessment. **p. 257**

Estoppel is the legal principle that precludes a party from asserting something contrary to what is implied by a previous action or statement of that party. **p. 53**

Ethics is, broadly stated, the theory of morality. **p. 14**

An **Examination for Discovery** is an oral cross-examination of a representative of an adverse party in a litigation. **p. 158**

An **Exclusion Clause** is a clause that purports to completely exclude the damages or remedies available to the innocent party upon the occurrence of specified events. **p. 126**

An **Exclusive Licence** grants the licensee (the party gaining the rights) the exclusive right to use the rights held by the licensor (the party granting the rights). **p. 122**

An **Excusable Delay** is a delay that is not attributable to the fault of any party and may or may not be a compensable delay. It entitles the parties to a contract to extend the time for completion of a contract and often includes force majeure. **p. 218**

An **Expert Witness** is a person with skill or expertise, based on either training or experience or both, who is hired to provide a professional opinion to the court or to an arbitration tribunal. **p. 162**

Express Authority, sometimes called actual authority, is the authority given by a principal to an agent often through a written or spoken contract between the principal and the agent.   **p. 68**

Express Terms are words, phrases, or conditions that have been discussed and agreed to by the parties to a contract.   **p. 44**

A **Fee Simple Right** or **Interest** means a right or interest in property that permits the holder to do almost anything with the property, including selling, leasing, occupying, or mortgaging the property.   **p. 24**

A **Fiduciary Duty** is a heightened duty to care for the interest of another party in priority to one's own interest, and to refrain from acting against the interest of the other party.   **p. 141**

A **Fixed-Price Contract**, sometimes called a stipulated sum contract, is a contract in which the exact amount that the contractor will be paid for the work is defined in the contract.   **p. 113**

A **Fixture** is personal property that becomes part of real property when it becomes affixed to the real property.   **p. 24**

Force Majeure generally means events beyond the control of the contracting parties, although most contracts specifically define the events to be considered *force majeure*.   **p. 218**

A *Force Majeure* **Clause** is a list of events that the parties to a contract agree are beyond their control and that provides for relief due to these events.   **p. 52**

Fraud, also known as deceit, is an intentional tort. It is often a fraudulent misrepresentation, because fraud in almost all cases involves a misstatement (misrepresentation) made by one party to deceive the other party.   **p. 140**

Frustration, or impossibility, is an event not foreseen at the time a contract was entered into that makes the performance of the contract either impossible or of no value.   **p. 52**

A **Fundamental Breach** is a breach of contract that goes to the root of the contract and deprives the innocent party of all or substantially all of the benefit of the contract.   **p. 61**

A **General Partner** is a partner in a limited partnership that operates the business of the limited partnership.   **p. 36**

The **Grantee** is the party that grants a right.   **p. 25**

A **Grievance** is a complaint by either the union on behalf of an employee or by the employer, stating that the other party has breached the collective agreement.   **p. 233**

A **Grubstake Agreement** was historically an agreement between a storeowner, who provided grub, and the prospector, so that the prospector did not starve while searching for a mineral find or strike. It now refers to any agreement with a geoscientist that provides necessary resources to complete a geotechnical investigation.   **p. 122**

Holdback, sometimes called retainage, is a percentage of contract value that must be withheld from each progress draw, not to be paid out until certain time periods have passed and certain conditions have been met.   **p. 209**

An **Impact**, also known as a disruption, is the effect of an event that detrimentally affects a project but that does not necessarily extend the time to complete the project.   **p. 216**

Implied Terms are those terms that have never been discussed or agreed to between the parties to a contract but which are taken for granted.   **p. 44**

Inadvertence means that one of the parties to the contract has unintentionally failed to carry out one of its promises or obligations.   **p. 58**

An **Indemnity** is an agreement by one party to bear the financial loss of another party for a specified event.   **p. 70**

An **Injunction** is a court order prohibiting a party from doing something.   **p. 59**

Insider Trading. See unlawful insider trading.   **p. 40**

Insurable Interest means that the insured benefits from the existence of or would be prejudiced by the loss of the insured property.   **p. 178**

An **Insurance Policy** is a contract of indemnity.   **p. 175**

The **Insured** is the party covered by an insurance policy.   **p. 175**

Intangible Property is all property other than real property.   **p. 23**

Intellectual Property is a set of rights, often referred to as a bundle of rights, in intangible property.   **p. 23**

An **International Trade Agreement** is an agreement between two countries that puts rules in place about the transfer of goods and services between the two countries.   **p. 110**

An **Interprovincial Trade Agreement** is an agreement between the federal government and the provinces that puts rules in place about the transfer of goods and services between the provinces.   **p. 110**

Interrogatories are a series of written questions sent by one party to another during a litigation, which must be answered under oath.   **p. 158**

An **Invitation** or **Invitation to Tender** is the same as a call for tenders.   **p. 88**

An **Invitation to Treat** is a request for offers that is not binding on the party making the invitation, except in the case of calls for tenders and requests for proposals.   **p. 45**

Joint Tenants hold rights in property concurrently such that if one of them dies, the remaining joint tenants automatically inherit the interest of the deceased.   **p. 25**

A **Joint Venture** is not a legal structure; rather, it is an organization where two or more joint venturers act together but retain their separate legal status.   **p. 34**

A **Jurisdictional Dispute** is a labour dispute in which two or more unions each claim that the work in question should properly be performed by its members.   **p. 230**

Just Cause refers to the legal acceptability of employment termination, such as when an employee has committed a fundamental breach of the employment contract. **p. 240**

A **Labour and Material Payment Bond** is a form of payment bond that is put in place to protect against liens by unpaid subcontractors, suppliers, and those working below them in the contractual chain. **p. 198**

Lateral Support means that the land and buildings on a property will always have the same or greater support and that a neighbour cannot undermine that support by digging in a manner that would have the effect of damaging the adjacent property. **p. 25**

A **Layoff** is the temporary suspension of employment. **p. 234**

A **Learned Intermediary** is a person positioned between a product manufacturer and the consumer, a person who has specialized technical knowledge and upon whom the law imposes a duty to warn the consumer. **p. 144**

A **Lease** provides a party with an exclusive right to occupy or possess all or part of the property for a set period of time. **p. 27**

A **Letter of Intent**, or a term sheet, is a document that sets out the basic terms of a business arrangement in advance of agreeing to all of the detailed contract terms. **p. 49**

Liability Insurance is insurance that protects an insured against claims made by third parties, such as a claim against a professional for errors and omissions. **p. 175**

A **Licensing Agreement** or **Licence** provides a grant of rights to a licensee to use a right held or owned by a licensor. **pp. 27, 113**

A **Lien** is a charge or claim against property. **p. 203**

A **Lien Bond** is a form of payment bond used as security to facilitate the discharge of a lien that has been filed against the land on which a project was constructed. **p. 199**

Limitation Clauses are contract clauses that purport to limit the damages or remedies available to the innocent party upon the occurrence of specified events. **p. 126**

A **Limitation Period** is a contractual or statutory requirement for a notice of a claim to be made or legal action to be commenced within a certain time period. **pp. 5, 175**

A **Limited Liability Partnership** is a partnership in which the liability of a limited partner is limited only to the liability of that partner. **p. 36**

Limited Partners are partners in a limited partnership whose liability is limited to their cash contribution to the limited partnership. **p. 36**

A **Limited Partnership** is a partnership in which the liability of each limited partner is limited only to the liability of that limited partner. **p. 36**

Liquidated Damages are genuine estimates of loss caused by breach of contract (mostly used for late performance of a contract) that are made before the contract is signed and before any breach of contract has occurred. They are defined in a clause in the contract. **p. 60**

Litigation is the use of the court system to resolve disputes. **p. 155**

Material Non-Disclosure is a breach of the duty of the insured to provide all relevant information to the insurer when purchasing an insurance policy. **p. 181**

The Measured Mile Approach is the valuation of a delay claim by using a portion of the contract unaffected by delays as a yardstick to compare productivity for the rest of the contract. **p. 223**

Mediation is an assisted negotiation process in which settlement discussions are facilitated by a neutral third party. **p. 161**

A **Misrepresentation** is a statement of fact made by one party that is untrue. **pp. 50, 139**

A **Mistake** is a misunderstanding with respect to the term of a contract. **p. 50**

Mitigation means that a party who has suffered an injury or loss must take reasonable steps to reduce or mitigate their own injury or loss. **p. 59**

Moral Principles are the standard of conduct required generally by society or specifically by an organization or group. **p. 14**

Moral Rights are rights that protect the artistic integrity of copyrighted work, such as the manner in which a painting is displayed. **p. 29**

A **Mortgage** is a form of security for a loan, which provides the lender with the right to be paid out of the value of the property, when the property is sold, in priority to the holder of the fee simple right. **p. 27**

A **Mutual Mistake** is a mistake made by both parties to a contract. **p. 50**

A **Negative Restrictive Covenant** is a requirement on a land title not to do something, such as not constructing a building. **p. 27**

Negligence is the failure to take reasonable care for another person. **p. 6**

Negotiation is discussion aimed at resolving a dispute, usually through compromise. **p. 161**

A **No-Damages-for-Delay** clause is a limitation clause designed to protect the owner from claims for damages associated with delay and acceleration; but it does not preclude the contractor from insisting upon an extension of time. **p. 220**

A **Notice of Claim Provision** is a clause that defines the process to be followed for giving notice of and information in respect of a claim under a contract. **p. 128**

Nuisance is the undue interference with the use and enjoyment of rights to land.   **p. 259**

An **Obligee** is the party to whom the obligations are owed under a bond.   **p. 191**

Occurrence Policies are insurance policies that cover claims in which the insured event (such as a house burning down or damage occurring to a building) occurred during the coverage period.   **p. 180**

An **Offer** is a proposal by an offeror to an offeree, containing the essential terms of a proposed contract. **p. 45**

Officers are individuals in a corporation who provide a closer operational direction for the business, such as the president, vice-president, and corporate secretary.   **p. 38**

An **Open Shop** is a workplace with no union membership requirement and in some places is synonymous with a non-union shop.   **p. 230**

An **Option Agreement** provides the optionee (the party purchasing the option) with the right to buy or not buy land or goods or rights from the optionor (the party granting the option).   **p. 122**

The **Parol Evidence Rule** is the rule of contractual interpretation that states that where a contract is entirely written, and its written language is clear and unambiguous, extrinsic evidence is not admissible to add to, vary, or contradict the written words.   **p. 66**

A **Partner** is one of the members of a partnership.   **p. 35**

A **Partnership** is a group of individuals or corporations who pool their resources with the goal of making profit.   **p. 34**

Patents are legal protections for inventions, which include any new and useful art, process, machine, manufacture, or composition of matter, or any new and useful improvement in any art, process, machine, manufacture, or composition of matter.   **p. 30**

A **Pay-If-Paid Clause** is a clause in a construction contract that shifts the risk of non-payment by the owner from the general contractor to the subcontractor by making payment to the subcontractor by the general contractor contingent on the general contractor being paid by the owner.   **p. 75**

A **Pay-When-Paid Clause** is a clause in a construction contract that applies to the timing of payment (not to whether payment must eventually be made). It does not allow the general contractor to refuse to pay the subcontractor if the owner never pays.   **p. 75**

A **Payment Bond** is a bond that guarantees the performance of a payment obligation.   **p. 198**

A **Penalty** is a sum of money included in a contract as punishment for breach of the contract rather than as compensation for the breach.   **p. 60**

A **Performance Bond** is a bond under which a surety guarantees the performance of a construction contract by a contractor.   **p. 194**

A **Performance Specification** is a specification that sets out the operating parameters that must be met by the final product.   **p. 79**

Performing Under Protest means that the protesting party gives the other party written notice that it will be performing certain work under a contract on the understanding that it is doing so without prejudice to its right to argue that it is entitled to additional payment for that work.   **p. 58**

Perils Covered are the events that trigger the obligation of the insurer to either indemnify, defend, or both.   **p. 175**

Personal Property is all property that is not real property.   **p. 23**

A **Phase I ESA** is the first part of an environmental site assessment, in which an assessor seeks to determine whether a particular property is or may be subject to actual or potential contamination.   **p. 257**

A **Phase II ESA** is the second part of an environmental site assessment, in which an assessor seeks to characterize and/or delineate the concentrations or quantities of substances of concern related to a site and compare those levels to criteria.   **p. 258**

A **Phase III ESA** is the third part of an environmental site assessment, which results in a detailed delineation of the contaminants, confirmatory sampling throughout the remediation process, and a measurement of the success of the remediation program.   **p. 258**

PIPEDA (Personal Information Protection and Electronic Documents Act) is the federal law governing privacy.   **p. 271**

The **Plaintiff** is the party in a court action who claims to have suffered injury or loss.   **p. 156**

Pleadings are documents filed in court in a lawsuit, or included in a trial record, including a statement of claim, statement of defence, reply, demand for particulars, interrogatories, and motions.   **p. 156**

A **Positive Restrictive Covenant** imposes a requirement on a land title to do something like build a building.   **p. 27**

The **Postal Acceptance Rule** is a contract rule that states that acceptance of the contract is deemed to have been communicated when the communication is placed into the mail system.   **p. 46**

Pre-Qualification is a process in which potential sellers are pre-qualified so as to limit the ultimate call for tenders or request for proposal to a limited number of qualified sellers.   **p. 89**

A **Principal** is a person who authorizes an agent to act on behalf of the principal. Under a bond, a principal is the party whose performance was guaranteed under the bond.   **pp. 68, 191**

A **Private Corporation**, also known as a closely held corporation, is a corporation in which all of the shares of the corporation are held by very few shareholders.   **p. 38**

Privileged Information is information revealed during one proceeding (such as arbitration) which cannot be used as evidence in a subsequent proceeding (such as a litigation).   **p. 161**

Privy to a contract means being a party to a contract.   **p. 44**

Procurement is the purchase of goods and/or services.   **p. 46**

Product Liability is the liability of the manufacturer to a consumer for a defective product.   **p. 143**

A *Profit à Prendre* is a combination of an easement and a licence.   **p. 27**

Property Insurance is insurance that protects an insured against loss or damage to property in which the insured has an interest as a result of certain causes, such as fire or theft.   **p. 175**

A **Proponent** is a potential seller that submits a detailed proposal responding to a request for proposals.   **p. 89**

A **Prospectus** is a written document containing specific financial and technical information about securities. It must be made available to prospective purchasers before sale of the securities can take place.   **p. 268**

A **Proximate Cause** is a cause that plays a significant part in producing the result, as compared with one which is too remote.   **p. 138**

A **Public Corporation** is a corporation in which the shares are publicly traded, generally on a stock exchange.   **p. 39**

A **Public-Private Partnership** is a project delivery method in which a public owner enters into a contract to have a project designed, constructed, financed, and operated on a long-term basis by a private company.   **p. 113**

A **Pure Economic Loss** is a loss where the only damages are monetary and where there are no personal injuries or property damage.   **p. 135**

Quantum Meruit means "the amount it is worth." It is the amount to be paid by one party to another when the contract that they were working under becomes inapplicable, and when a court finds that one party has been unjustly enriched at the other's expense.   **p. 54**

The **Rand Formula** is a term in a collective agreement that states that employees are required to pay union dues but are not required to join the union.   **p. 230**

Real Property generally means anything that is land and anything attached to land.   **p. 23**

A **Rejection** is an express or implied refusal to accept an offer.   **p. 45**

Remoteness is the lack of a connection between a wrong and an injury or loss.   **p. 59**

A **Repudiation** of a contract occurs when one party to the contract, either by word or by conduct, lets the other party know that it does not intend to perform its obligations.   **p. 62**

A **Request for Proposals** is a process in which potential sellers are invited to propose a solution to a particular problem, but where the details of the potential solution are less well developed than in a call for tenders.   **p. 89**

A **Request for Qualifications** is a process used to choose a project participant based solely on the qualifications of the respondents, rather than on other criteria, such as price.   **p. 88**

A **Request for Quotation** is generally an informal process whereby separate requests are made with no formal closing time or tender closing conditions.   **p. 88**

A **Request for Standing Offers** is a process in which the buyer prearranges prices, terms, and conditions with sellers for frequently ordered goods or services.   **p. 89**

Rescission means the cancellation of a contract.   **p. 139**

A **Restrictive Covenant** in an employment contract forbids a former employee from working for a competitor, usually for a specified period of time, within a specified geographic area, or within an area of business.   **pp. 27, 239**

Resultant Damage is damage to property in addition to the property containing the faulty workmanship, material, or design.   **p. 185**

Retainage is the same as holdback.   **p. 209**

A **Right of Way** is an easement that grants a right to cross land.   **p. 27**

Right to Title is the exclusive right of a professional regulatory authority and its members to use a particular title.   **p. 8**

Riparian Rights provide rights to use water on or adjacent to a property.   **p. 25**

Scope of Practice is the exclusive right of a professional regulatory authority and its members to practice in their particular field.   **p. 8**

The **Scope of Work** is the definition of goods or services to be supplied pursuant to a contract.   **p. 88**

Securities in the context of securities legislation are publicly traded shares, bonds, and other investment devices.   **p. 268**

Self-regulation means that a profession is provided with statutory authority to govern itself.   **p. 8**

Self-retention is the same as *deductible*.   **p. 175**

Shareholders own a corporation in proportion to their shareholdings.   **p. 38**

A **Simple Breach** is a breach of contract that does not entitle the innocent party to treat the contract as ended or permit the innocent party to stop performing their part of the contract. **p. 61**

Sole Proprietorship is the legal term used to describe an individual carrying on business. **p. 34**

Specifications describe the detailed work requirements, including the quality and standard of materials and workmanship. **p. 78**

Specific Performance is a court order that requires one party to perform specific acts. **p. 59**

Standard of Care is an element of negligence and is the level of skill and care required of a person. **p. 136**

A **Standard Specification** is a specification that describes in detail all of the individual components of the final product. **p. 79**

A **Statement of Claim** is a pleading that commences a litigation in some Canadian provinces. A statement of claim describes the names and addresses of the parties, the allegations that form the basis of the claim, and the nature of the relief sought by the plaintiff in a litigation. **p. 156**

A **Statement of Defence** is a pleading setting out the legal basis for denying a statement of claim by a defendant in a litigation. **p. 156**

A **Statutory Declaration** in a construction law context is a sworn statement declaring that to the best of the knowledge of the declarant, all of the financial obligations of the subcontractor with respect to the project have been met, including payments to workers, subcontractors, suppliers, and assessments of all authorities having jurisdiction, such as workers' compensation premiums, and taxes. **p. 213**

A **Stipulated Sum Contract** is the same as a fixed-price contract. **p. 113**

Strict Liability is a doctrine of tort law under which the defendant may be held liable to the plaintiff regardless of whether the defendant was negligent. **p. 142**

A **Strict Liability Offence** is an offence for which the prosecution need only prove that the act occurred (such as contamination of land in breach of an environmental statute). **p. 262**

Subrogation allows one party who pays for a loss suffered by another party (such as an insurer and an insured) to assume the rights of that other party for the purpose of recovering the loss from a third party. **pp. 177, 192**

A **Surety** is a guarantor under a bond. **p. 191**

A **Surface Lease** is a real property right that provides access to a surface holder's property to permit subsurface access. **p. 25**

Tangible Personal Property, also known as a chattel, is personal property that is also tangible property. **p. 23**

Tangible Property is any property that has physical attributes. **p. 23**

A **Target Price Contract** is a construction contract in which the cost is fixed as long as the cost of construction exceeds a certain amount (the target); but savings below the target are shared in accordance with some previously agreed proportion between the owner and the contractor. **p. 116**

Tenancy in Common is a form of joint ownership in which each of the holders of a right to property owns a stated portion of the right; and if one of them dies, the heirs of the deceased tenant-in-common inherit his or her interest. **p. 25**

A **Tender** is a bid. **p. 46**

Tendering is bidding. **p. 92**

A **Termination Clause** lists the acts by parties to a contract that entitle the other party to terminate the contract. **p. 62**

A **Third Party** is a party to a litigation that is added by the defendant, rather than the plaintiff. **p. 157**

A **Third Party Claim** is a pleading setting out the claim of a defendant against a third party. **p. 157**

A **Tort** is a breach of a duty to care for another party where the breach causes injury or loss to that party, independent of whether the two parties involved do not have a contract, for which the law provides a remedy. **pp. 6, 131**

The **Total Cost Approach** is the valuation of a delay claim by calculating the difference between the expected and actual cost of a project. **p. 223**

Trademarks are protected logos or words that represent a company's goods or services. **p. 30**

A **Trade Secret** is confidential information contained in a document, product, formula, or patent. **p. 238**

Trespass is the unauthorized entry onto the land of another person. **p. 142**

A **Trust** is a legal doctrine that separates the legal ownership of a right or of a property from the beneficial interest or ownership. **p. 199**

A **Trust Beneficiary** is a person for whose benefit trust property or rights must be used. **p. 199**

The **Trust Provision** of a construction lien statute is a clause that creates a trust, such that funds received by a contractor or subcontractor are designated as trust funds, obliging the contractor to use those funds to pay subcontractors, workers, and suppliers working under the contractor. **p. 207**

A **Trustee** is a person appointed under the terms of a trust to have legal ownership of the trust property or rights, but who is required by law to act on behalf of and in the best interests of the trust beneficiaries. **p. 199**

An **Unconscionable Contract** is one that is so unfair, oppressive, or one-sided that it would be offensive for the court to enforce it. **p. 51**

A **Union Shop** is a workplace in which employees must become union members after they are hired.   **p. 230**

A **Unit Price Contract** is a contract that requires the owner to pay a stipulated amount for each unit or quantity of work performed.   **pp. 113, 117**

Unlawful Insider Trading is the offence of using privileged, non-public information to trade on securities or commodities markets in contravention of the law.   **p. 40**

Unpatented Land is land for which there is no certificate of title or registry entry.   **p. 26**

Vicarious Liability is the liability of a party for the acts or omissions of another party. It is usually used in the context of an employer being liable for the acts or omissions of its employee.   **p. 240**

A **Voidable Contract** is a contract that can be terminated or ended by a party that is not in breach of the contract; however, that party can also choose to continue with the contract.   **p. 45**

A **Volunteer** is someone who pays money to another without any legal obligation to do so.   **p. 197**

A **Waiver** of rights occurs when a party, by word or conduct, does not enforce its own contractual rights.   **p. 52**

A **Waiver of Subrogation** is a contract clause eliminating the insurer's right to subrogation.   **p. 177**

A **Warranty** is a contractual promise to repair defects in the goods and services for a specific period of time after the goods and services were performed or provided.   **p. 124**

A **Wrap-up Policy** is a liability insurance policy that insures all of the direct participants in a construction project, including the consultants, the owner, and all subcontractors.   **p. 187**

Writ is the pleading that commences a litigation some Canadian provinces. A writ is usually accompanied by a statement of claim.   **p. 156**

SELECTED REFERENCES

CHAPTER 3

Andrews, G.C., and J.D. Kemper. Canadian Professional Engineering Practice and Ethics, 2nd ed. (Toronto: Nelson Thomson Learning, 1999)

Guidelines for Professional Excellence. (Association of Professional Engineers and Geoscientists of British Columbia, 1994)

Mantell, M. Ethics and Professionalism in Engineering. (New York: MacMIllan, 1964)

CHAPTER 4

DiCastri, V. The Law of Vendor and Purchaser, 3rd ed. (Toronto: Carswell, 1988) [looseleaf]

Harris, L. Canadian Copyright Law, 3rd ed. (Toronto: McGraw-Hill Ryerson, 2001)

Henderson, G.F., et al. (ed.). Copyright and Confidential Information Law of Canada. (Scarborough ON: Carswell, 1994)

———. Patent Law of Canada. (Scarborough ON: Carswell, 1994)

Hughes, R.T. Hughes on Trade Marks. (Toronto: Butterworths, 1984) [looseleaf]

Kratz, M.P.J. Canada's Intellectual Property Law in a Nutshell. (Toronto: Carswell, 1998)

———. Protecting Copyright and Industrial Design. (Scarborough ON: Carswell, 1994)

Oosterhoff, A.H., and W.B. Rayner. Anger and Honsberger: Law of Real Property, 2nd ed. (Aurora ON: Canada Law Book, 1985)

Salvatore, B., et al. Agreements of Purchase and Sale. (Toronto: Butterworths, 1996)

Sinclair, A.M. Introduction to Real Property Law. (Toronto: Butterworths, 1982)

Tamaro, N. 2000 Annotated Copyright Act. (Toronto: Carswell, 2000)

CHAPTER 5

Fraser, W.K. Fraser's Handbook on Company Law, 8th ed. (Scarborough ON: Carswell, 1994)

Hansell, C. Directors and Officers in Canada: Law and Practice. (Toronto: Carswell, 1999) [looseleaf]

Peterson, D.H. Shareholder Remedies in Canada. (Toronto: Butterworths, 1989) [looseleaf]

Phillips, G. Personal Remedies for Corporate Injuries. (Scarborough ON: Carswell, 1992)

Sutherland, H., and D.B. Horsley. Company Law of Canada, 6th ed. (Scarborough ON: Carswell, 1993)

VanDuzer, J.A. The Law of Partnerships and Corporations. (Concord ON: Irwin Law, 1997). [Available from QUICKLAW by subscription]

CHAPTER 6

Fridman, G.H.L. The Law of Contract in Canada, 4th ed. (Toronto: Carswell, 1999)

Goldsmith, I., and T.G. Heintzman. Goldsmith on Canadian Building Contracts, 4th ed. (Toronto: Carswell, 1999)

Waddams, S.M. The Law of Contracts, 4th ed. (Toronto: Canada Law Book Inc., 1999)

CHAPTER 7

Snell, E.H.T. Snell's Equity, 31st ed. (London: Sweet & Maxwell, 2004)

Waddams, S.M. The Law of Damages, 3rd ed. (Aurora ON: Canada Law Book, 1997)

CHAPTER 9

Bowstead, W., F.M.B. Reynolds, and B.J. Davenport. Bowstead on Agency. (London: Sweet & Maxwell, 1976)

Cushman, R.F., and S.D. Butler (eds.). Construction Change Order Claims. (New York: Aspen Law & Business, 1994)

CHAPTER 10

Kirsch, H.J., and L. Roth. Kirsh and Roth: The Annotated Construction Contract. (Toronto: Canada Law Book Inc., 1997)

CHAPTER 11

Worthington, R.C. Purchasing Law Handbook, 2nd ed. (Markham ON: Butterworths LexisNexis, 2005) [looseleaf]

CHAPTER 12

Jackson, R. M., and J. L. Powell. Professional Negligence, 5th ed. (London: Sweet & Maxwell, 2002)

Linden, A.M. Canadian Tort Law, 7th ed. (Toronto: Butterworths, 2001)

CHAPTER 14

Bazerman, M.A., and M.A. Neale. Negotiating Rationally. (New York: The Free Press, 1993)

Brandenburger, A., and B.J. Nalebuff. Co-opetition. (New York: Doubleday, 1996)

Fisher, R., W. Ury, and B. Patton. Getting to Yes, 2nd ed. (New York: Penguin, 1991)

CHAPTER 16

Brown, C. Insurance Law in Canada, 2nd ed. (Scarborough ON: Carswell, 1999)

———. Introduction to Insurance Law. (Toronto: Buttersworth, 2003)

Canadian Cases on the Law of Insurance [case law reporter]. (Toronto: Carswell, 1983)

Hilliker, G.G. Liability Insurance Law in Canada, 2nd ed. (Toronto: Butterworths, 1996)

Insurance Law Reporter. (Toronto: CCH Canadian, 1934)

McNairn, C.H.H. Consolidated Insurance Companies Act of Canada Regulations and Guidelines. (Scarborough ON: Carswell, 1993)

Norwood, D., and J. P. Weir. Norwood on Life Insurance Law in Canada, 3rd ed. (Scarborough ON: Carswell, 2002)

CHAPTER 18

Bristow, D.I., *et al.* Construction, Builders' and Mechanics' Liens in Canada, 7th ed. (Toronto: Carswell, 2005)

British Columbia Builders Liens Practice Manual. (Vancouver: Continuing Legal Education Society of British Columbia, updated annually)

Kirsh, H. Kirsh's Guide to Construction Liens in Ontario, 2nd ed. (Toronto: Butterworths, 1995)

CHAPTER 19

Bramble, B.B., and M.T. Callahan. Construction Delay Claims, 3rd ed. (Baltimore: Aspen, 2000)

Cushman, R.F., *et al.* Proving and Pricing Construction Claims/With 2003 Cumulative Supplement. (Baltimore MD: Aspen, 2000)

CHAPTER 20

Adams, G.W. Canadian Labour Law, 2nd ed. (Aurora ON: Canada Law Book, 1993) [looseleaf]

Carter, D., G. England, B. Etherington and G. Trudeau. Labour Law in Canada, 5th ed. (Toronto: Butterworths, 2002)

MacNeil, M. *et al.* Trade Union Law in Canada. (Aurora ON: Canada Law Book, 1994) [looseleaf]

Sack, J., and E. Poskanzer. Labour Law Terms: A Dictionary of Canadian Labour Law. (Toronto: Lancaster House, 1984)

Sanderson, J.P. The Art of Collective Bargaining, 2nd ed. (Aurora ON: Canada Law Book 1989)

CHAPTER 21

Ball, S.R. Canadian Employment Law. (Aurora ON: Canada Law Book, 1998)

England, G., *et al.* Employment Law in Canada. (Markham ON: Butterworths, 1998) [looseleaf]

Harris, D. Wrongful Dismissal, 3rd ed. (Don Mills ON: R. DeBoo, 1984) [looseleaf]

Levitt, H.A. The Law of Dismissal in Canada, 2nd ed. (Aurora ON: Canada Law Book, 1992)

Sproat, J.R. Employment Law Manual: Wrongful Dismissal, Human Rights and Employment Standards. (Scarborough ON: Carswell, 1995)

CHAPTER 23

Faieta, M.D. Environmental Harm: Civil Actions and Compensation. (Toronto: Butterworths, 1996).

Hughes, E.L., *et al.* (ed.). Environmental Law and Policy, 3rd ed. (Toronto: E. Montgomery, 2003)

Pardy, B. Environmental Law: A Guide to Concepts. (Toronto: Butterworths, 1996)

Saxe, D. Environmental Offences: Corporate Responsibility and Executive Liability. (Aurora ON: Canada Law book, 1990)

CHAPTER 24

Borrows, J., and Rotman, L. Aboriginal Legal Issues: Cases, Materials and Commentary, 2nd ed. (Toronto: Butterworths, 2003)

Elliott, D.W., (ed.). Law and Aboriginal Peoples of Canada, 4th ed. (North York ON: Captus Press, 2000)

Woodward, J. Consolidated Native Law Statutes, Regulations, and Treaties. (Scarborough ON: Carswell, 1998)

CHAPTER 25

Findlay, P.G. Securities Law and Practice, 3rd ed. (Toronto: Carswell, 2003) [looseleaf]

Gillen, M.R. Securities Regulation in Canada, 2nd ed. (Scarborough ON: Carswell, 1998)

Johnston, D.L. Canadian Securities Regulation, 2nd ed. (Markham, ON: Butterworths, 1998)

CHAPTER 27

Gahtan, A.M. Internet Law: A Practical Guide for Legal and Business Professionals. (Scarborough ON: Carswell, 1998)

INDEX

A

Aboriginal law
  Aboriginal participation in projects, 266
  contracts with Band Councils, 266
  duty to consult, 266
  nature of reserve property, 266
  overview, 265
acceleration, 216, 220
acceleration claim, 220, 221–223
acceptance, 44, 45–46
accidents, 250–251
actual authority, 68–69, 69f
Aeronautics Act, 40
agency, 68–70
agent, 68–70
Agreement on Internal Trade (AIT), 110
agreements to agree, 49–50
air rights, 26
air space parcel, 26
Alberta's Code of Ethics, 19
all-risk policies, 183–186, 184f
alliance agreement, 119
allocation of risk, 5
alternate dispute resolution (ADR), 159
amendment, 52
American Arbitration Association (AAA), 160
American Association of Petroleum Geologists (AAPG), 121
American Consulting Engineers Council, 115, 119
American Institute of Architects (AIA), 114, 115
answer, 156n
anticipatory breach, 62
apparent authority, 68–69, 69f
appeal, 92
arbitration, 155, 159–161
arbitrator, 159

architects
  breach of duty, 136
  code of ethics, 17
  definition, 10
  dual role, 20
  duties, 16
  see also ethics
  duty of care, 133–134
  examination component, 11
  experience requirement, 11
  regulation of. *See* self-regulation
Architects Act (Ont.), 10
architectural services agreements, 121
articles of incorporation, 38
Associated Owner's & Developers (AOD), 115
Association of Consulting Engineering Companies (ACEC), 9, 121
Association of Engineers and Geoscientists, 4
authority, 68–70, 79–80

B

balance of probabilities test, 15, 17, 17n, 163n
Band Councils, 265, 266
Bank Act, 40
bankability, 128–129
bankable contract, 128–129
Bankruptcy Act, 40
bar chart, 217, 217f
BCCA 200, 76, 76n
bid, 46
bid bond, 92, 191, 193–194
bid chopping, 100n
bid chiseling, 100n
bid depositories, 92–93, 101–102
bid shop, 95
bid shopping, 98–100
bidders, 84
bidding process
  bid bond, 92

bid depositories, 92–93, 101–102
bid shopping, 95, 98–100
choosing the best process and delivery method, 105–108
consultant's role, 98
contract award, 97–98
contract formation, 93–95, 94f
described, 92–93
mistakes, 100–101
preparation of bid documents, 95–96
receipt and examination of bids, 97
stages, 95 98
subcontract formation, 74, 102–105
submission of bids, 96–97
tendering law, 99f
tips for defining, 90f
trade agreements, effects of, 108–110
binding mediation, 162
boiler insurance, 176, 176n
bona fide concern, 89
bonding company, 191
bonds
  bid bonds, 92, 191, 193–194
  and contract of indemnity, 191
  described, 191
  fidelity bonds, 191
  indemnities, 192–193
  labour and material payment bond, 198, 199
  lien bond, 199
  obligee, 191
  one-tier payment bonds, 76
  payment bonds, 76, 191, 198–199
  performance bonds, 191, 194–198, 195f
  primary obligations, 192
  principal, 191

roles and responsibilities, 192
subrogation, 192–193
surety, 191
surety recourses, 192–193
types of, 191
bonus clause, 60, 124
Bre-X scandal, 140, 269
breach of contract, 57–62
  anticipatory breach, 62
  conditions for, 57–59
  contract termination, 61–62
  damages, 59–61
  described, 57
  disagreement, 58–59
  fundamental breach, 61
  inability, 58
  inadvertence, 58
  lack of profit, 59
  performing under protest, 58–59
  repudiation, 62
  simple breach, 61
  termination clauses, 62
breach of duty, 132, 136–137
breach of ethics, 14
breach of trust, 207
British Columbia's Code of Ethics, 17, 19
builder's lien. *See* construction liens
Builders' Lien Act (B.C.), 204
builders' risk policies, 176, 183–186, 184*f*
building code, 71, 72, 151
bundle, 28
business interruption insurance, 176, 176*n*
business name, 34
business organization
  advantages and disadvantages, 42*t*
  business name, 34
  company, 34
  corporation, 34, 37–41
  forms of, 34
  joint venture, 34
  partnership, 34, 35–36
  sole proprietorship, 34, 34–35

voiding a contract, 34
"but for" test, 137–138

C

call for tenders, 84, 88
"call-up", 89
Canada Business Corporations Act, 37
Canada Health and Safety Regulations (CHSR), 249
Canada-Korea Agreement on the Procurement of Telecommunications Agreement, 110
Canada Labour Code (CLC), 226*n*, 249
Canada Pension Plan, 40
Canada Revenue Agency (CRA), 206
Canada Transportation Act, 249
Canadian Charter of Rights and Freedoms, 1, 3, 244–245
Canadian Constitution, 1–2
Canadian Construction Association (CCA), 115
Canadian Construction Document Committee 2 Contract (CCDC 2), 115
Canadian Construction Document Committee (CCDC), 114
Canadian Council of Professional Engineers (CCPE), 8
Canadian Council of Professional Geoscientists (CCPG), 9, 10
Canadian court system, 2–3, 3*f*
Canadian Environmental Assessment Act (CEAA), 260, 263
Canadian Environmental Assessment Agency, 263
Canadian Environmental Protection Act (CEPA), 260
Canadian Geotechnical Society (CGS), 9
Canadian International Trade Tribunal (CITT), 110
Canadian Portland Cement Association (CPCA), 9
Canadian Securities Administrators (CSA), 269

Canadian Society of Mechanical Engineering (CSME), 9
Canadian Standards Association (CSA), 257
canons of contract construction
  all parts of contract, 65
  commercial purpose, 65
  context, 65
  contra preferentem, 65
  contract as a whole, 65
  described, 65
  plain and ordinary meaning, 65
  restriction by express provisions, 65
  special meaning, 65
capacity of corporation, 38
case citations, 7
case law, 7
cause of action, 203
CCA 5, 118
CCA 14, 118
CCDC 2, 115
CCDC 3, 116
CCDC 4, 117
CCDC labour and material payment bond, 199
Certificate of Authorization, 37
certificate of pending legislation, 204
certificates of title, 26
change directive, 221
change orders
  defined, 49, 54
  described, 71–72
  extras for design negligence, 72
  impact costs, 72–73
  timing and pricing of changes, 73, 74*f*
changes, 123–125
Charter of Rights and Freedoms, 1, 3, 244–245
chattels, 23, 24, 28
Cicero, 140
Civil Code, 4
civil law, 4
claims, 4–5
claims consultants, 216
claims-made policy, 180–181
clear title, 26

client, duty to, 18
client dissatisfaction, 166, 169–171
client's interest, 18, 163, 168
closed shop, 230
closely held corporation, 38
codes of ethics, 14, 16–17
codes of professional practice, 14
codes and standards, 151–153
collective agreements, 233
commercial agreements, 65
commercial general liability (CGL) policies, 176, 186
common employers, 231
common interests, 251
common law, 4, 256, 259–260, 269–270
common law presumptions, 167–168
communications system policy (CSP), 277
communications system risk management, 277
company, 34
comparative negligence statutes, 71
compensable delay, 218–219
competence, 240
Competition Act, 1
complaint, 156n
compliance audits, 258
concession, 129
concurrent delays, 219
concurrent liability, 148
condominiums, 28
confidential information, 238–239
confidentiality, 16, 238–239, 257
confidentiality agreement, 21, 238–239
conflict of interest
    client's interest, 16, 18
    consultants, and extra payment claims, 20
    for corporations, 39
    employer's interest, 19
    insured and insurer, 182–183
    meaning of, 14
conflict of laws, 274
consequences, 14–16

consequential damages, 60, 60n, 126
consideration, 6, 44, 46–47, 46f, 49
Constitution, 1–2
construction contract models
    see also project delivery processes
    comparison of methods, 106t
    construction management, 109
    described, 105
    design-bid-build process, 105–107, 106f, 107t
    design-build model, 107, 107t, 108f
    design-build-operate finance-model, 109
    design-build-operate model, 108
    public-private partnerships, 108
    responsibilities in each method, 85t
    traditional, 105
construction contracts
    alliance agreement, 119
    cases, 53
    construction management contracts, 118
    cost-plus contract, 113, 115–116, 117f
    cost reimbursable contract, 113
    described, 6, 113
    design-build contract, 113, 118–119
    fixed-price contract, 113, 114–115
    guaranteed maximum price (GMP) contracts, 116
    licensing agreement, 113
    lump sum contract, 114–115
    public-private partnerships, 113, 120
    stipulated price contract, 114–115
    stipulated sum contract, 113
    target price contract, 116
    unit price contract, 113, 117

construction lien statute, 203
construction liens
    claimants, 204–205
    described, 27, 202–203
    holdback, 209–212
    legislation, 203
    making a claim, 203–204
    proving a claim, 203–204
    risk to the contractor and owner, 213
    substitute lien security, 205–206
    trust provisions, 207–208
construction management, 109, 109f, 109t
construction management contracts, 118
construction projects, 5, 6
Construction Specifications Canada (CSC), 9
constructive acceleration, 220
constructive changes, 54
constructive dismissal, 242
constructive trust, 19
consult, duty to, 266
consultants
    claims consultants, 216
    delay claims consultant, 222
    dual role, 20
    role of, in bidding process, 98
    self-interest, 20
consulting engineering, 6
Consumer Products Warranties Act, 40
contamination, 258–260
contestability procedures, 92
context, 65
contingent fees, 163
continuing professional development (CPD), 11
contra proferentem, 65
contract administration
    authority, 79–80
    field reviews, 80–82
    timeliness, 80
"Contract for Geoscience Services", 121
contract interpretation. *See* interpretation of contracts

contract issues
  agency, 68–70
  authority, 68–70, 79–80
  change orders, 71–74, 74f
  codes and standards,
    151–153
  concurrent liability, 148
  contract administration,
    79–82
  described, 68
  drawings, 78–79
  extras for design negligence,
    72
  impact costs, 72–73
  indemnities, 70–71
  joint and several liability, 150
  limitation periods, 148–150
  performance under protest,
    73
  specifications, 78–79
  subcontract issues. See sub-
    contract issues
  timing and pricing of
    changes, 73
  unforeseen conditions,
    76–78
  vicarious liability, 151
contract law, 6
contract of indemnity, 191
contract of suretyship, 191
  see also bonds
contract time, 123
contracting process
  see also procurement process
  bidders, 84
  call for tenders, 84
  delivery processes, 85, 85t
  delivery methods, 85t, 85–86
  delivery systems, 84–86
  payment methods, 84–85,
    85t
  project participant selection.
    See procurement process
  scope of work, 88
  tendering process. See bid-
    ding process
  transfer of risk and obliga-
    tion, 86, 87f
contractor-caused delay, 218

contractors
  breach of duty, 136
  duty of care to, 20
  independent contractor,
    242–243
contractors' all risks insurance,
  183
contracts
  see also contract issues; stan-
    dard clauses
  acceptance, 44, 45–46
  administration. See contract
    administration
  agreements to agree, 49–50
  amendment, 52
  award of, 97–98
  with Band Councils, 266
  bankable contract, 128–129
  bid, 46
  bonus clause, 60
  breach. See breach of con-
    tract
  change order, 54
  consideration, 44, 46–47,
    46f, 49
  construction contracts. See
    construction contracts
  constructive changes, 54
  defined, 6, 44
  drafting, 66–67
  duress, 51
  electronic contracts, 275–276
  employment contract. See
    employment contract
  enforceability, 44, 44n
  enforceable contract, 44
  estoppel, 53
  express terms, 44
  financing, 128–129
  formation, 44, 45–46, 93–95,
    94f
  frustration, 52
  fundamental principles of
    contract law, 6–7
  geoscience agreements, 113,
    122
  grubstake agreement, 122
  health and safety law, 252
  implied terms, 44

  impossibility, 52
  of indemnity, 191
  insurance, 176
  interpretation. See interpre-
    tation of contracts
  invitation to treat, 45, 45n,
    93
  letter of intent, 49–50
  licensing agreement, 113,
    122
  misrepresentation, 50–51
  mistake, 50
  notice provisions, 218
  offer, 44, 45–46, 93
  option agreement, 122
  Postal Acceptance Rule, 46
  privy to, 44
  process. See contracting
    process
  procurement, 46
  professional service agree-
    ments, 114, 120–122
  purpose of, 44, 166
  quantum meruit, 53–54
  quasi-contract, 53–54
  rejection, 45
  repudiation, 62
  revocation of offer, 45
  standard clauses. See stan-
    dard clauses
  standard form contracts. See
    standard form contracts
  target price contract, 116
  tender, 46
  termination, 61–62
  unconscionable contract,
    51–52
  voidable contract, 45
  voiding a contract, 50–52
  waiver, 52–53
  as a whole, 65
contractual context, 64
contributory negligence, 150
cooperation, 182–183
copyright
  defined, 23, 29
  Internet law, 274
  moral rights, 29
  plans and specifications,
    29–30

principal rights, 29
protection, 29
requirements to gain protection, 29
term of protection, 29
Copyright Act, 29
corporate indemnity, 40
corporate seal, 49
corporation
    articles of incorporation, 38
    capacity, 38
    closely held corporation, 38
    conflicts of interest, 39
    control, 38
    creation of, 37
    debt, 38
    defined, 34, 37
    described, 37
    directors, 38, 39–41
    due diligence, 40
    equity, 38
    fiduciary duty, 39
    governmental liabilities, 39–40
    insider trading, 40–41
    officers, 38, 39–41
    organization, 38
    private, 38
    public, 39
    separate legal entity, 37–38
    shareholders, 38, 39–41
correction of deficiencies, 60, 116
cost-plus contract, 113, 115–116, 117f
cost reimbursable contract, 113
costs of litigation, 157
counter-offer, 45
counterclaims, 157
Court of Appeal, 2
court system, 2–3, 3f
court-appointed masters and referees, 162
coverage period, 175
craft, 230
creation of law, 3–7
credits, 72
Criminal Code, 15, 248–249, 261–262
criminal conduct, 15–16

critical path, 217
critical path method (CPM), 217

D

damage clause, 124
damages
    bonus clause, 60
    breach of contract, 59–61, 139
    consequential damages, 60, 60n, 126
    correction of deficiencies, 60, 116
    declaratory order, 59
    injunction, 59
    limitations on recovery of, 59
    liquidated damages, 60, 60n
    liquidated damages clauses, 124
    meaning of, 59
    misrepresentation, 50
    mitigation, 59
    remoteness, 59–60
    resultant damage, 185
    specific performance, 59
    standard clauses, 124
    standard measure, 60
    withdrawal of subcontractor bid, 104–105
debt, 38
debt financing, 38
deceit, 140
declaratory order, 59
deductible, 175
defamation, 274
Defence Production Act, 40
defences
    bid bonds, 193–194
    duty to defend, 177
    performance bonds, 197–198
    revocability of the bid, 194
defendants, 6–7, 156
delay
    acceleration, 216, 220
    acceleration claim, 220, 221–223
    compensable delay, 218–219
    concurrent delays, 219

contractor-caused delay, 218
    defined, 216
    excusable delay, 218–219
    force majeure, 218
    heads of damage, 221–222
    law relating to claims, 216
    measured mile approach, 223
    no-damages-for-delay clause, 75, 220
    notice provisions, 218
    overview, 216
    proving, 221–223
    required documentation, 222–223
    scheduling principles, 217
    total cost approach, 223
delay claims consultant, 222
delayed opening insurance, 176, 176n
dependent contractor, 228
derivative actions, 39
design-bid-build model, 105–107, 106f, 107t
design-build contract, 113, 118–119
Design-Build Institute of America (DBIA), 114, 118–119
design-build model, 107, 107t, 108f
design-build-operate finance-model, 109
design-build-operate model, 108
Digital Millennium Copyright Act, 274
diminution of value, 60–61
directors, 38, 39–41
disability, 244
disagreement, 58–59
disciplinary consequences, 15–16
discipline, 12
disclaimer clauses, 96
disclaimers, 171–172
disclosure guidelines, 269
disclosure of material facts, 140
disclosure requirements, 268–269
discovery, 158–159
discovery of documents, 158

discrimination, 244
dispute resolution
    alternate dispute resolution (ADR), 159
    arbitration, 155, 159–161
    dispute resolution provision, 127–128
    litigation, 155, 155–159
    mediation, 155, 161–162
    negotiation, 155, 161
    other methods, 162
    types of, 155
dispute resolution provision, 127–128
disputes
    client dissatisfaction, 169–171
    common law presumptions, 167–168
    described, 4–5
    disclaimers, 171–172
    jurisdictional disputes, 229–230
    problem solving, 174
    record keeping, 173–174
    resolution of. *See* dispute resolution
disruption, 216
division of powers, 1
domain names, 275
double jeopardy, 15
drafting contracts, 66–67
drawings, 78–79
dual role of consultant, 20
due diligence, 40, 262
duress, 45, 51
duty of care
    architects, 133–134
    described, 132–133
    engineers, 133–134
    general rule, 132–133
    pure economic loss, 135
    reducing risk of negligence-claims, 135–136
    standard of care, 136–137
duty of competence, 240
duty of fidelity, 238–239
duty of good faith, 181–182
    see also good faith obligations
duty to consult, 266

duty to cooperate, 182–183
duty to defend, 177
duty to indemnify, 177
duty to mitigate, 241
duty to the client, 16, 18
duty to the employer, 16, 18–19
duty to the profession, 16, 19–20
duty to the public, 16, 17–18
duty to warn, 143

E

easement, 27
economic loss, 135
electronic contracts, 275–276
email, 277
employee, 228–229, 242–243
employer, duty to, 18–19
employer's interest, 18–19
employment contract
    confidential information, 238–239
    duty of competence, 240
    duty of fidelity, 238–239
    duty to mitigate, 241
    implied terms, 237–242
    notice, 240–242
    restrictive covenants, 239
    severance, 240–242
    termination, 240–242
    trade secrets, 238–239
employment law
    see also employment contract; labour law
    and *Charter of Rights and Freedoms*, 244–245
    constructive dismissal, 242
    employment standards legislation, 243–244
    human rights, 244
    independent contractor *vs* employee, 242–243
    minimum wages, 243
    notice, 243
    overview, 226, 237
    termination, 244, 245*t*
employment standards legislation, 243–244
enforceability, 44, 44*n*
enforceable contract, 44

enforcement, 12
engineer-in-training (EIT), 9
engineering service agreements, 121
engineers
    breach of duty, 136
    code of ethics, 16–17
    duties, 16
    see also ethics
    duty of care, 133–134
    Engineers Canada definition, 9–10, 10–11
    examination component, 11
    experience requirement, 11
    regulation. *See* self-regulation
Engineers Canada, 8, 9, 9–10, 10–11, 16
Engineers Canada Model Code of Ethics, 16
entitlement, 221
environmental assessment process, 261
environmental audits, 258
environmental cleanup, 262
environmental contamination, 259–260
environmental law
    environmental assessment process, 261
    environmental audits, 258
    environmental site assessments (ESAs), 257, 257–258
    governmental regulation, 260–262
    nuisance, 259–260
    overview, 256–257
    remedies for private landowners, 258–260
environmental legislation
    common legislative concepts, 261
    environmental cleanup, 262
    environmental offences, 261–262
    federal environmental legislation, 260
    governmental liabilities, 39

jurisdiction, 260–261
provincial environmental legislation, 260–261
Environmental Management Act, 260
environmental offences, 261–262
environmental site assessments (ESAs), 257, 257–258
equitable remedies, 53*n*
equity, 38
equity investment, 38
errors and omissions insurance, 187
estate in fee simple, 24
estoppel, 53
ethics
breach of ethics, 14
codes of ethics, 14, 16–17
criminal conduct, 15–16
duties, 16
duty to the client, 18
duty to the employer, 18–19
duty to the profession, 19–20
duty to the public, 17–18
and health and safety law, 253
and the law, 14–16
legal liability, 15–16
overview, 14
Evaluation Society, 92
Evidence Acts, 173
examination for discovery, 158–159
Excise Tax Act, 40
exclusion clause, 125–126
excusable delay, 218–219
expert witness, 18, 20, 162–163
Export and Import Permits Act, 40
express authority, 68–69
express terms, 44
extras, 71
extras for design negligence, 72

F

face value of bond, 194
factual matrix, 66
fairness, 89, 91–92
fairness commissioner, 92
fairness monitor, 91
Federal Courts, 2
federal jurisdiction, 1
federal environmental legislation, 260
federal privacy legislation, 271–272
fee simple, 24–25
fidelity, 238
fidelity bonds, 191
fiduciary duty, 36, 39, 141
fiduciary relationship, 16*n*, 141
field reviews, 80–82
financing the project, 128–129
Fisheries Act, 260
fixed-price contract, 113, 114–115
fixture, 24
float, 217
force majeure clause, 52, 218, 220
foreclosure action, 203*n*
formation of contract, 44, 45–46
fraud, 15, 37–38, 140–141
fraudulent misrepresentation, 50–51, 139, 149–141
freehold interest, 24
frustration, 45, 52
fundamental breach, 61

G

gas, 25
general conditions, 74*n*
general contractor, 74, 118
general partner, 36
geophysical trespass, 142
geoscience agreements, 113, 122
geoscience service agreements, 121–122
geoscientists
code of ethics, 17
definition, 10
duties, 16
see also ethics
examination component, 11
regulation. *See* self-regulation
trespass, 142

good faith obligations, 45, 89–90
goods, 28
Government of British Columbia, 120
governmental liabilities, 39–40
Governor General, 2
grantee, 25
grievance, 233, 234
grievance arbitration, 233
grubstake agreement, 122
guaranteed maximum price (GMP) contracts, 116

H

harassment, 244
heads of damage, 221–222
health and safety law
accident occurs, 250–251
Canadian Labour Code (CLC), 249
common interests, 251
contracts, 252
Criminal Code, 248–249
ethical considerations, 253
federal and provincial law, 248–249
fundamental rights of workers, 247
legislation, 247
other safety statutes, 249
overview, 247–248
provincial statutes, 249
regulators, 251
responsibility for health and safety, 249–250
safety, importance of, 248
torts, 252–253
workers' compensation legislation, 134, 252–253
hidden conditions, 76–78
holdback, 209–212
human rights, 244
hybrid methods, 89

I

impact, 216
impact claims
acceleration, 216

described, 221
heads of damage, 221–222
impact, 216
law relating to, 216
measured mile approach, 223
overview, 216
proving, 221–223
required documentation, 222–223
scheduling principles, 217
total cost approach, 223
impact costs, 72–73, 221
implied authority, 68
implied terms, 44
impossibility, 45, 52
in privity, 44
inability, 58
inadvertence, 58
Income Tax Act, 40, 206
incorporation by reference, 76
indemnification clause, 70–71, 125
indemnities, 70–71, 192–193
indemnity clause, 70–71, 125
independent contractor, 228, 242–243
indices, 67, 67*n*
industrial design, 23, 31
Industrial Design Act, 31
infringement of intellectual property rights, 31
inherent jurisdiction, 2
injunction, 59
insider trading, 40–41, 268
insolvency dispute, 5
inspection of property, 158
Institute of Civil Engineering (U.K.), 9
Institute of Civil Engineers (ICE), 115
Institute of Electrical and Electronics Engineers (IEEE), 9
insubordination, 242
insurable interest, 178–179
insurance
    conflict between insured and insurer, 182–183
    duty of good faith, 181–182
    duty to cooperate, 182–183

duty to defend, 177
duty to indemnify, 177
insurable interest, 178–179
liability insurance, 15, 36, 40, 175
material non-disclosure, 181–182
operating without insurance, 176–177
prejudice to third parties, 181–182
property insurance, 175
subrogation, 177–178
insurance policies
    all-risk policies, 183–186
    builders' risk policies, 176, 183–186, 184*f*
    claims-made policy, 180–181
    commercial general liability (CGL) policies, 176, 186
    contractual indemnity, 70
    coverage period, 175
    deductible, 175
    elements of, 175
    errors and omissions insurance, 187
    limitation periods, 175
    malpractice insurance, 186
    occurrence policies, 180–181
    overview, 175–176
    perils covered, 175
    professionals and, 176
    self-retention, 175
    types of, 183–187
    wrap-up policy, 187
insured, 175
intangible property, 23
integrated circuits, 31
Integrated Circuits Topography Act, 31
intellectual property
    copyright, 29–30, 274
    described, 23, 28–29
    industrial design, 31
    infringement and remedies, 31
    integrated circuits, 31
    patents, 30
    trademarks, 30–31, 275
intention of the parties, 64

intentional torts, 132*f*
interest arbitration, 233
interest in fee simple, 24
International Federation of Consulting Engineers (FIDIC), 114, 115
international law, 5–6
international trade agreement, 108–110
Internet law
    communications system policy (CSP), 277
    communications system risk management, 277
    copyright, 274
    defamation, 274
    domain names, 275
    electronic contracts, 275–276
    email, 277
    jurisdiction, 274
    negligent advice, 274
    overview, 273
    privacy, 275
    securities regulation, 275
    security, 275
    torts, 274
    trademarks, 275
    websites, 276
interpretation of contracts
    all parts of contract, 65
    canons of contract construction, 64, 65
    commercial purpose, 65
    context, 65
    contra preferentem, 65
    contract as a whole, 65
    golden rule, 65
    intention of the parties, 64
    introduction of additional evidence, 65–66
    legal principles, 64
    parol evidence rule, 66
    plain and ordinary meaning, 65
    restriction by express provisions, 65
    special meaning, 65
interprovincial trade agreement, 110
interrogatories, 158

invitation to tenders, 88
invitation to treat, 45, 45n, 93

J

joint and several liability, 150
joint tenants, 25
joint venture, 34
judgment-proof, 176
jurisdiction
    conflict of laws, 274
    environmental legislation,
      260–261
    internet law, 274
just cause, 240

K

Kyoto Accord, 110, 260

L

labour and material payment
    bond, 76, 198, 199
labour law
    see also unions
    collective agreements, 233
    common employers,
      231–233
    employee, definition of,
      228–229
    layoffs, 234, 240
    overview, 226–227
    seniority, 234
    successor employers, 231
labour standards legislation, 39,
    40
lack of profit, 59
latent defect, 149
lateral support, 25
law
    Aboriginal law. *See* Aborigi-
      nal law
    case citations, 7
    civil law, 4
    common law, 4, 256,
      259–260, 269–270
    common law presumptions,
      167–168
    contract law, 6

creation of, 3–7
delay and impact claims, 216
employment law. *See* em-
    ployment law
environmental law. *See* envi-
    ronmental law
ethics and, 14–16
health and safety law. *See*
    health and safety law
international law, 5–6
Internet law. *See* Internet law
labour law. *See* labour law
privacy law. *See* privacy law
Quebec law, 4
securities law. *See* securities
    law
statute law, 3–4
subject areas and principles,
    6–7
layoffs, 234, 240
learned intermediary, 144
lease, 27
legal consequences, 14–16
legal liability, 15–16
letter of assurance, 12
letter of intent, 49–50
letter of interest, 89
letters patent, 38
liability
    concurrent liability, 148
    contributory negligence, 150
    corporate indemnity, 40
    due diligence, 40, 262
    exclusion, limitation, or
      waiver clauses, 125–126
    governmental liabilities,
      39–40
    joint and several liability, 150
    limitation of liability clause,
      120
    product liability, 143–144
    securities and, 269
    strict liability, 142, 143
    vicarious liability, 71, 151,
      240
liability insurance
    claims-made policy, 180–181
    commercial general liability
      (CGL) policies, 186
    duty to defend, 177

    errors and omissions insur-
      ance, 187
    governmental liabilities, 40
    limited liability partnership,
      36
    malpractice insurance, 186
    purpose of, 15, 175
    vicarious liability, 151
    wrap-up policy, 187
liability policy. *See* liability
    insurance
libel, 274
licence, 27
licensing agreement, 113, 122
lien, 203.
    see also construction liens
lien bond, 199
limitation clause, 126, 220
limitation of liability clause, 120
limitation periods, 5, 148–150,
    175
limited liability partnership, 36
limited partners, 36
limited partnership, 36
liquidated damages, 60, 60n
liquidated damages clauses, 124
lis pendens, 204
litigation
    advantages, 156
    costs, 157
    counterclaims, 157
    disadvantages, 156, 159
    discovery, 158–159
    meaning of, 155
    pleadings, 156–157
    statement of claim, 156
    statement of defence, 156
    technical evidence, 159
    third-party claims, 157
lockouts, 230–231
loss caused by breach, 139
lump sum contract, 114–115

M

malpractice insurance, 186
mandatory arbitration provi-
    sion, 128, 155, 161
mandatory binding arbitration,
    159

marine and transportation insurance, 176, 176n
material non-disclosure, 181–182
measured mile approach, 223
mechanics liens. See construction liens
mediation, 155, 161–162
member-in-training (MIT), 9
memoranda of association, 38
minerals, 25
mines, 25
mini-trial, 162
minimum wage provision, 243
mining law options, 122
Ministry of Competition, Science, and Enterprise, 120
Ministry of Labour, 249
misrepresentation, 50–51, 139–140, 274
mistake, 50, 100–101
mitigation, 59, 237
mode of acceptance, 45
Model Code of Ethics, 16–17
moral principles, 14
moral rights, 29
mortgage, 27
multiple holdback system, 209, 210
mutual mistake, 50

N

NAFTA, 6, 11, 110, 260
named subcontractor, 102–104
Napoleonic Code, 4
National Building Code, 249
National Council of Architectural Registration Boards (NCARB), 11
National Council of Examiners for Engineering and Surveying (NCEES), 11
National Instruments (NI), 269, 270
negative restrictive covenant, 27
negligence
    breach of duty, 136–137
    contributory negligence, 150
    described, 6–7
development of law of, 131
duty of care, 132, 132–136
duty to the public, 17–18
elements of, 131–132
expert witness, 18
extras for design negligence, 72
loss caused by breach, 139
misrepresentation, 139–140
negligence claim, 6–7
proximate cause, 137–138
reducing risk of negligence claims, 135
standard of care, 136–137
negligent advice, 274
negligent misrepresentation, 139–140
negotiation, 45, 155, 161
network diagrams, 217
no-damages-for-delay clause, 75, 220
non-union shop, 230
nonbinding arbitration, 159
North American Free Trade Agreement (NAFTA), 6, 11, 110, 260
notice of claim provision, 127–128
notice of termination, 240–242, 243
notice provisions, 218
nuisance, 143, 259–260

O

objective standard, 64
obligation of good faith, 45
obligations of professional members, 11
obligee, 191
occupational health and safety (OH&S) legislation, 247, 248–251
    see also health and safety law
occurrence policies, 180–181
offer, 44, 45–46, 93
offer and acceptance, 6
officers, 38, 39–41
oil, 25
one-tier payment bonds, 76
Ontario Labour Relations Act, 233n
open shop, 230
openness, 90–91
operating without insurance, 176–177
option agreement, 122
organizational structures. See business organization
"orphan" sites, 262
ostensible authority, 68
owner, duty of care to, 20
owner's agent, 20, 118
ownership, 24

P

P3, 84, 120, 129
    see also public-private partnerships
parol evidence rule, 66
partner, 35
partnering, 162
partnering model, 105
partnership
    described, 34, 35
    fiduciary duty, 36
    formation of, 35–36
    limited liability partnership, 36
    limited partnership, 36
Partnership Acts, 35–36
partnership agreement, 35–36
patent, 23, 30
Patent Act, 30
patent defect, 149
pay-if-paid clause, 75
pay-when-paid clause, 75
payment bond, 76, 191, 198–199
penalty, 60
penalty clauses, 124
PEO Code of Ethics, 16–17
perfecting the lien, 203–204
performance bond, 191, 194–198, 195f
performance specification, 79
performance under protest, 58–59
perils covered, 175
Personal Information Protection Act (PIPA), 272

Personal Information Protection and Electronic Documents Act (PIPEDA), 271–272, 275
personal property, 23
Phase ESA, 257–258
Phase II ESA, 258
Phase III ESA, 258
picketing, 231
"piercing the corporate veil", 37–38
PIPA (*Personal Information Protection Act*), 272
PIPEDA (*Personal Information Protection and Electronic Documents Act*), 271–272, 275
plaintiff, 7, 156
plans (copyright), 29–30
pleadings, 156–157
positive restrictive covenant, 27
Postal Acceptance Rule, 46
pre-qualification, 89
prejudice to third parties, 181–182
presumptions, 167–168
primary obligations, 192
prime contract, 76
prime contractor, 74, 103–105, 194, 252
principal, 68, 191
principal rights
    copyright, 29
    industrial design, 31
    integrated circuits, 31
    patents, 30
    trademarks, 31
privacy law
    federal legislation, 271
    Internet activities, 275
    overview, 271
    PIPEDA, 271–272
    provincial legislation, 272
private corporation, 38
private landowners, remedies for, 258–260
privilege clauses, 96
privileged, 161
privy, 44
problem solving, 174
    see also dispute resolution
procedural issues, 89–92

procurement, 46
procurement process
    see also contracting process
    cost of fairness and transparency, 91–92
    delivery systems, types of, 88–89
    fairness and good faith, 89–90
    hybrid methods, 89
    invitation to tenders, 88
    letter of interest, 89
    openness or transparency, 90–91
    pre-qualification, 89
    procedural issues, 89–92
    request for proposals (RFPs), 88, 89
    request for qualifications, 88
    request for quotation, 88
    request for standing offers, 89
    scope of work, 88
    tendering process. *See* bidding process
    tips for implementing, 91f
product liability, 143–144
productivity tracking, 223
the profession, duty to, 19–20
professional engineers. *See* engineers
Professional Engineers Ontario Code of Ethics, 16–17
professional evaluators, 92
professional regulation. *See* self-regulation
professional seal, 12
professional service agreements, 114, 120–122
profit à prendre, 27
project delivery
    see also construction contract models
    common risks and obligations, 86f
    comparison, 106t
    described, 84–86, 85t
    project participant selection. *See* procurement process

responsibilities in each method, 85t
transfer of risk and obligation, 86, 87f
types of delivery systems, 88–89
project finance, 128–129
promissory estoppel, 46n, 98
property
    boundaries, 25–26, 142
    chattels, 23, 24, 28
    fixture, 24
    intangible property, 23
    intellectual property. *See* intellectual property
    personal property, 23
    real property. *See* real property
    tangible property, 23, 186
    trespass, 142–143
    types of, 24f
property boundaries, 25–26, 142
property insurance
    all-risk policies, 183–186, 184f
    builders' risk policies, 183–186, 184f
    purpose of, 175, 183
property law
    chattels, 28
    intellectual property, 28–31
    overview, 23–24
    real property, 24–28
property policy. *See* property insurance
proponent, 89
prospectus, 268–269
Provincial Court, 2–3
provincial jurisdiction, 1
provincial labour standards acts, 40
provincial OH&S statutes, 249
provincial environmental legislation, 260–261
provincial privacy legislation, 272
provincial regulatory bodies, 8–9
proximate cause, 137–138
the public, duty to, 17–18
public corporation, 39

public-private partnerships, 84, 108, 109*f*, 113, 120
Public Private Partnerships Best Practices Guide, 120
pure economic loss, 135

Q

quantum meruit, 53–54
quasi-contract, 53–54
quasi-criminal penalties, 207
Quebec law, 4

R

Rand formula, 230
real property, 24–28
  air rights, 26
  air space parcel, 26
  clear title, 26
  condominiums or stratas, 28
  fee simple, 24–25
  grantee, 25
  joint tenants, 25
  meaning of, 23, 24
  mines, minerals, oil and gas, 25
  property boundaries, 25–26
  registration of real property rights, 26–27
  rights running with the land, 27
  riparian rights, 25–26
  surface lease, 25
  tenancy in common, 25
  unpatented, 26
real property rights
  construction liens, 27
  easement, 27
  lease, 27
  licence, 27
  mortgage, 27
  profit à prendre, 27
  restrictive covenant, 27
  right of way, 27
reasonable notice, 241
record keeping, 173–174
registration
  professional, 10
  real property rights, 26–27

regulation of the professions. *See* self-regulation
regulatory bodies, 8–9, 11
  see also self-regulation
rejection, 45
remedies
  construction liens. *See* construction lien
  damages. *See* damages
  declaratory order, 59
  environmental contamination, 259–260
  equitable remedies, 53*n*
  for infringement, 31
  injunction, 59
  for private landowners, 258–260
  specific performance, 59
  termination of employment, 244, 245*t*
remoteness, 59–60
repudiation, 62
request for proposals (RFPs), 88, 89
request for qualifications, 88
request for quotation, 88
request for standing offers, 89
requirements to gain protection
  copyright, 29
  industrial design, 31
  integrated circuits, 31
  patents, 30
  trademarks, 31
requisition, 89
rescission, 139–140
reserve property, 266
restriction by express provisions, 65
restrictive covenant, 27, 239
resultant damage, 185
retainage, 209–212
revocation of offer, 45
rework, 116
right in fee simple, 24
right of way, 27
right to title, 8, 9
right-to-work, 230
riparian rights, 25–26
risk
  allocation of, 5

assessment, 166–167
common law presumptions, 167–168
communications system risk management, 277
construction liens, 213
contracts, and allocation of, 166
disclaimers, 171–172
email, 277
in project delivery, 86*f*
shifting risk, 168–169
transfer of, 86, 87*f*
Royal Architectural Institute of Canada (RAIC), 9, 10, 114
Royal Architectural Institute of Canada (RAIC) Document 6, 121, 149
Royal Architectural Institute of Canada (RAIC) Document 7, 121
run with the land, 26

S

safety. *See* health and safety law
sale of goods statute, 28, 124
Samuels, Brian, 74*n*, 102*n*
Saskatchewan Code of Ethics, 18
schedule, 123, 217
schedule interpretation and analysis, 222
scheduling principles, 217
scope of practise, 8, 9–11
scope of the work, 123
scope of work, 88
seals, 49
secondary activity, 231
securities, 268
securities law
  common law, 269–270
  information disclosure requirements, 268–269
  and the Internet, 275
  overview, 268
  prospectus, 268–269
  statutory liability, 269–270
  technical disclosure guidelines, 269

self-regulation
continual professional development (CPD), 11
discipline, 12
duty to the profession, 19–20
enforcement, 12
ethics. *See* ethics
letters of assurance, 12
meaning of, 8
obligations of professional members, 11
professional seals, 12
registration, 10
regulatory bodies, 8–9, 11
right to title, 8, 9
scope of practise, 8, 9–11
technical associations, 9
self-retention, 175
seniority, 234
separate legal entity, 37–38
settlement conference, 162
severance, 241–242
sexual harassment, 244
shareholders, 38, 39–41
simple breach, 61
single holdback system, 209
slander, 274
small claims courts, 2, 157n
sole proprietorship, 34, 34–35
special meaning, 65
specific performance, 59
specifications, 78–79
specifications (copyright), 29–30
standard clauses
bonuses, 124
change, 123–125
contract time, 123
damages, 124
disclaimers, 171–172
dispute resolution provision, 127–128
exclusion clause, 125–126
force majeure clause, 52
indemnification, 70–71, 125
limitation clause, 126, 220
liquidated damages clauses, 124
mandatory arbitration provision, 128, 155, 161

no-damages-for-delay clause, 220
notice of claim provision, 127–128
penalty clauses, 124
scope of the work, 123
termination, 125
waiver clause, 125–126
waiver of subrogation clause, 177
warranty, 124–125
standard form contracts
architectural service agreements, 121
change order clauses, 72
construction management contract, 118
cost-plus contract, 113, 115–116, 117f
described, 114
design-build contract, 113, 118–119
electronic contracts, 275–276
engineering service agreements, 121
geoscience service agreements, 121–122
incorporation by reference, 76
indemnity clauses, 71
insurance, 176
stipulated price contract, 114–115
unit price contract, 113, 117
standard of care, 136–137, 163
standard specification, 79
standards, 151–153
standards of proof, 15
statement of claim, 156
statement of defence, 156
statute law, 3–4
statutory declaration, 213
stipulated price contract, 114–115
stipulated sum contract, 113
stratas, 28
strict liability doctrine, 142, 143, 144
strict liability offences, 262
subcontract formation

holding subcontractor to a price, 104
named subcontractor, 102–104
quantification of damages, 104–105
withdrawal of subcontractor bid, 104–105
subcontract issues
described, 74
formation of subcontract, 74
incorporation by reference, 76
one-tier payment bonds, 76
pay-if-paid clauses, 75
subcontractors, 74
subjective standard, 64
submission of bids, 96–97
subrogation, 177–178, 192–193
substitute lien security, 205–206
subtrades, 74
successor employers, 231
summary jury trials, 162
superior courts, 2
Supreme Court of Canada, 2
supreme courts, 2
surety, 191, 192–193
surety recourses, 192–193
suretyship agreement, 191
surface lease, 25

T

tangible personal property, 23, 28
tangible property, 23, 186
target price contract, 116
tax treaties, 6
technical associations, 9
technical disclosure guidelines, 268–269
technical evidence, 159, 160
tenancy in common, 25
tender, 46
tendering, 88
tendering law, 99f
tendering process. *See* bidding process
term of protection
copyright, 29

industrial design, 31
integrated circuits, 31
patents, 30
trademarks, 31
termination clause, 62, 125
termination of contract
    anticipatory breach, 62
    fundamental breach, 61
    repudiation, 62
    rescission, 139–140
    simple breach, 61
    termination clauses, 62
termination of employment,
    240–242, 244, 245t
third party, 157
third party claims, 157
timeliness, 80
TN visa, 6
torrens system, 26
tort issues
    codes and standards,
      151–153
    concurrent liability, 148
    joint and several liability, 150
    limitation periods, 148–150
    vicarious liability, 71, 151,
      240
tort standard, 80
torts
    deceit, 140
    defamation, 274
    defined, 6, 131
    duty to warn, 143
    fiduciary duty, 141
    fraud, 140–141
    and health and safety law,
      252–253
    intentional, 132f
    Internet law, 274
    misrepresentation, 139–140,
      274
    negligence. See negligence
    nuisance, 143
    obligations, 6
    product liability, 143–144
    strict liability, 142, 143
    trespass, 142–143
    types of, 132f
    unintentional torts, 131, 132f

and workers' compensation
    legislation, 134, 252–253
total cost approach, 223
trade agreements, 108–110
trade contractors, 74
Trade-Marks Act, 30
trade secrets, 238–239
trade unions. See unions
trademarks, 23, 30–31, 275
transfer of risk and obligation,
    86, 87f
transfer pricing, 115
transparency, 90–91, 91–92
Transportation of Dangerous
    Goods Act, 260
treaties, 5–6
trespass, 142–143
trial-level court, 2
trust, 199, 207
trust beneficiary, 199
trust provision, 207
trustee, 199
two-tier bonds, 76n, 199n

U

unconscionability, 45
unconscionable contract, 51–52
under protest, 58–59, 58n, 73
unforeseen conditions, 76–78
Uniform Electronic Commerce
    Act, 276
unintentional torts, 131
union shop, 230
unions
    see also labour law
    closed shop, 230
    development of, 227
    employee resistance, 229
    jurisdictional disputes,
      229–230
    open shop, 230
    picketing, 231
    Rand formula, 230
    representation, 229–229
    right-to-work, 230
    secondary activity, 231
    union security, 230
    union shop, 230
    work stoppages, 230–231

unit price contract, 113, 117
unlawful insider trading, 40–41
unpatented, 26
unwritten codes and standards,
    151–153
U.S. National Council of Archi-
    tectural Registration Boards
    (NCARB), 11

V

vicarious liability, 71, 151, 240
vmail, 277
voidable contract, 45
voiding a contract
    described, 50
    duress, 51
    frustration, 52
    misrepresentation, 50–51
    mistake, 50
    unconscionability, 51–52
voluntary binding arbitration,
    159

W

waiver, 52–53
waiver clause, 125–126
waiver of subrogation, 177
warn, duty to, 143
warranty, 124–125
websites, 276
withdrawal of subcontractor bid,
    104–105
work stoppages, 230–231
Workers Compensation Act
    (Ont.), 134, 253
workers' compensation board
    (WCB), 247, 252
workers' compensation legisla-
    tion, 134, 252–253
Workmen's Compensation Act
    (Sask.), 253
workplace accidents, 250–251
Worksafe New Brunswick, 249
World Trade Organization
    (WTO) Agreement on Gov-
    ernment Procurement, 110
wrap-up policy, 187
writ, 156